FOOD SCIENCE AND TECHNOLOGY

HANDBOOK OF CHICKPEAS

NUTRITIONAL VALUE, HEALTH BENEFITS AND MANAGEMENT

Food Science and Technology

Additional books and e-books in this series can be found on Nova's website under the Series tab.

Nutrition and Diet Research Progress

Additional books and e-books in this series can be found on Nova's website under the Series tab.

FOOD SCIENCE AND TECHNOLOGY

HANDBOOK OF CHICKPEAS

NUTRITIONAL VALUE, HEALTH BENEFITS AND MANAGEMENT

ALBERT T. LUND
AND
NOAH D. SCHULTZ
EDITORS

nova science publishers
New York

Library of Congress Cataloging-in-Publication Data

ISBN: 978-1-53616-374-2
Library of Congress Control Number:2019950483

Published by Nova Science Publishers, Inc. † New York

CONTENTS

PREFACE

Legumes are an important source of proteins, carbohydrates, vitamins, minerals, and dietary fiber. Chickpea is one of the most consumed legumes in the world. It is an affordable food product for people who cannot get animal protein, giving them a nutritious substitute in their diet. As such, Handbook of Chickpeas: Nutritional Value, Health Benefits and Management discusses the current information regarding the nutritional value of chickpea.

Following this, the authors highlight the nutritional value of domesticated Kabuli chickpeas used in different forms, as well as the impact of different factors which regulate the nutritional value.

The authors then explore the nutritional value, health benefits, and uses of desi chickpeas in comparison to Kabuli chickpeas.

Current information is presented regarding the amino acid profile and nutritional protein quality in raw and processed chickpea seeds and flours, the health or nutraceutical effects of chickpea protein isolates, hydrolysates or bioactive peptides and the generation of selenized proteins during sprouting and their antioxidant and anticancer potential.

The demand for gluten-free products is increasing since the prevalence of gluten-related disorders is rising. As such, chickpea is studied in the context of replacing wheat/gluten in some types of gluten-free food to improve the quality of the gluten-free diet.

Chickpea is also studied as a dairy-product substitute since it represents a good source of protein, iron, and fiber.

Chickpeas are composed of interesting amounts of protein, fibers, iron, zinc, and others, arousing interest by vegetarians. Therefore, the use of chickpeas is studied as an important ingredient in a vegetarian diet to contribute to nutritional quality.

The authors present the way in which legume proteins can be used to fortify some foods to increase nutritional value and balance amino acid content. The most important of these foods are bakery products.

Current information related to starch digestion and glycemic index of raw/processed chickpea flours and starch isolates, undigested carbohydrates content in raw/processed chickpea and health effects of chickpea digestible and undigestible carbohydrates is also presented.

Chickpea seeds are vulnerable, both in the field and in storage conditions to attack by several insect pests of economic importance, whereby infested grains lose their viability. Therefore, basic steps in order to control insects in warehouse or processing plants that must be followed are: monitoring for infestation detection, identification of the problem and insect control measures.

Traditionally, preferred control options are carbamate and pyrethroid sprays as soon as the threshold has been reached, as insects grow rapidly and a few days delay in spraying can result in major crop damage and increased difficulty in control.

The final study proposes that chickpea protease inhibitor concentrates exerted anti-genotoxic effects on LNCaP cells, supporting previous findings that Bowman-Birk inhibitors and protease inhibitors have a protective effect on oxidative damage. Chickpea may therefore play a role in prostate cancer prevention, however, further research is needed to fully understand the molecular mechanisms involved.

Chapter 1 – Legumes are an important source of proteins, carbohydrates, vitamins, minerals, and dietary fiber. They are an economical source of good quality proteins, being, in general, the sulfur-containing amino acids the limiting ones. Recently, legumes have gained much attention because of the health benefits associated with their consumption; even if they have antinutritional factors that have been linked to a reduction of the assimilation of nutrients, these can be reduced by processing the legumes, and it has been reported that these factors can also provide health benefits to the consumer. Chickpea is one of the most consumed legumes in the world. It is an affordable food product for people who can not get animal protein, giving them a nutritious substitute in their diet. In general, its protein has a better value than the protein of other pulses. Carbohydrates are the major component in chickpea which, being a pulse, has a low glycemic index. The dietary fiber in chickpea is mostly insoluble and has been linked with prebiotic effects, as well as with the decrease of the risk of certain diseases. Chickpea has a higher lipids content than many other legumes, but most of them are unsaturated fatty acids, which have been related to several health benefits. Studies have shown that not only due to its nutritional composition but also because of its bioactive compounds chickpea has a protective effect against several diseases such as cancer, cardiovascular diseases, and diabetes, to mention some, as well as an antioxidative effect. Since now researchers are looking for food that besides nourishing people can help to heal certain health conditions, a particular interest has emerged for the study of the bioactive compounds of chickpea and their influence on human health. This chapter will discuss the current information about the nutritional value of chickpea, as well as the health benefits given by its components.

Chapter 2 – Chickpea (*Cicer arietinum* L.) is a key dietary component due to its cheap and inexpensive protein, dietary fibres, vitamins and minerals along with other significant functional properties. Chickpea is one of the oldest cultivated edible legumes with high demand in the modern World. The consumption of chickpea is increasing steadily due to its demand especially in the developing and under-developed countries. Chickpeas provide wide taste and texture variety to cereal-based diet and ensures a balanced diet for both a vegetarian and non-vegetarian population. Chickpea is an herbaceous, annual legume and can be classified into three main groups of desi, kabuli, and pea on the basis of seed structure and consumption. However, the major cultivated chickpea types are Kabuli, and desi types. Kabuli varieties are preferred for consumption as whole seeds in different forms, whereas desi types are typically processed into flour and splitted form. Turkey is the major producer of both types of kabuli and desi chickpea. Turkey is also a major contributor of kabuli type chickpea and is ranked 5th after India, Australia, Myammar and Ethiopia and is ranked 1st in the Mediterranean region in terms of production. In Turkey, chickpea is consumed in different ways from raw to cooked or roasted. The production and consumption of desi chickpea is limited and confined to a specific area of Turkey. The nutritional value and their effects on human health varies for both types and affected by number of factors. This study highlights the nutritional values of domesticated Turkish chickpeas used in different forms and also highlights the impact of different factors which regulate the nutritional values.

Chapter 3 – Desi type chickpeas (*Cicer arietinum* L.) or microsperma are small sized, dark colored and angular shaped chickpeas with high nutritive and medicinal values. Desi chickpeas are considered as most primitive chickpeas due to its resemblance to the wild ancestor of chickpea. The desi chickpeas are commonly consumed and cultivated in India and Pakistan but also show its presence in other countries like Turkey, Mexico, Iran and some African countries. Desi chickpeas are used in different forms such as cooked, boiled or roasted. Whereas, *dhal* (splitted dehulled chickpeas) and *besin* (chickpea flour) are very popular in India and Pakistan for making different foods like *helva* and *chapatti* etc. Desi chickpeas are highly rich in dietary fibers, proteins, vitamins, minerals, carbohydrates and other compounds which are essential for dietary program. Apart from its nutritional values, desi chickpeas are considered as healthy food for curing diabetes, cardiovascular risks, cancer and skin diseases. However, the knowledge about desi chickpea consumption and its advantages are less known. This study enlightens the nutritional values, health benefits, and its uses in different forms along with its comparison with Kabuli chickpeas.

Chapter 4 – Chickpea (*Cicer arietinum L.*) is the third most relevant food legume produced worldwide. Its nutritional quality has been considered superior compared to other pulses and its consumption provides several proven health benefits. This legume crop is usually consumed as an alternative to more expensive animal proteins in developing countries. Chickpeas are a significant source of high-quality protein (12-30%)

with a good balance of essential amino acids, high protein digestibility and relatively low levels of antinutritional factors. The limiting amino acids are sulphur-containing and therefore can be easily complemented by consumption of cereal-based foods. The amino acid composition, protein quality and functional properties of chickpea proteins are affected by processing methods, such as decortication, soaking, germination, fermentation, boiling, among others, which remove or inactivate antinutritional factors and improve protein digestibility. Chickpea proteins are mainly composed of globulins (56%), glutelins (18%), albumins (12%) and prolamins (3%). Chickpea proteins isolates have been employed as food ingredients to improve both nutritional and functional properties. Besides the nutritional and functional properties of proteins, they may also possess other biological properties. These properties are attributed to active peptides encoded in protein molecules that can be released during gastrointestinal digestion or by controlled hydrolyses using exogenous proteases. The enzymatic hydrolysis of chickpea proteins is an important tool for obtaining hydrolysates and bioactive peptides with antioxidant, antifungal, angiotensin I-converting enzyme (ACE) inhibitory (hypotensive), hypocholesterolemic and anticarcinogenic activities. Moreover, the production of selenoproteins during germination has been recently proposed as alternative to diminish oxidative stress and improve health. This is due to the synthesis of selenoproteins, which are used to synthesize Glutathione Peroxide that protects the human system against oxidative stress and cancer.

This chapter describes current information of the following 1) amino acid profile and nutritional protein quality in raw and processed chickpea seeds and flours, 2) health or nutraceutical effects of chickpea protein isolates, hydrolysates or bioactive peptides and 3) generation of selenized proteins during sprouting and their antioxidant and anticancer potential.

Chapter 5 – The gluten-free diet (GFD) is arising as a food trend due to gluten-related disorders (GRD) and also due to people who believe that they genuinely benefit from GFD. However, the GFD tend to be costly and socially restrictive since the lack of gluten in products can lead to food products without adequacy on nutritional, sensorial, and technological quality. Therefore, some studies have been attempted to develop tastier and healthier gluten-free products to help people who need to follow the GFD. In this context, chickpea - as a naturally gluten-free product, and also high in protein, iron, and fiber - has been studied to replace wheat/gluten in some types of gluten-free food to improve the quality of GFD.

Chapter 6 – The adverse reactions to dairy products are prevalent worldwide in the population. Dairy products reactions include allergies and lactose intolerance, reaching almost 80% of the population. There is also the demand for dairy-free products by vegans. Therefore, the market of the dairy-free product tends to continue increasing. There are some milk substitutes based on cereals, legumes, and nuts or modified cow milk, as lactose-free dairy products.

Some of these products present limitations to replace dairy products because of their allergenic potential; some of them present high glycemic or low protein content; and others are not entirely accepted by consumers.

In this context, chickpea has been studied as a dairy-product substitute since it represents a good source of protein, iron, and fiber.

Chapter 7 – Vegetarianism has become an increasingly common practice around the world. A vegetarian diet excludes any foods containing meat, fish, and poultry. Since these foods are sources of vital nutrients such as protein, iron, calcium, zinc, Vitamin B12, and magnesium, adequate nutritional replacements in the diet are highlighted. Chickpeas (*Cicer arietinum*) present in their composition interesting amounts of protein, fibers, iron, zinc, and others, arousing interest by vegetarians. Therefore, the use of chickpeas has been studied as an important ingredient on a vegetarian diet to contribute to nutritional quality.

Chapter 8 – Legumes are very important economic crops and have a high protein content up to 40% in some species. Legumes proteins can be used to fortify some foods to increase its nutritional value and balance amino acid content. The most important of these foods are bakery products.

The main ingredient of bakery products is cereal flour, especially wheat, which contains functional proteins that contribute to obtaining a good final product through the formation of the gluten network. Nutritionally, wheat protein is rich in sulfur amino acids, such as cysteine and methionine, which contribute significantly to formation of sulfur bonds in the gluten network, but on the other hand, poor in lysine (one of essential amino acids), which caused that cereal proteins not completed nutritionally. So many researchers tend to fortify cereal proteins with other sources proteins rich in these amino acids. Soy protein, chickpeas, peas and sweet or bitter lupine were used to balance the amino acids in the final product.

Chickpea (*Cicer arietinum L.*) is one of the most important legumes after beans and peas in many countries of the world such as India, America, Australia, Asia, Europe, East Africa and Middle East. Syria and Turkey are among the most important producing countries. Chickpea is a crop cultivated from ancient times and in different agricultural lands and climatic areas, which gives it high agricultural productivity. There are two main types of chickpeas, Desi seeds which are small size, brown, green or black color and Kabuli seeds are large size, light colors and have low fiber content. Chickpea protein is rich in lysine and poor in methionine content, making it an ideal model for the fortification of cereal proteins for balance of essential amino acids.

The rheological properties of the dough play a very important role in predicting the characteristics of the final product. The gluten network is the main component of the wheat flour dough, which is formed primarily from the gliadin and glutenin, which give the dough its ideal structure. Therefore, when wheat flour is fortified with any other component, an interfere of this component with the gluten network occurs, which affects its rheological properties. This can be measured by instruments of rheological properties, as well as studying the properties of resulting bread to determine the optimum addition which improve the nutritional value of wheat bread dough with optimum technological characteristics.

Chapter 9 – Pulses are good sources of protein, slow digesting starch, insoluble and soluble dietary fibers, oligosaccharides and relevant antioxidants mainly of phenolic nature. In addition, they commonly contain less than 10% lipids rich in polyunsaturated fatty acids. These nutrients and phytochemicals have been related to relevant health benefits especially in terms of prevention of chronic diseases as diabetes, cardiovascular and cancers. Chickpea is the third most produced pulse worldwide, after common beans and peas. This crop is classified as Desi or Kabuli. The first yield seeds with thick and pigmented seed coats and the second thin and cream-white seed coat pigmentation. Carbohydrates in chickpea constitute more than 60% of the total seed weight and are divided into digestible and non-digestible. The amounts of carbohydrate fractions vary according to the class (Desi or Kabuli genotypes). Starch is the main carbohydrate and its content varies from 41 to 50%. Unlike cereal grains, the chickpea starch is rich in slowly digestible (SDS) and resistant (RS) fractions. In chickpea seeds, RS is 20-30 % of total starch and the rest is comprised of SDS and rapidly digestible starch (RDS). As a result, the i*n vitro* starch digestibility and glycemic index in chickpea foods is significantly lower compared to cereal starches favoring diabetic patients because they release glucose slowly into the blood stream and leftover starch readily fermented in the hind gut. Besides RS, the non-digestible carbohydrates are oligosaccharides, insoluble dietary fiber and soluble dietary fiber rich in β-glucans. The oligosaccharides reported in chickpea are raffinose, stachyose, verbascose and ciceritol. Chickpea non-digestible carbohydrates play an important role in cholesterolemia, glucose homeostasis, cancer prevention and the modulation of microbiota composition. Hence, this chapter aims to review and analyze current information related to: 1) starch digestion and glycemic index of raw/processed chickpea flours and starch isolates, 2) undigested carbohydrates content in raw/processed chickpea and 3) health effects of chickpea digestible and undigestible carbohydrates.

Chapter 10 – Chickpea (*Cicer arietinum* L.), is one of the most important grain legumes, major source of protein and forms an integral part of diet. However, major constraints to chickpea during different stages of crop growth include pod borer *Helicoverpa armigera* (Hubner) (Lepidoptera: Noctuidae), cutworm (*Agrotis ipsilon* Hufnagel), leaf feeding caterpillar (*Spodoptere exigua* Hubner), leaf miner (*Liriomyza cicerina* Rondani) and aphid (*Aphis craccivora* Koch). Small larvae of *H. armigera* are primarily foliage feeders, while large caterpillars feed on pods and foliage, but have not any preference for flowers which show no serious yield loss incurred if pod borers are tolerated during flowering stage. During pods formation, the result is a sharp decline in chickpea production due to increasing losses of even up to 100%. In rare situations, control may be warranted during the vegetative and flowering stages, when pest pressure is extreme. If using controlling products that are only effective against small larvae, it may be necessary to apply a spray during flowering to control a population of small larvae to prevent them causing damage as large larvae during pod-set and pod-fill. A variety of predatory and parasitic insects, spiders, birds, bats, rodents and diseases attack pod borer at different stages of its lifecycle. Traditionally, preferred control options are carbamates and pyrethroids to spray as soon as the threshold has been reached as insect grows rapidly and a few days delay in spraying can result in major crop damage and increased difficulty in control. Check crop regularly from flowering onwards to identify when crop is most susceptible, familiar with the economic thresholds by using them as a basis for control decision and time to control pest most effectively. Conserve populations of predatory and parasitic insects by using selective insecticides to reduce infestations below economic threshold if natural enemies are not disrupted by broad-spectrum sprays. Manage the local population with trap cropping, cultivating to destroy overwintering pupae, destroying weed hosts within the crop and surrounding areas and adopting effective chemical control strategy. Following the implementation of integrated pest management, pod borer incidence is significantly reduced, a two to six fold increase in grain yield obtained and there is also a two to five fold increase in net income.

Chapter 11 – Chickpea (*Cicer arietinum* L.), (Fabales: Fabaceae), is one of the most popular vegetables in several areas of the world having an important source of protein particularly for vegetarian population. The chickpea seeds are vulnerable, both in the field and storage conditions to attack by several insect pests of economic importance, whereby infested grains lose their quality and viability. Five insect pests such as *Callosobruchus chinensis* (L.), *Tribolium castaneum* (H.), *Trogoderma granarium* (E.), *Rhizopertha dominica* (F.), and *Sitophilus oryzae* (L.), are mainly observed in stored chickpea. The losses due to insects activity during storage are physical loss, losses in carbohydrates and proteins, nutritional losses, and contamination of products with uric acid, fragments and fecal matter. Bruchids or pulse beetles (Coleoptera: Bruchidae) belonging to the genus *Callosobruchus* (*C. chinensis* Linnaeus, *C. maculates* Fabricius and *C. analis* Fabricius) are major storage insect pests of chickpea and cause severe damage to stored grains. Their adults and grubs feed on the grain by making a small hole and white eggs are seen on damaged seed. Owing to the bad smell created by the attack of the pest, the damaged grains do not remain fit for humans as well as animals consumptions. Contrasting to cereal grains, chickpea cannot be merely treated with protectants to prevent insect infestations in storage. Therefore, basic steps in order to control insects in warehouse or processing plants that must be followed are monitoring for infestation detection, identification of the problem and insect control measures. Insect control measures can be any one or a combination of subsequent methods like, physical and mechanical methods, chemical methods, sanitation, exclusion and Integrated Pest Management (IPM). Sun drying is the most common and preferred method not only to kill the insects, but it also reduces grain moisture to the limits of safe storage. Adoption of meticulous hygiene and aeration cooling to manage storage temperature and moisture are crucial to prevent insect damage and from downgrading of stored chickpea. Genetic resistance in grain is considered the most desirable control method since it is more cost effective and environment friendly than the use of chemicals. Most of the physical methods suggested when may not give total insects control then application of chemicals sometimes becomes essential for complete pests mortality and prevention of their growth.

Chapter 12 – Epidemiological studies support a role for diets rich in legumes and pulses in prostate cancer prevention. A protease inhibitor from soybean (Bowman-Birk Inhibitor; BBI) has been shown to exert anti-cancer properties in several *in vitro* and *in vivo* carcinogenesis model systems, including those of prostate cancer.

The anti-cancer effects of protease inhibitors from other legumes have been poorly investigated. BBI and other protease inhibitor concentrates (PICs) have previously been shown to decrease oxidative damage by inhibiting free radical production. This mechanism may account for BBI's potent anti-inflammatory effects and consequently BBI's anti-cancer properties.

The ability of PICs to modulate oxidatively induced DNA damage is not fully elucidated and the aim of the current study was to determine if a PIC from chickpea exerts anti-genotoxic effects on LNCaP and DU 145 human prostate cancer cells *in vitro*.

Cells were treated with PIC from chickpea (0.05, 0.1, 0.3, 1 and 3 mg/ml) for 24 hours and the Comet assay was used to assess the effect on DNA strand breaks induced by a H_2O_2 challenge (75 μM, 5 min, 4°C). Purified BBI was used as a reference. The genotoxicity of PIC was assessed using the Comet assay following exposure to PBS. The toxicity of chickpea PIC was assessed in LNCaP cells using the MTT assay.

Chickpea PIC (1, 3mg/ml) reduced H_2O_2-induced DNA damage in LNCaP cells by up to 15.4% ($P < 0.01$). Similar effects were apparent with BBI (0.1, 0.3 mg/ml).

Of note, the tested PIC showed no genotoxicity to LNCaP cells. Neither the chickpea PIC nor BBI exerted any significant anti-genotoxic effects in hydrogen peroxide challenged DU 145 cells, while no genotoxic activity was induced.

Pre-incubation with PIC (3 mg/ml) also significantly modulated the level of expression of genes linked with detoxifying or DNA repair properties.

In conclusion, chickpea PIC exerted anti-genotoxic effects on LNCaP cells, supporting previous findings that BBI and protease inhibitors have a protective effect on oxidative damage. Chickpea PIC may therefore play a role in prostate cancer prevention, however, further research is needed to fully understand the molecular mechanisms involved.

In: Handbook of Chickpeas
Editors: Albert T. Lund and Noah D. Schultz
ISBN: 978-1-53616-374-2

Chapter 1

CHICKPEAS: NUTRITIONAL VALUE AND HEALTH BENEFITS

Brenda Karely Bon Padilla[1], MD, Cuauhtémoc Reyes Moreno[1,2], PhD, and Roberto Gutiérrez Dorado[1,2,*], PhD

[1]Programa Regional de Posgrado en Biotecnología, Facultad de Ciencias Químico Biológicas (FCQB), Universidad Autónoma de Sinaloa (UAS), Cd Universitaria, Culiacán, Sinaloa, México

[2]Programa de Posgrado en Ciencia y Tecnología de Alimentos, FCQB, UAS, Cd Universitaria, Culiacán, Sinaloa, México

ABSTRACT

Legumes are an important source of proteins, carbohydrates, vitamins, minerals, and dietary fiber. They are an economical source of good quality proteins, being, in general, the sulfur-containing amino acids the limiting ones. Recently, legumes have gained much attention because of the health benefits associated with their consumption; even if they have antinutritional factors that have been linked to a reduction of the assimilation of nutrients, these can be reduced by processing the legumes, and it has been reported that these factors can also provide health benefits to the consumer. Chickpea is one of the most consumed legumes in the world. It is an affordable food product for people who can not get animal protein, giving them a nutritious substitute in their diet. In general, its protein has a better value than the protein of other pulses. Carbohydrates are the major component in chickpea which, being a pulse, has a low glycemic index. The dietary fiber in chickpea is mostly insoluble and has been linked with prebiotic effects, as well as with

* Corresponding Author's Email: robe399@hotmail.com.

the decrease of the risk of certain diseases. Chickpea has a higher lipids content than many other legumes, but most of them are unsaturated fatty acids, which have been related to several health benefits. Studies have shown that not only due to its nutritional composition but also because of its bioactive compounds chickpea has a protective effect against several diseases such as cancer, cardiovascular diseases, and diabetes, to mention some, as well as an antioxidative effect. Since now researchers are looking for food that besides nourishing people can help to heal certain health conditions, a particular interest has emerged for the study of the bioactive compounds of chickpea and their influence on human health. This chapter will discuss the current information about the nutritional value of chickpea, as well as the health benefits given by its components.

Keywords: chickpea, nutritional value, nutraceutical properties, bioactive compounds

INTRODUCTION

Legumes are known as a nutritious food that is high in protein, dietary fiber, resistant starch, vitamins, and minerals (Udahogora 2012). Since there are places with limited meat consumption, pulses have served to replace it (Hall et al. 2016).

Chickpea (*Cicer arietinum* L) is one of the most consumed legumes worldwide (Sánchez-Chino et al. 2015). It has its origins in Turkey in 5450 B.C. (Maesen and Vander 1972) and is cultivated in several countries like India, Pakistan, Mexico, the Mediterranean regions, among others (Asif et al. 2013). Two types of chickpeas are cultivated in the world: desi chickpea (colored seed, smaller, with a thick seed coat) and kabuli chickpea (white or beige seed color, bigger, with a thin seed coat) (Chavan et al. 1987; Wood et al. 2008). Distinctions between each type will be made through the chapter when considered necessary.

Chickpea is reported to have a better nutritional value than many other pulses (Asif et al. 2013). Chickpea is rich in protein and carbohydrates, constituting together about 80% of its total dry weight. It is also a rich source of vitamins and minerals (Jukanti et al. 2012), and relatively high in oil, but its content comprises mainly polyunsaturated fatty acids (Wood and Grusak 2007), which have been related with some positive health effects (Jandacek 2017).

Thanks to its many components, chickpea has shown to have several health benefits regarding the treatment of many diseases. Besides its mentioned content in macronutrients, it possesses a lot of bioactive compounds that have demonstrated to have beneficial health effects against some diseases like cancer, diabetes, cardiovascular diseases, obesity, among others. Therefore, chickpea seeds can be considered as a source of natural nutraceuticals. Among the nutraceutical chemical components present in these seeds there are phenolic compounds, bioactive peptides, dietary fiber, and resistant starch. Even when antinutritional factors can affect the nutritional quality of chickpea, some have also shown to exhibit favorable bioactive properties (Gupta et al. 2016). This chapter will discuss the health benefits of chickpea, given by its nutritional and bioactive compounds.

NUTRITIONAL VALUE

Chemical Composition

Chickpea chemical composition can vary depending on the type (desi or kabuli) and on the diverse growing conditions, cultivation and agronomic practices. Chickpea chemical composition is summarised in Table 1.

Protein

Chickpea has been one of the principal sources of protein in human nutrition among pulses, especially in developing countries (Wood and Grusak 2007; Gupta et al. 2016).

Chickpea has a nutritional value of protein superior of soybean, peas, lentils, cowpeas, mung beans, black gram, dry beans, faba bean and pigeonpea (Asif et al. 2013). Its protein content is of 17-31% (Milán-Carrillo et al. 2000, 2007; Reyes-Moreno et al. 2004; Wood and Grusak 2007; Garzón-Tiznado et al. 2012; Sharma et al. 2013; Domínguez-Arispuro et al. 2017). This content is similar to the one of pea and lentil (Hall et al. 2016). The major protein fraction of chickpea are globulins, representing 42-57% of the total protein content, being the principal one the legumin, followed by glutelins (10-18%), albumins (12-16%) and prolamins (0.5-3%) (Singh and Jambunathan 1982; da Silva et al. 2001; Rachwa-Rosiak et al. 2013).

A biochemical parameter used to measure the nutritional quality of foods is the protein digestibility (Ratnavathi 2019). Chickpea has been reported to have a protein digestibility *in vitro* of 65.3-92.3% (Sotelo et al. 1987; Chitra et al. 1995; El-Adawy 2002; Milán-Carrillo 2000, 2007; Angulo-Bejarano et al. 2008) and *in vivo* of 81.1-82.7% (apparent) and 83.7-84.9% (true) (Milán-Carrillo et al. 2000; Reyes-Moreno et al. 2001, 2002, 2004; Angulo-Bejarano et al. 2008; Reyes-Bastidas et al. 2010;); whereas casein, the main protein of milk, presents a protein digestibility *in vitro* of 90% and *in vivo* of 90.6% (apparent) and 93.8% (true) (Milán-Carrillo et al. 2000). A study has shown that proteins of chickpea are looser in structure (in comparison with sorghum proteins) and, therefore, easier to access by our body (Omima et al. 2010).

The amino acid content of foods is important as an indicator of the nutritional value of those (Rachwa-Rosiak et al. 2015). Chickpea is reported to have an essential amino acid score of 83-87% (Milán-Carrillo et al. 2007; Angulo-Bejarano et al. 2008). Grela et al. (2017) reported that chickpea showed the highest essential amino acid index among the studied legumes (including broad bean, common bean, pea, lupine and grasspea), with an essential amino acids content of 34-41 g/100g of protein.

Chickpea amino acid composition includes proline, glutamic acid, lysine, and aspartic acid, among others (Gupta et al. 2006). The limiting amino acids of chickpea, as well as of most legumes, are reported to be the sulfur-containing ones, methionine and cysteine (FAO/WHO 1991; Milán-Carrillo et al. 2007; Angulo-Bejarano et al. 2008; Rachwa-Rosiak et al. 2015); however, a more detailed study (Milán-Noris 2017) pointed out that the sulfur-containing amino acids are the limiting ones of kabuli chickpea, whereas the limiting amino acid of desi chickpea can be either tryptophan or threonine.

Table 1. Chickpea chemical composition

Component (%)	Desi	References	Kabuli	References
Protein	16.7-30.57	Wood and Grusak 2007; Sharma et al. 2013; Garzón-Tiznado et al. 2012; Domínguez-Arispuro et al. 2017	12.6-31.0	Reyes-Moreno et al. 2004; Milán-Carrillo et al. 2000, 2007; Wood and Grusak 2007; Sharma et al. 2013
Lipids	2.60-7.42	Wood and Grusak 2007; Sharma et al. 2013; Garzón-Tiznado et al. 2012; Domínguez-Arispuro et al. 2017	3.1-8.83	Reyes-Moreno et al. 2004; Milán-Carrillo et al. 2000, 2007; Wood and Grusak 2007; Sharma et al. 2013
Ash	2.04-4.2	Wood and Grusak 2007; Sharma et al. 2013; Garzón-Tiznado et al. 2012;	2.0-3.9	Reyes-Moreno et al. 2004; Milán-Carrillo et al. 2000, 2007; Wood and Grusak 2007; Sharma et al. 2013
Carbohydrates	50.4-70.85	Wood and Grusak 2007; Garzón-Tiznado et al. 2012	54.27-71.14	Reyes-Moreno et al. 2004; Milán-Carrillo et al. 2000, 2007; Wood and Grusak 2007; Sharma et al. 2013
Total starch	32.0-56.3	Veena et al. 1995; Wood and Grusak 2007; Demirkesen-Bicak et al. 2018	30.0-57.2	Wood and Grusak 2007; Aguilera et al. 2009
Resistant starch	3.39-10.3	Veena et al. 1995; Wood and Grusak 2007; Demirkesen-Bicak et al. 2018	0.31-19.3	Wood and Grusak 2007; Aguilera et al. 2009; Sánchez-Magaña et al. 2014
Dietary fiber	18.4-29.29	Wood and Grusak 2007; Domínguez-Arispuro et al. 2017	10.6-24.64	Wood and Grusak 2007; Sánchez-Magaña et al. 2014
SDF	0.00-1.00	Wood and Grusak 2007; Domínguez-Arispuro et al. 2017	~5.60	Sánchez-Magaña et al. 2014
IDF	13.9-28.29	Wood and Grusak 2007; Domínguez-Arispuro et al. 2017	~19.04	Sánchez-Magaña et al. 2014

SDF: Soluble dietary fiber; IDF: Insoluble dietary fiber.

Lipids

Legumes can be classified according to their lipid content in two groups: low lipid content legumes (1-6%), where the chickpea is included along with other legumes such as lentil (1-2%) and bean (2-3%); and high lipid content legumes, such as soybeans (18%) and peanuts (50%) (Sánchez-Chino et al. 2015; Hall et al. 2016).

Chickpea lipid content varies from 2.6 to 8.8% (Milán-Carrillo et al. 2000, 2007; Reyes-Moreno et al. 2004; Wood and Grusak 2007; Garzón-Tiznado et al. 2012; Sharma et al. 2013; Domínguez-Arispuro et al. 2017). This content comprises mostly polyunsaturated fatty acids (62-67%), followed by monounsaturated (19-26%) and saturated fatty acids (12-14%). The main fatty acid in chickpea is the linoleic acid (LA), an essential fatty acid that represents 46-62% in desi chickpea oil and 16-56% in kabuli chickpea oil, followed by the oleic acid (OA), representing 18-23% in desi chickpea oil and 19-32% in kabuli chickpea oil (Wood and Grusak 2007). Consumption of LA has been related to a decrease in the risk of atherosclerosis, cardiovascular diseases and inflammation (Jandacek 2017). OA also has shown to possess antiinflammatory effects and has been related to a lower risk of developing cancer and cardiovascular diseases (Sales-Campos et al. 2013).

Chickpea oil also contains tocopherols, tocotrienols and sterols (Akihisa et al. 1992). Chickpea contains a higher amount of tocopherols (1.78 mg/100g) than most other pulses (0.26-0.82 mg/100g) (Kalogeropoulos et al. 2010). Tocopherols exhibit a great antioxidant activity; this helps the consumer due to the health benefits it provides and it can aid against oxidative rancidity, improving chickpea's shelf life. β-sitosterol is the main sterol in chickpea, representing about 73-76% in chickpea oil. β-sitosterol has shown to provide health benefits; some of them are discussed in the following sections (Jukanti et al. 2012).

Carbohydrates

As most legumes, carbohydrates are the major nutritional component in chickpea, representing about 58-73% of it (Milán-Carrillo et al. 2000, 2007; Reyes-Moreno et al. 2004; Wood and Grusak 2007; Garzón-Tiznado et al. 2012; Sharma et al. 2013), being this content a little higher than the content of pea and lentil (57-60%) (Venkidasamy et al. 2019). Chickpea carbohydrates are of low glycemic index, so after its consumption, the blood glucose rise produced is low (Clark et al. 2018).

Carbohydrates can be either available or unavailable. The available ones are digested in the small intestine by enzymes; these are the monosaccharides and the disaccharides. The unavailable ones are not digested in the small intestine; these are the oligosaccharides, resistant starch, pectins, fiber, among others (Jukanti et al. 2012).

It is reported (Sánchez-Mata et al. 1998) that chickpea has a monosaccharides concentration of 0.05% galactose, 0.11% ribose, 0.25% fructoses, and 0.7% glucose. The most abundant disaccharides on chickpea are maltose (0.6%) and sucrose (1-2%). Chickpea contains 0.4-2.8% and 1.2-3.9% of oligosaccharides on types desi and kabuli, respectively, being the most important ones: raffinose, stachyose, verbascose and ciceritol, being present in quantities up to 2.2%, 6.5%, 0.4% and 3.5%, respectively (Wood and Grusak 2007).

Starch content in pulses can go from as little as 2% in lupin to as high as 58% in mung bean (Hall et al. 2016). Total starch content in chickpea varies from 32.0 to 57.2%; 34-36.7% of it is available starch and up to 19.3% is resistant starch (Veena et al. 1995;

Wood and Grusak 2007; Aguilera et al. 2009; Sánchez-Magaña et al. 2014; Demirkesen-Bicak et al. 2018), which acts like dietary fiber, since it can not be digested and it reaches the large intestine performing the beneficial effects the dietary fiber does (Aguilera et al. 2009; Demirkesen-Bicak et al. 2018).

Dietary Fiber

Dietary fiber includes polysaccharides, lignin, among others (Jukanti et al. 2012). In legumes, most of the fiber is present in the seed coat (Singh 1984), thereby, desi chickpea has a higher dietary fiber content (18-29%) (Wood and Grusak 2007; Domínguez-Arispuro et al. 2017) than kabuli chickpea (10-24%) (Wood and Grusak 2007; Sánchez-Magaña et al. 2014). Chickpea dietary fiber content is higher than the content in legumes such as peas (10.7-14.8%) and lentils (13.1-14.7%) (Hall et al. 2016).

Dietary fiber can be classified in soluble and insoluble (Anderson et al. 2009). In pulses, most of the dietary fiber is insoluble (85-93% of the total dietary fiber) (Hall et al. 2016). The soluble fiber is fermented in the colon by human microflora, producing short-chain fatty acids (SCFA), like acetate, propionate, and butyrate, and playing a prebiotic effect (Anderson et al. 2009; Slavin 2013); it helps to decrease serum cholesterol and to reduce the risk of having a heart attack and of developing colon cancer (Hefnawy 2011). The insoluble fiber has a bulking action, helping to the prevention or the relieving of constipation; it can be fermented just to a limited extent (Anderson et al. 2009).

Dietary fiber regular consumption (14g/1000 kcal) has been related to the reduction of the incidence of diseases like diabetes, obesity, heart diseases, and gastrointestinal disorders (Anderson et al. 2009).

Vitamins and Minerals

Vitamins are required in small quantities that are reached through a diet that ideally incorporates all food groups (Jukanti et al. 2012). Chickpea contains several water-soluble vitamins, such as B-complex vitamins, including thiamin (0.028-0.40 mg/100g), riboflavin (0.15-0.26 mg/100g), niacin (1.22-1.72 mg/100g), pantothenic acid (1.02-1.72 mg/100g), pyridoxine (0.30-0.55 mg/100g), folic acid (150.0-557.0 mg/100g), cyanocobalamin (0.535 mg/100g) and biotin, as well as 1.34-6.0 mg/100g of ascorbic acid (vitamin C) (Chavan et al. 1987; Jukanti et al. 2012). It also contains liposoluble vitamins, like retinol (vitamin A), 115.4 mg/100g of calciferol (vitamin D), 0.820-22.0 mg/100g of α-tocopherol (vitamin E) and 9.0-120.0 mg/100g of phylloquinone (vitamin K) (Jukanti et al. 2012). Vitamins are necessary for energy metabolism and metabolic processes, and their deficit provokes the development of diseases such as pellagra, scurvy, rickets, and beriberi (Huskisson et al. 2007).

Minerals, as well as vitamins, are required for metabolic processes, and their deficiency comprises a wide range of health conditions (FAO/WHO 2001). Chickpea is rich in minerals like calcium (93-213 mg/100g), magnesium (176 mg/100g), potassium

(988-1333 mg/100g), phosphorus (268-429 mg/100g), iron (3.1-7.8 mg/100g), zinc (4.1mg/100g), among others (Jukanti et al. 2012).

ANTINUTRITIONAL FACTORS

Antinutritional factors are secondary metabolites of plants which exhibit a high biological activity. They are named "antinutritional factors" since they can reduce the assimilation of some nutrients or cause digestion problems and even be toxic for individuals (Muzquiz 2000; Gemede and Ratta 2014). The majority of them, being heat-labile ones, can be removed by diverse thermal treatments, while other kinds of antinutritional factors can be removed by processes like soaking, dehulling and bioprocessing (germination, fermentation) (Patterson et al. 2016; Venkidasamy et al. 2019).

Despite the troubles caused by their consumption, it is now known the health benefits that antinutritional factors possess (Muzquiz 2000). Next, we will discuss some of the principal antinutritional factors that are present in chickpea seeds.

Protein Antinutritional Factors

Protein antinutritional factors are heat labile due to their nature (Giri and Kachole 2004; Gemede and Ratta 2014); therefore, thermal treatments can cause their inactivation or elimination, such as boiling, pressure cooking, extrusion cooking, microwave heating and infrared heating (Patterson et al. 2017). It has been observed that soaking chickpeas in water produces a 12% reduction in trypsin inhibitor activity (Frias et al. 2000). Bioprocesses like germination (Alonso et al. 2000; Shimelis and Rakshit 2007) and fermentation (Cuadrado et al. 2002; Martin–Cabrejas et al. 2004; Abu-Salem and Abou-Arab 2011) have also been used effectively to reduce protein antinutritional factors in legumes (Patterson et al. 2017).

Protease Inhibitors

Protease inhibitors can suppress the activity of the proteolytic enzymes trypsin and chymotrypsin, binding to them and interfering with the digestion of proteins. Protease inhibitors are resistant to pepsin (a digestive enzyme), and to the acidic pH of the stomach (Jukanti et al. 2012).

Protease inhibitors belong to two families: the Kunitz and the Bowman-Birk family. The first one can inhibit trypsin activity and consist of a single chain of polypeptides with two disulfide bridges; the second one can inhibit trypsin and chymotrypsin and also consist of a single chain of polypeptides, but with eleven disulfide bridges (Jukanti et al.

2012; Gemede and Ratta 2014). These inhibitors have shown to cause pancreatic hypertrophy and growth inhibition, as well as hypersecretion of digestive enzymes (Muzquiz and Wood 2017).

The content of trypsin inhibitor in chickpea is higher (11.90 Trypsin inhibited unit (TIU)/mg protein) (El-Adawy 2002) than in other Indian pulses, but lower than in soybean (34.7-122.6 TIU/mg protein) (Mohan et al. 2016).

Despite the antinutritional effects of protease inhibitors, the health-promoting value of them has been recognized, especially as anticarcinogenic agents (Muzquiz and Wood 2007; Gemede and Ratta 2014).

Lectins

Lectins are glycoproteins that can bind to carbohydrates reversibly on the surface of erythrocytes, agglutinating (they are also called "haemagglutinins"). Lectins can interfere with nutrient absorption because they can interact with the enterocytes of the intestinal mucosa, causing lesions. Different types of lectins exhibit varying degrees of toxicity (Gemede and Ratta 2014).

Grant et al. (1983) studied lectins of 15 legumes and found that the ones of chickpea showed the lowest haemagglutinin activities, and were non-toxic. Chickpea lectin is reported to have an haemagglutinin activity of 2.73-6.22 hemagglutinin unit (HU)/mg sample (El-Adawy 2002; Shi et al. 2018), while other legumes, like lentil, bean and soybean have greater values (11, 88 and 693 HU/mg, respectively) (Shi et al. 2018).

Even when lectins have shown to be toxic, they also have proved to have some health benefits, like the stimulation of the gut function, reduction of the tumor growth and mitigation of obesity (Muzquiz and Wood 2007).

Non-Protein Antinutritional Factors

Phytic Acid

Phytic acid is the principal source of phosphorus in legumes (Urbano et al. 2000). In chickpea, the content of phytic acid is of 1.21% (El-Adawy 2002), which is a minor content than in other pulses, like white and black beans (7.8% and 9.11%) and black gram (11.24-11.47%) (Mohan et al. 2016). Phytic acid can work in a wide pH region as a negatively charged ion, binding to important cations and forming insoluble complexes, making them undigestable (Jukanti et al. 2012; Gemede and Ratta 2014). It can also form complexes with proteins and inhibit the action of digestive enzymes such as α-amylase, lipase, pepsin, trypsin and chymotrypsin, reducing the digestibility of nutrients; as well as bind with starch through phosphate linkages (Muzquiz and Wood 2007).

Phytic acid is soluble in water, so it can be removed by soaking, which is a step in the preparation of legumes for their consumption (Patterson et al. 2017). Infrared heating

(Arntfield et al. 2001) and germination (Alajaji and El-Adawy 2006; Shimelis and Rakshit 2007; Bilgicli et al. 2008; Bains et al. 2014) also have proved to reduce the phytic acid content in legumes (Patterson et al. 2017). Lactic acid fermentation (Chitra et al. 1996) and solid state fermentation (Abu-Salem and Abou-Arab 2011; Reyes-Moreno et al. 2000) help to reduce the phytic acid content in legumes thanks to the action of the enzyme phytase, produced by the microorganisms used to carried on the fermentation processes (Patterson et al. 2017).

Saponins

Saponins give the pulses a bitter taste (Jukanti et al. 2012), their content in chickpea is higher (3.6 g/kg) than in other legumes such as peas (2.5 g/kg) and black gram (2.3 g/kg), but lower than in most other pulses (Mohan et al. 2016). Their name is given due to their ability to form foams when shaken in water. They are a diverse group of compounds, consisting of nonpolar aglycones joined with one or more moieties of simple sugars (Muzquiz and Wood 2007; Gemede and Ratta 2014).

Saponins in legumes can be reduced by soaking them in water or by bioprocesses like germination (Sharma and Sehgal 1992 a, b; Shimelis and Rakshit 2007; Gemede and Ratta 2014).

The huge diversity of saponins makes them have a broad range of properties/toxic effects. Saponins are known to reduce the nutrient uptake due to the inhibition of digestive enzymes activities; also, some of them possess hemolytic activity (Gemede and Ratta 2014). Even when they have some antinutritional effects, saponins have proved to possess hypocholesterolemic, anticarcinogenic, immunostimulatory and antimicrobial properties, and they can also prevent cardiovascular diseases, peptic ulcer, and osteoporosis (Muzquiz and Wood 2007; Gemede and Ratta 2014). Saponins in chickpea, being mainly soyasaponin IV (Ruiz et al. 1996) and soyasaponin βg (Milán-Noris 2017), have been reported to prevent lipid peroxidation of DNA and proteins; they also possess an antifungal activity and can inhibit human immunodeficiency virus *in vivo* (Thompson 1993; Muzquiz and Wood 2007).

α-Amylase Inhibitors

α-amylase is an enzyme secreted by the pancreas and salivary cells. This enzyme acts hydrolyzing starch and glycogen, cleaving α-D-(1-4) glycosidic bonds, contributing to their digestion and yielding oligosaccharides with different lengths with α-limit dextrins, initiating the starch breakdown (de Sales et al. 2012).

Since carbohydrates are an energy source, the inhibition of their metabolism to glucose (like the inhibition of key enzymes, such as α-amylase) can be seen as something detrimental. However, this inhibition has been used as a strategy for the development of drugs against diabetes, such as acarbose, a competitive inhibitor of the enzyme (de Sales et al. 2012), minimizing the digestion of starch and avoiding its complete degradation and absorption (Tormo et al. 2004). The inhibition of α-amylase has proved to help to maintain the glucose levels to a lower level on diabetic patients (Da Ros et al. 2005; Preuss 2009).

Currently, α-amylase inhibitors from natural sources are being studied due to their hypoglycemic activity, being the inhibitors from common bean the ones that have gained more attention (Muzquiz and Wood 2007). Nevertheless, other legumes, like chickpea, have also demonstrated to possess compounds with good values of α-amylase inhibition activity (Singh et al. 1982; Sreerama et al. 2012; Sánchez-Magaña et al. 2014), some of which will be discussed ahead.

Oligosaccharides

Pulses contain a very high concentration of oligosaccharides. Oligosaccharides cannot be absorbed or hydrolyzed by the digestive system of humans, so they are fermented by colonic microflora, producing gases (Jukanti et al. 2012). The most important oligosaccharides in chickpea are α-galactosides, representing about 62% of the total sugar content (Sánchez-Mata et al. 1998). Even when they can cause indigestion, oligosaccharides can act as prebiotics, since they pass into the lower gut mostly undigested, acting as dietary fiber (Mitsuoka 1996; Martínez-Villaluenga et al. 2008; Zhang et al. 2017).

Phenolic Compounds

Phenolic compounds are secondary metabolites of plants; they present a wide variety of structures and possess bioactive properties (Rice-Evans et al. 1997; Cheynier 2012). They derive from the shikimate that comes either from the pathway of the phenylpropanoid or the polyketide and have one or more phenolic rings with one or more hydroxyl substituents (Cheynier 2012; Vuolo et al. 2019).

Phenolic compounds are classified in many classes; the most important ones found in the human diet are phenolic acids, flavonoids and tannins (Minatel et al. 2016; Vuolo et al. 2019). Phenolic acids possess at least one aromatic ring with one hydrogen substituted by a hydroxyl group. They can consist of hydroxybenzoic acids (general structure C6-C1) or hydroxycinnamic acids (general structure C6-C3) (Vuolo et al. 2019).

The total phenolic content (TPC) in desi chickpea is 0.746-1.6 mg gallic acid equivalents (GAE)/g (Garzón-Tiznado et al. 2012, 2013; Heiras-Palazuelos et al. 2012; Domínguez-Arispuro et al. 2017) and 0.57 mg GAE/g in kabuli chickpea (Heiras-Palazuelos et al. 2012). The main phenolic acids in chickpea are sinapic acid, gallic acid, ρ-hydroxybenzoic acid, ferulic acid and caffeic acid (Kalogeropoulos et al. 2010; Domínguez-Arispuro et al. 2017; Grela et al. 2017; Quintero-Soto et al. 2018).

Flavonoids are polyphenols with a general structure C6-C3-C6, in which the two aromatic rings, named A and B (of phenolic nature) are joined by a heterocyclic ring (C) of three carbons (Tsao 2010; Birt and Jeffery 2013). The diversity of compounds within the flavonoids classes are due to the variations in the C-ring, resulting in subgroups such as flavonols, flavones, flavanones, flavanols (catechins), isoflavones, flavanonols, and anthocyanidins (Birt and Jeffery 2013).

The total flavonoid content in desi chickpea is 0.553-1.890 mg catechin equivalents (CAE)/g (Garzón-Tiznado et al. 2012, 2013; Heiras-Palazuelos et al. 2012) and in kabuli chickpea is 2.040 mg CAE/g (Heiras-Palazuelos et al. 2012). Biochanin A, as well as its glycosides, have been reported as the major flavonoids in chickpea (Aguilera et al. 2009; Wu et al. 2012; Gao et al. 2017; Milán-Noris et al. 2018). It also contains significant amounts of genistein hexoside, catechin, daidzein, formononetin and quercetin (Gao et al. 2017; Grela et al. 2017; Milán-Noris et al. 2018; Quintero-Soto et al. 2018). Isoflavones are known to possess activity against breast cancer and cardiovascular diseases (Venkidasamy et al. 2019).

Phenolic compounds give the foods an astringent flavor and can reduce the availability of some minerals, like zinc (Shahidi 1997). Polyphenols can interact with dietary proteins forming complexes and provoking protein precipitation, causing detrimental effects in human health. Digestive enzymes, such as proteases (pepsin, trypsin, chymotrypsin, elastase) have been successfully inhibited by polyphenols (Velickovic and Stanic-Vucinic 2018). Other digestive enzymes, such as lipase, α-amylase, and α-glucosidase also have shown to be affected by phenolic compounds, with the consequent decrease of the uptake of nutrients. The *in vitro* protein digestibility also has been observed to be reduced, as an effect of the ingestion of a polyphenol compound rich food (Martínez-González et al. 2017; Velickovic and Stanic-Vucinic 2018).

Since the seed coat of legumes has a higher proportion of phenolic compounds, dehulling them provokes a considerable decrease of the TPC (Patterson et al. 2017), as observed when dehulling kabuli and cream-colored desi chickpeas, which suffered a 11-16% reduction of TPC, whereas black and red desi chickpeas presented a 75% loss (Sotelo et al. 1987). Soaking in water also has proved to be effective in reducing the TPC of chickpeas, being observed a more significant loss in desi chickpea (30-70%) than in kabuli chickpea (0-45%) after the process (Khandelwal et al. 2007; Han and Baik 2008; Segev et al. 2010; Patterson et al. 2017).

As for the effect of processes like extrusion and germination, there have been several reports. There are studies where extrusion cooking has shown to reduce polyphenols content in pulses (Melcion and van der Poel 1993), as well as the germination process (Alonso et al. 1998; Khandelwal et al. 2007). However, both processes, as well as fermentation, carried out properly, are recently being used to enhance the TPC. In chickpea, extrusion (Garzón-Tiznado et al. 2013), germination (Domínguez-Arispuro et al. 2017; Milán-Noris et al. 2018) and fermentation (Sánchez-Magaña et al. 2014) had been used successfully in desi and kabuli chickpea to obtain increases of 9%, 97-861% and 278% of TPC, respectively.

The strategy to use processes to increase the TPC of legumes, despite this being an antinutritional factor, is because even when they have some antinutritional properties, the beneficial effects they possess are even greater. They are known to be potent antioxidants that prevent biomolecules (proteins, lipids, DNA) from undergoing oxidative damage (Vuolo et al. 2019). They can donate an electron or hydrogen to neutralize free radicals, or chelate metal cations (Tsao 2010; Minatel et al. 2016). Mainly thanks to this feature, phenolic compounds are believed to have properties against many diseases (Velickovic and Stanic-Vucinic 2018), as a diet rich in these compounds has demonstrated to reduce the incidence in chronic degenerative diseases such as cardiovascular diseases, arteriosclerosis, diabetes, cancer, among others (Minatel et al. 2016), some of which will be discussed ahead.

Tannins

Tannins are polyphenolic secondary metabolites that contain groups which can bind to proteins and other macromolecules (Muzquiz et al. 2000). They are categorized into: (1) hydrolyzable and (2) condensed (also named proanthocyanidins) according to their capacity to fractionate hydrolytically (Muzquiz and Wood 2007). Due to their ability to form complexes, they reduce the digestibility of proteins; they have also shown to inhibit the absorption of iron (Jukanti et al. 2012; Gemede and Ratta 2014).

Tannins are heat stable; in legumes, they are maily contained in the seed coat (Muzquiz and Wood 2007; Gemede and Ratta 2014), so removing it could be a strategy to decrease the tannin content of legumes, as discussed before (Patterson et al. 2017). Irradiation has shown to be effective in reducing the tannin content in legumes, showing a reduction of 28.1% at a dose of 10 kGy (El-Niely 2007). Solid state fermentation (tempeh production) also has proved to be a useful strategy for tannins reduction in chickpea; this is specially attributed to the soaking, dehulling and discarding of the cooking water steps of the process (Reyes-Moreno et al. 2000).

In particular, hydrolyzable tannins are the ones that give the seed coat its color, whereas condensed tannins are reported to be the responsible for forming complexes with dietary protein and iron (Muzquiz and Wood 2007).

Chickpea has a tannin content of 3.78-4.85 mg/g, of which only <0.04-0.1% are condensed tannins (El-Adawy 2002; Muzquiz and Wood 2007). These values are lower than the ones of other pulses such as lentil (8.72 mg/g) and blue lupine (5.02 mg/g) (Grela et al. 2017), and did not have an impact on iron absorption when compared with tannins of different sources (tea, coffee and wine) (Muzquiz and Wood 2007).

HEALTH BENEFITS

Antioxidant Activity

Legumes have a variety of phytochemicals (phenolic acids, flavonoids, carotenoids, phytosterols, phytohaemagglutinins) that perform through different mechanisms of action, exhibiting various bioactive activities, such as anticarcinogenic, antiangiogenic, antimutagenic, immunostimulation, modulation of lipid metabolism, inhibition of the oxidation of lipoprotein and suppression of platelet aggregation. All these activities have been associated with the antioxidant properties of the phytochemicals. In plants, the phenolic compounds that exhibit antioxidant activity are flavonoids, phenolic acids, tannins, stilbenoids, and lignans (Grela et al. 2017).

Polyphenols can convert the potentially damaging reactive oxygen species into nontoxic species thanks to their capacity to donate H-atoms, decrease oxygen concentration, intercept singlet oxygen, prevent the first-chain initiation by scavenging initial radicals, bind metal ions catalyst and decompose primary products of oxidation to nonradical species (Zia-ul-Haq et al. 2008).

Several authors have reported the antioxidant activity by polyphenolic compounds of different kinds of chickpeas. It is observed that the genotypes with the higher activities were the desi ones. It has been reported that when measured by DPPH, the total antioxidant activity of desi chickpea is 0.52-16.50 μmol TE/g, and 0.93-2.43 for kabuli chickpea. By the FRAP method, there is a reported antioxidant activity of 0.41-11.81 μmol TE/g for desi chickpea, and 0.46-1.72 μmol TE/g for kabuli chickpea (Quintero-Soto et al. 2018). By the ABTS method, the total antioxidant activity is 3.26-58.71 μmol TE/g for desi chickpea (Domínguez-Arispuro et al. 2017; Quintero-Soto et al. 2018), whereas for kabuli chickpea is 2.78-32.98 μmol TE/g (Sánchez-Magaña et al. 2014; Quintero-Soto et al. 2018). When measured by the ORAC method, the total antioxidant activity of TPC for desi chickpea is 43.9-66.2 μmol TE/g (Garzón-Tiznado et al. 2012, 2013; Heiras-Palazuelos et al. 2012) and 44.43 μmol TE/g for kabuli chickpea (Heiras-Palazuelos et al. 2012). These values are similar to the ones reported for common bean (55.09 μmol TE/g by ABTS and 64.02 μmol TE/g by ORAC) (Guzmán-Uriarte et al. 2013).

Bioactive peptides are present in food in their natural form or can be generated after the digestion of food (by hydrolysis of proteins in the digestive tract) or by controlled protein hydrolysis with enzymes. These peptides commonly contain a few amino acid residues; they cross the gastrointestinal epithelial barrier and reach the blood vessels, exhibiting their beneficial effect in the target organs (Yust et al. 2003).

Bioactive peptides perform various nutraceutical functions, such as anticarcinogenic/anticancer, antioxidant, antihypertensive, antiobesity, and antimicrobial (Xue et al. 2017).

Antioxidant peptides typically contain 3-20 amino acid residues, and the activity of the peptide is influenced by its amino acid parts, its sequence and structure; that is determined by the protein source, the enzyme(s) and the hydrolysis process used to obtain the peptides (Pihlanto-Leppälä 2000; Kou et al. 2013). The presence of the amino acids histidine, tryptophan and phenylalanine in the peptide sequence has demonstrated to have a hydrogen donor function (Torres-Fuentes et al. 2015).

Medina-Godoy et al. (2012) obtained chickpea peptides after hydrolysis (either with alcalase, papain or pancreatin) and those peptides showed antioxidant activity by the DPPH method of 18.17-46.76 μmol TE/μg, and by the ABTS method of 27.73-95.61 μmol TE/μg. Torres-Fuentes et al. (2015) found the peptides sequences ALEPDHR, TETWNPNHPEL, FVPH and SAEHGSLH in chickpea hydrolysates, which displayed antioxidant activity and belonged to the legumin fraction. The peptide RQSHFANAQP (named CPe-III), purified from chickpea albumin hydrolysis also seemed to show antioxidant effects (Kou et al. 2013), and the antioxidant peptide NRYHE isolated from chickpea hydrolysate was found to exhibit a more significant inhibition on lipid peroxidation than α-tocopherol (Zhang et al. 2011).

Polysaccharides also have been reported to exhibit antioxidant activity. Although the mechanism of action has not been fully elucidated, it has been related with the presence of pectins and pectic acids and their reduction power and free radical scavenging activity (Urias-Orona et al. 2010). Myint et al. (2017) reported that rats fed with chickpea husk had an improved antioxidative status due to the polyphenolic components contained in the husk. Pectin obtained from chickpea husk which mainly had galacturonic acid showed a dose-dependent free radical scavenging activity, exhibiting a 29% scavenging rate on DPPH radicals at 1.0 mg/mL (Urias-Orona et al. 2010).

Ye et al. (2016) reported that three chickpea polysaccharides showed ABTS radical scavenging activities (at a concentration of 2.5 mg/mL) between 57.96 and 93.47%. Those activities were higher than the activity of water-soluble polysaccharides from wheat bran (43.8%), at a concentration of 5 mg/mL (Yan et al. 2015).

Dyslipidemias and Obesity

It has been observed that people with a high intake of soluble fiber manifest a reduction in total serum cholesterol and low-density lipoprotein cholesterol (LDL-C). High fiber intake can also provide benefits against obesity because it can aid to reach a state of satiety faster that last longer due to its bulking action and because it takes longer to digest (Bouchenak and Lamri-Senhadji 2013).

Fiber, soluble and insoluble, can bind to bile acids in the intestine, preventing them from being reabsorbed. Thereby, bile acids production increase with a decrement of hepatic cholesterol, so there is an uptake of blood cholesterol and a decrease of serum cholesterol levels. On the other hand, SCFA production by soluble fiber fermentation by colon microflora helps to reduce the synthesis of cholesterol in the liver (Abeysekara et al. 2012).

It was reported (Pittaway et al. 2008) that regular consumption of chickpeas (728g/week minimum) led to a lower serum total cholesterol and low-density lipoprotein cholesterol (7.7 mg/dL and 7.3 mg/dL, respectively) and that chickpea dietary fiber had the most significant single effect.

In all the experiments where rats were fed with diverse diets, testing the difference between a chickpea/chickpea bran diet and either a basic purified diet (Hamid and Kalsoom 2017), a wheat and corn bran (Muzzafar et al. 1989) or a purified diet with cellulose/corn starch (Myint et al. 2017), it was reported that the chickpea/chickpea bran diet was better to reduce the serum cholesterol level and the liver and heart cholesterol concentrations, and to increase the cecal levels of butyrate and total SCFA.

The ratio of monounsaturated fatty acids (MUFAs) to saturated fatty acids (SFAs), plays an important role in obesity, which is the result of a high-fat diet (Shinohara et al. 2016). Chickpea has a higher amount of fat compared with many other pulses, but most of it is unsaturated fat. LA and OA, two unsaturated fatty acids, constitute about 50–60% of chickpea fat. Intake of unsaturated fatty acids has beneficial effects on serum lipids (Jukanti et al. 2012). Chickpea also possesses phytosterols, being the most important one the β-sitosterol, which helps to decrease serum cholesterol levels and the incidence of cardiovascular diseases (Jukanti et al. 2012).

Pittaway et al. (2006) reported a higher reduction of total serum cholesterol (-3.9%) and LDL-C (-4.6%) levels in humans after a chickpea-supplemented diet in comparison with a wheat supplemented diet; this was attributed to the differences in the polyunsaturated fatty acids and the dietary fiber content of each diet.

Chickpea contains ferulic and ρ-coumaric acids. These have been shown to reduce blood lipid levels in rats (Sharma 1980, 1984).

It has been reported (Nakagawa et al. 1976, Sharma 1979) that biochanin A and formononetin, isolated from chickpea, possess hypolipidemic properties in rats. Chickpea ethanol extracts treated fibroblast 3T3-L1 cells showed a reduction, as well as inactivation of enzymes participating in lipogenesis, having a suppressive effect on the storage of fat, with the consequent decrease in the lipid content of them (Shinohara et al. 2016).

The intake of food rich in saponins has been reported to reduce plasma cholesterol by 16–24% (Thompson 1993). This effect may be due to their capacity to inhibit pancreatic lipase activity, which also could help against obesity, or by their ability to bind to dietary cholesterol or bile acids, increasing their excretion through feces (Jukanti et al. 2012).

To prevent or treat diseases related with conditions such as overweight and obesity, the development of natural inhibitors of digestive enzymes has been a strategy since they can inhibit the hydrolysis and therefore the absorption of carbohydrates and lipids from the diet. Pancreatic lipase is a key enzyme necessary for the absorption of dietary triglycerides in the small intestine. Chickpea protein hydrolysates (with pepsin, chyme and pancreatin) showed to inhibit lipase with an IC_{50} value of 9.74±1.09 μg/mL (Ercan and Nehir El 2016).

Yang et al. (2007) reported that a chickpea diet normalized body weight, adipose tissue and hepatic lipase activity of rats; this was attributed to a synergistic effect of the isoflavones and the unsaturated fatty acids in chickpeas composition.

Chickpea peptides were obtained by hydrolysis with alcalase, and the sequence "VFVRN" was identified. The peptides inhibited the activities of enzymes such as fatty acid synthetase (FAS) and HMG-CoA reductase (HMGR) and were found to decrease the body weight, adipose tissue size, total cholesterol (TC), total triglyceride (TG), LDL-C, atherogenic index (AI) in serum and liver TC and TG of mice. On the other side, there was an increase in serum high-density lipoprotein cholesterol (HDL-C), bile acids in liver and feces, fecal TC and TG (Shi et al. 2019).

Similar results were obtained when hydrolyzing with albumin (Xue et al. 2012), where the peptide RQSHFANAQP (named CPe-III) was discovered (Xue et al. 2017). Since the peptide also had antioxidant activity, it could inhibit the oxidation of LDL-C, which is the principal cause of arteriosclerosis (Xue et al. 2012, 2017).

Hypertension and Cardiovascular Diseases

Hypertension is an important risk factor for the development of cardiovascular diseases (CVD). Angiotensin I-converting enzyme (ACE; peptidyldipeptide hydrolase, EC 3.4.15.1), hydrolyzes the decapeptide angiotensin-I to obtain angiotensin-II (an octapeptide), a strong vasoconstrictor; its inhibition has become a strategy to control high blood pressure (Barbana and Boye 2010).

Polyunsaturated fatty acids, such as LA, are important for prostaglandins (PG) production. PG have a function in blood pressure lowering control and the smoothing of muscle construction. Phytosterols, such as β-sitosterol, can also help to reduce blood pressure (Jukanti et al. 2012).

The IC_{50} value can be defined as the concentration required to inhibit 50% of the activity of a particular enzyme. Therefore, smaller values mean greater potential (Guzmán-Uriarte et al. 2013; Rochín-Medina et al. 2015).

Protein hydrolysates obtained from chickpea by digestion with different enzymes (such as pepsin, trypsin, chymotrypsin, alcalase, flavorzyme, papain, or a combination of those), have shown to possess an ACE inhibitory activity of IC_{50} = 140-673 μg/mL. Some authors (Yust et al. 2003; Medina-Godoy et al. 2012) have also hydrolyzed the globulin/legumin fraction of chickpea with different enzymes (including alcalase, papain and pancreatin) and reported an IC_{50} of 0.016 μg/mL-0.18 mg/mL, as well as some peptides with antihypertensive activity (IC_{50} = 0.011-0.021 mg/mL), being MDFLI the one that showed the highest inhibition. Pedroche et al. (2002) purified four peptides (after digestion with alcalase and flavourzyme) with ACE-inhibitory activity and discovered that the inhibitory mechanism was either competitive (IC_{50} = 0.108 and 0.117 mg/mL) or uncompetitive (IC_{50} = 0.103 and 0.105 mg/mL). The IC_{50} value of chickpea hydrolysate (16.340 μg/mL, obtained by pancreatin) (Medina-Godoy et al. 2012) is better than the ones reported for common bean (Guzmán-Uriarte et al. 2013) and black bean (Rochín-Medina et al. 2015) (79.22 and 93.53 μg/mL, respectively obtained by sequential digestion by pepsin and pancreatin), whereas the IC_{50} for a common drug for the treatment of hypertension, captopril, is 0.00193-0.00206 μg/mL (Chen et al. 2013).

The differences in the ACE-inhibitory activity in the different fractions in chickpea can be due to different causes, such as the sample and the protein extraction procedure, the enzyme(s) utilized to obtain the protein hydrolysates, and the variables in the process of digestion (pH, time, temperature, substrate/enzyme ratio), which causes a difference in the peptide mixture composition or the method used for the determination of the ACE-inhibitory activity (Barbana and Boye 2010). The best ACE-inhibitory peptides contain hydrophobic amino acids such as Pro, or positively charged amino acids such as Lys or Arg at the C-terminal position and present short sequences from 2 to 12 amino acids (García et al. 2013).

Phenolic compounds from chickpea also provide ACE-inhibitory activity, showing an IC_{50} of 0.061-0.152 mg/mL (Sreerama et al. 2012; Mamilla and Mishra 2017). Phenolic compounds inhibit ACE by the interaction with the zinc ion at the active center of the enzyme; this is stabilized by interactions with other amino acids at the active center and increases with the increment of hydroxyl groups. It is suggested that the free hydroxyl groups are important structural moieties to chelate the zinc ions, thus inactivate the ACE-I activity (Sreerama et al. 2012; Al Shukor et al. 2013).

Diabetes

It has been reported that a legume-based diet can contribute to decreasing fasting blood and plasma glucose, fasting insulin, plasma insulin and homeostatic model assessment (HOMA), and improves glycemic control and insulin resistance (Nestel et al. 2004; Yang et al. 2007; Jenkins et al. 2012; Hosseinpour-Niazi et al. 2015; Zafar et al. 2017). Pulses have a low glycemic index, generating a low rise in blood glucose after their consumption, so they can be beneficial for diabetes management since there is also a reduction in the insulin response, producing better concentrations of blood glucose and lipid concentrations in the individuals (Trinidad et al. 2010; Clark et al. 2018). Resistant starch increased consumption is associated with better tolerance to glucose and insulin sensitivity (Jukanti et al. 2012).

To breakdown the carbohydrates we consume, the action of hydrolytic enzymes is necessary; α-amylase and α-glucosidase are ones of the enzymes involved in this process, and their inhibition is a strategy for type II diabetes management (Prathapan et al. 2011).

Chickpea protein hydrolysates (with pepsin, chyme and pancreatin) showed an IC_{50} of 2885 μg/mL against α-amylase and 167 μg/mL against α-glucosidase (Ercan and Nehir El 2016). Phenolic compounds have shown an inhibitory index of 3.56 against α-amylase and 2.94 against α-glucosidase (Sánchez-Magaña et al. 2014), and an IC_{50} of 108.3 μg/mL and 92.2 μg/mL, respectively (Sreerama et al. 2012). Those values are better than the ones obtained for cowpea against α-amylase (159.6 μg/mL) (Sreerama et al. 2012), while acarbose possesses an IC_{50} of 6.9 μg/mL against α-amylase and 91.0 μg/mL against α-glucosidase (Yilmazer-Musa et al. 2015).

It is reported that isoflavones and anthocyanins can also attenuate insulin resistance and improve insulin secretion (Clark et al. 2018). Gao et al. (2015) observed that isoflavones of chickpea stimulated glucose uptake through the down-regulation of glucose transporter-4 mRNA expression in a dose-dependent manner and inhibited lipid accumulation, which can improve glucose utilization and insulin sensitivity. Shen et al. (2016) suggested that biochanin A (the major isoflavone in chickpea) has better potential than other isoflavones as an anti-diabetic agent.

Cancer

Studies have reported diverse chemopreventive effects exhibited by legumes, including the induction of apoptosis, the regulation and arrest of the cell cycle, the suppression of the proliferation and the mitigation of inflammation, the reparation of the DNA, the increase of the activity of the enzyme glutathione S transferase (GST), the inhibition of the cytochrome P450 and the contribution of the synthesis of SCFA (Sánchez-Chino et al. 2017).

There has been observed an inverse association between legume intake levels and colorectal cancer risk (Wang et al. 2013). A study (Murillo et al. 2004) showed a 64% suppression of azoxymethane-induced aberrant crypt foci for rats fed with 10% chickpea flour, suggesting that chickpea possesses bioactive compounds that can inhibit the formation of pre-cancerous lesions, reducing the risk of developing colon cancer.

Butyrate is a SCFA that has demonstrated to inhibit cell proliferation and promote apoptosis (Cao et al. 2019); the production of this SCFA can be stimulated after the consumption of a high fiber diet, which is reachable with a chickpea diet (200 g/day) (Jukanti et al. 2012).

Bowman-Birk inhibitors (BBI) are reported to prevent or inhibit carcinogenic and inflammatory processes either *in vitro* or *in vivo* in the gastrointestinal tract (Clemente et al. 2011). This protease inhibitor has shown activity against gastric adenocarcinoma cell

GS (IC_{50} = 3.23 mg/mL), colorectal adenocarcinoma cell SW480 (IC_{50} = 5.71 mg/mL) and prostate carcinoma cell DU145 (non-dose-dependent inhibition) (Xu and Chang 2009). A protease inhibitor concentrate obtained from chickpea showed inhibition of the viability of breast cancer cells MDA-MB-231 and prostate cancer cells PC-3 and LNCaP (Magee et al. 2013). A study (Raicht et al. 1980) showed that to add β-sitosterol in the diet of rats was effective to decrease the colonic tumors induced by N-methyl-N-nitrosourea. Foods rich in saponins also have demonstrated to inhibit azoxymethane-induced lesions in rat colon (Koratkar and Rao 1997).

There has been shown that Asian people, who regularly consume an isoflavone-rich diet, have a minor rate of breast, colorectal, and prostate cancer than western people (HaiRong et al. 2013). Some studies (Yanagihara et al. 1993; Dixon 2004) reported that biochanin A inhibited the growth of stomach cancer cells *in vitro* and reduced the tumor growth in mice. Chickpea polyphenolic extracts also have demonstrated activity against breast cancer (Gowen et al. 2007; Ma et al. 2013; Domínguez-Arispuro 2017), and colorectal cancer cells proliferation, as well as suppression of epithelial tumor growth (Girón-Calle et al. 2004).

The level of estrogens is extremely higher in breast cancer tissues than in plasma. The enzyme aromatase converts androgens into estrogens, so its inhibition is a valuable strategy to treat hormone-dependent breast cancer, eliminating the stimulating effect of estrogen on tumor growth (Dawood et al. 2018; Zhang et al. 2018). Chickpea total extract has shown to inhibit the aromatase enzyme with an IC_{50} of 234 μg/mL, whereas the butanol-fractionated extract of chickpea showed an IC_{50} of 75.3 μg/mL (Dawood et al. 2018).

Zhang et al. (2018) found 11 compounds from chickpea extract as potent human aromatase inhibitors, identifying biochanin A-7-*O* glucopyranoside, pratensein and biochanin A as the ones with the highest inhibition rate (62.22, 60.74, and 59.54%, respectively).

Lectins promote apoptosis and regulate the expression of apoptotic genes in cancer cells (Gautam et al. 2018). Gupta et al. (2018b) reported that chickpea lectin-induced apoptosis in treated breast cancer cells (MCF-7) and retained less cell viability (47.3%) than the untreated cells (95%). Chickpea lectins also have shown effects against other human cancer cell lines such as hepatic cancer cells (HepG2), endometrial cancer cells (Ishikawa) and breast cancer cells (MCF-7 and MDA-MB-231) with IC_{50} values of 44.20-61.8, 46.67-54.4, 37.5-53.58 and 37.47-44.2 μg/mL, respectively (Gautam et al. 2018; Gupta et al. 2018a).

It has been reported that chickpea albumin hydrolysate can increment the tumor inhibition rate and decrement the tumor size (Xue et al. 2012). A peptide from this hydrolysate, with the sequence RQSHFANAQP, named CPe-III, showed a high inhibition against breast cancer cells (MCF-7 and MDA-MB-231) with IC_{50} values of 2.38 and 1.50 μmol/mL. The inhibition of the proliferation was due to the increase of the p53 protein (a key tumor suppressor protein), which increased by the stimulation of the increasing concentration of CPe-III (Xue et al. 2015).

It is reported (Álvarez et al. 2016; Sánchez-Chino et al. 2018) that when administered chickpea protein hydrolysates to mice with azoxymethane-induced colon cancer, reduced carcinogenesis was observed, which was probably related with the decrease of oxidized lipids and proteins stimulated by the hydrolysates.

Other Health Benefits

It has been reported that chickpea phytosterols have various properties besides the ones previously described, such as antiulcerative, antibacterial, antifungal, and antiinflammatory (Jukanti et al. 2012).

Lectins from chickpea have been reported to exhibit antifungal activity (Gupta et al. 2018a). Proteins from chickpea have also shown antifungal properties against *Candida krusei*, *Candida tropicalis*, and *Candida parapsilosis* (Kumar et al. 2014).

Chickpea polyphenolic extracts possess antiinflammatory effects, modulating the activity of some proinflammatory markers such as tumor necrosis factor alpha (TNF-α), interleukin-6 (IL6), nuclear factor kappa beta (NF-κβ) and ciclooxygenase-2 (COX-2) (Hermsdorff et al. 2010; Mekky et al. 2016; Wahby et al. 2017).

On the other hand, chickpea contains soluble and insoluble polysaccharides, galactooligosaccharides and resistant starch. Besides their prebiotic function (Mitsuoka 1996; Martínez-Villaluenga et al. 2008; Anderson et al. 2009; Slavin 2013; Zhang et al. 2017), their consumption has been associated with an improved immune function, increasing the production of antiinflammatory cytokines and decreasing the proinflammatory cytokines (Vulevic et al. 2008).

It is reported that adding ciceritol, the most abundant oligosaccharide in chickpeas, to human colonic microflora can function as a prebiotic, enhancing the growth of *Lactobacillus–Enterococcus* group and *Bifidobacterium* spp., and inhibiting the growth of *Bacteroides–Prevotella*, *Clostridium histolyticum* and *Eubacterium–Clostridium* groups. Also, the addition of ciceritol can aid in the production of SCFA (Zhang et al. 2017).

Biochanin A has demonstrated to help against Alzheimer's disease in mice. It has a neuroprotective effect on dopaminergic neurons, protecting them against the lipopolysaccharide-induced neurotoxicity, inhibiting the activation of microglia and NADPH-oxidase induced by the lipopolysaccharide (LPS) administration, attenuating the oxidative stress and increasing the activities of antioxidant enzymes (Wang et al. 2015).

A large amount of aluminum has been observed accumulated in the brain of subjects suffering from Alzheimer's disease (Jason and Timothy 2011), which causes chronic inflammation of nervous tissue cells, resulting in massive neurodegeneration and cognitive dysfunction. Wahby et al. (2017) showed that pre- and co-administration of a chickpea phenolic extract in rats attenuated the inflammation caused by the administration of $AlCl_3$. On the other hand, estrogen delays the neurodegeneration and mitigates the cytotoxicity, oxidative injury and neuronal degeneration (Howard 2002; Mateos et al. 2012); due to structure similarities, phytoestrogens can interact with estrogen receptors (ER) and regulate estrogenic responses (Amjad 2015).

The non-organic phase of chickpea dichloromethane extract showed anticonvulsant effects in mice with seizures induced by pentylenetetrazole (Sardari et al. 2015). After performing a screening, it was found that the extract and its fractions possessed a significant quantity of alkaloids. A compound of the non-organic fraction was isolated, and it contained two hydrocarbons with the functional groups -COOH and -NH, as well as similarities with valproic acid, an antiepileptic drug (Bialer et al. 2007).

Conclusion

The potential of chickpea to be used as a nutritional and nutraceutical food is presented towards this chapter. Chickpea is a good source of protein, carbohydrates, dietary fiber and unsaturated fatty acids. Chickpea antinutritional factors can be removed easily by processing; nevertheless, some of them have shown important health benefits. Studies have shown that diverse chickpea components had exhibited several positive bioactivities that can reduce the risk of developing several chronic degenerative diseases. The acknowledge of this could persuade to a higher consumption and to a more extensive research of this grain, so scientific community could find the most effective ways to increase chickpea nutritional value and bioactive compounds, as well as to elucidate the mechanism(s) of action on which each chickpea component acts and, maybe, to discover more health benefits given by them.

References

Abeysekara, S., Chilibeck, P. D., Vatanparast, H. and Zello, G. A. (2012). A pulse-based diet is effective for reducing total and LDL-cholesterol in older adults. *British Journal of Nutrition*, 108: S103–S110.

Abu-Salem, F. M. and Abou-Arab, E. A. (2011). Physico-chemical properties of tempeh produced from chickpea seeds. *Journal of American Science*, 7: 107-118.

Aguilera, Y., Esteban, R. M., Benítez, V., Molla, E. and Martín-Cabrejas, M. A. (2009). Starch, functional properties, and microstructural characteristics in chickpea and lentil as affected by thermal processing. *Journal of Agricultural and Food Chemistry*, 57: 10682–10688.

Akihisa, T., Nishismura, Y., Nakamura, N., Roy, K., Gosh, P., Thakur, S. and Tamura, T. (1992). Sterols of *Cajanus cajan* and three others leguminosae seed. *Phytochemistry*, 31: 1765-1768.

Al Shukor, N., Van Camp, J., Gonzales, G. B., Staljanssens, D., Struijs, K., Zotti, M. J., Raes, K. and Smagghe, G. (2013). Angiotensin-converting enzyme inhibitory effects by plant phenolic compounds: A study of structure activity relationships. *Journal of Agricultural and Food Chemistry*, 61(48): 11832-11839.

Alajaji, S. A. and El-Adawy, T. A. (2006). Nutritional composition of chickpea (*Cicer arietinum* L.) as affected by microwave cooking and other traditional cooking methods. *Journal of Food Composition and Analysis*, 19: 806-812.

Alonso, R, Orue, E., and Marzo, F. (1998). Effect of extrusion and conventional processing methods on protein and antinutritional factor contents in peas seeds. *Food Chemistry*, 63: 505-512.

Alonso, R., Aguirre A. and Marzo, F. (2000). Effects of extrusion and traditional processing methods on antinutrients and *in vitro* digestibility of protein and starch in faba and kidney beans. *Food Chemistry*, 68: 159-165.

Álvarez, G. I., Muñoz, C. L. S., Pérez, G. A., Madrigal, E., Jiménez, M. C., Dávila, O. G., Villa, T. S., Sánchez, C. X. (2016). Chemopreventive effect of chickpea (*Cicer arietinum*) protein hydrolysate on colon cancer in mouse. *Toxicology Letters*, 259: S93.

Amjad, F. (2015). Structural and functional diversity of estrogen receptor ligands. *Current Topics in Medicinal Chemistry,* 15(14): 1372–1384.

Anderson, J. W., Baird, P., Davis Jr R. H., Ferreri, S., Knudtson, M., Koraym, A., Waters, V. and Williams, C. L. (2009). Health benefits of dietary fiber. *Nutrition Reviews,* 67(4): 188–205.

Angulo-Bejarano, P. I., Verdugo-Montoya, N. M., Cuevas-Rodriguez, E. O., Milán-Carrillo, J., Mora-Escobedo, R., López-Valenzuela, J. A., Garzón-Tiznado, J. A. and Reyes-Moreno, C. (2008). Tempeh flour from chickpea (*Cicer arietinum* L.). Nutritional and physicochemical properties. *Food Chemistry*, 106: 106-112.

Arntfield, S. D., Scanlon, M. G., Malcolmson, L. J., Watts, B. M. Cenkowski, S., Ryland, D. and Savoie, V. (2001). Reduction in lentil cooking time using micronization. Comparison of 2 micronization temperatures. *Journal of Food Science*, 66: 500-505.

Asif, M., Rooney, L. W. Ali, R. and Riaz, M. N. (2013). Application and opportunities of pulses in food system: A review. *Critical Reviews in Food Science and Nutrition*, 53(11): 1168-79.

Bains, K., Uppal, V., and Kaur, H. (2014). Optimization of germination time and heat treatments for enhanced availability of minerals from leguminous sprouts. *Journal of Food Science and Technology*, 51: 1016-1020.

Barbana, C. and Boye, J. I. (2010). Angiotensin I-converting enzyme inhibitory activity of chickpea and pea protein hydrolysates. *Food Research International*, 43: 1642–1649.

Bialer, M., Johannessen, S. I., Kupferberg, H. J., Levy, R. H., Perucca, E. and Tomson, D. (2007). Progress report on new antiepileptic drugs: a summary of the eight. Eilat conferences (EILAT VIII). *Epilepsy Research*, 73: 1-52.

Bilgicli, N. (2008). Effects of cooking and drying processes on physical, chemical and sensory properties of legume base bulgur. *Journal of Food Processing and Preservation*, 33:590-604.

Birt, D. F. and Jeffery, E. (2013). Flavonoids. *Advances in Nutrition*, 4: 576–577.

Bouchenak, M. and Lamri-Senhadji, M. (2013). Nutritional Quality of legumes, and their role in cardiometabolic risk prevention: A review. *Journal of Medicinal Food*, 16: 1–14.

Cao, M., Zhang, Z., Han, S. and Lu, X. (2019). Butyrate inhibits the proliferation and induces the apoptosis of colorectal cancer HCT116 cells via the deactivation of mTOR/S6K1 signaling mediated partly by SIRT1 downregulation. *Molecular Medicine Reports*, 19 (5): 3941-3947.

Chavan, J. K., Kadam, S. S., Salunkhe, D. K. and Beuchat, L. R. (1987). Biochemistry and technology of chickpea (*Cicer arietinum* L.) seeds. *Critical Reviews in Food Science and Nutrition*, 25(2): 107-158.

Chen, J., Wang, Y., Ye, R., Wu, Y. and Xia, W. (2015). Comparison of analytical methods to assay inhibitors of angiotensin I-converting enzyme. *Food Chemistry*, 141(4):3329-34.

Cheynier, V. (2012). Phenolic compounds: from plants to foods. *Phytochemistry Reviews*, 11: 153–177.

Chitra, U., Singh, U., and Rao, P. V. (1996). Phytic acid, *in vitro* protein digestibility, dietary fiber and minerals of pulses as influenced by processing methods. *Plant Foods for Human Nutrition*, 49:307-316.

Clark, J. L., Taylor, C. G. and Zahradka, P. (2018). Rebelling against the (Insulin) resistance: A review of the proposed insulin-sensitizing actions of soybeans, chickpeas, and their bioactive compounds. *Nutrients*, 10(4): 434.

Clemente, A., Sonnante, G. and Domoney, C. (2011). Bowman-Birk inhibitors from legumes and human gastrointestinal health: Current status and perspectives. *Current Protein and Peptide Science*, 12: 358-373.

Cuadrado, C., Hajos, G., Burbano, C., Pedrosa, M. M., Ayet, G., Muzquiz, M. Pusztai, A. and Gelencser, E. (2002). Effect of natural fermentation of the lectin of lentil measured by immunological methods. *Food and Agricultural Immunology*, 14:41-49.

Da Ros, R., Assaloni, R. and Ceriello, A. (2005). Postprandial hyperglycemia and diabetic complications. *Recenti progressi in medicina*, 96(9): 436-444.

da Silva, M. A., Neves, V. A., and Lourenço, E. J. (2001). Protein fractions and major globulin from chickpea (*Cicer arietinum* L.). *Alimentose Nutrição*, 12: 131–149.

Dawood, H. M., Ibrahim, R. S., Shawky, E., Hammoda, H. M. and Metwally, A. M. (2018). Integrated *in silico-in vitro* strategy for screening of some traditional Egyptian plants for human aromatase inhibitors. *Journal of Ethnopharmacology*, 224(5): 359-372.

de Sales, P. M., Monteiro de Souza, P., Simeoni, L. A. de Oliveira Magalhães, P. and Silveira, D. (2012). α-amylase inhibitors: A review of raw material and isolated compounds from plant source. *Journal of Pharmacy and Pharmaceutical Sciences*, 15(1): 141-183.

Demirkesen-Bicak, H., Tacer-Caba, Z. and Nilufer-Erdil, D. (2018). Pullulanase treatments to increase resistant starch content of black chickpea (*Cicer arietinum* L.) starch and the effects on starch properties. *International Journal of Biological Macromolecules*, 11: 505-513.

Dixon, R. A. (2004). Phytoestrogens. *Annual Review of Plant Biology*, 55: 225–261.

Domínguez-Arispuro, D. M., Cuevas-Rodríguez, E. O., Milán-Carrillo, J., León-López, L., Gutiérrez-Dorado, R. and Reyes-Moreno, C. (2017). Optimal germination condition impacts on the antioxidant activity and phenolic acids profile in pigmented desi chickpea (*Cicer arietinum* L.) seeds. *Journal of Food Science and Technology*, 55(2): 638-647.

Domínguez-Arispuro, D. M. (2016). “Desi Chickpea (*Cicer arietinum* L) Bioprocessing to Increase Antioxidant Potential/Inhibition of Breast Cancer Cells Proliferation.” PhD diss., Universidad Autónoma de Sinaloa.

El-Adawy, T. A. (2002). Nutritional composition and antinutritional factors of chickpeas (*Cicer arietinum* L.) undergoing different cooking methods and germination. *Plant Foods for Human Nutrition*, 57: 83–97.

El-Niely, H. F. G. (2007). Effect of radiation processing on antinutrients, *in-vitro* protein digestibility and protein efficiency ratio bioassay of legume seeds. *Radiation Physics and Chemistry*, 76: 1050-1057.

Ercan, P. and Nehir El, S. (2016). Inhibitory effects of chickpea and *Tribulus terrestris* on lipase, α-amylase and α-glucosidase. *Food Chemistry*, 205:163-9.

FAO/WHO. (1991). *Protein quality evaluation, food and agricultural organization of the United Nations.* Rome: FAO/WHO.

Food and Agriculture Organization of the United Nations/World Health Organization (FAO/WHO). (2001). Human Vitamin and Mineral Requirements. Report of a joint FAO/WHO expert consultation Bangkok, Thailand.

Frias, J., Vidal-Verde, C., Sotomayor, C., Diaz-Pollan, C., and Urbano, G. (2000). Influence of processing on available carbohydrate content and antinutritional factors of chickpeas. *European Food Research and Technology*, 210: 340-345.

Gao, Y., Yao, Y., Zhu, Y. and Ren, G. (2015a). Isoflavones contents and composition in chickpeas (*Cicer arietinum* L.) sprouts germinated under different conditions. *Journal of Agricultural and Food Chemistry*, 63(10): 2701-7.

Gao, Y., Yao, Y., Zhu, Y. and Ren, G. (2015b). Isoflavones in chickpeas inhibit adipocyte differentiation and prevent insulin resistance in 3T3-L1 cells. *Journal of Agricultural and Food Chemistry*, 63: 9696–9703.

García, M. C., Puchalska, P., Esteve, C. and Marina, M. L. (2013). Vegetable foods: A cheap source of proteins and peptides with antihypertensive, antioxidant, and other less occurrence bioactivities. *Talanta*, 106: 328–349.

Garzón-Tiznado, J. A., Ochoa-Lugo, M. I., Heiras-Palazuelos, M., Domínguez-Arispuro, D. M., Cuevas-Rodríguez, E. O., Gutiérrez-Dorado, R., Milán-Carrillo, J. and Reyes-Moreno C. (2012). Acceptability properties and antioxidant potential of desi chickpea (*Cicer arietinum* L.) cultivars. *Food and Nutrition Sciences*, 3: 1281-1289.

Garzón-Tiznado, J., Heiras-Palazuelos, M., Espinoza-Moreno, R., Cano-Campos, M., Delgado-Vargas, F., Gutiérrez-Dorado, R., Milán-Carrillo, J. and Reyes-Moreno C. (2013). Antioxidant and antimutagenic activities of optimized extruded desi chickpea (*Cicer arietinum* L) flours. *Journal of Pharmacy and Nutrition Sciences*, 3: 38-47.

Gautam, A. K., Gupta, N., Narvekar, D. T., Bhadkariya, R. and Bhagyawant, S. S. (2018). Characterization of chickpea (*Cicer arietinum* L.) lectin for biological activity. *Physiology and Molecular Biology of Plants*, 24 (3): 389–397.

Gemede, H. F. and Ratta. (2014). Antinutritional factors in plant foods: Potential health benefits and adverse effects. *International Journal of Nutrition and Food Sciences*, 3(4): 284-289.

Girón-Calle, J., Vioque, J., del Mar Yust, M., Pedroche, J., Alaiz, M. and Millán, F. (2004). Effect of chickpea aqueous extracts, organic extracts and protein concentrates on cell proliferation. *Journal of Medicinal Food*, 7(2): 122–129.

Gowen, A., Abu-Ghannam, N., Frias, J. and Oliveira, J. (2007). Modelling the water absorption process in chickpeas (*Cicer arietinum* L.)—The effect of blanching pre-treatment on water intake and texture kinetics. *Journal of Food Engineering*, 78: 810–819.

Grant, G., Murray, S., Gravells, E., Duguid, T. J., Bardocz, S. and Pusztai, A. (1998). The body metabolism of mammals can be adversely affected by enzyme inhibitors from legume seeds. In: *Proceedings of the Third International Workshop on Antinutritional Factors in Legume Seeds and Rapeseed*, edited by: Jansman, A. J. M., Hill, G. D., Huisman, J. and van der Poel, A. F. B. 233-237. Wageningen, The Netherlands.

Grela, E. R., Kiczorowska, B., Samolińska, W., Matras, J., Kiczorowski, P., Rybiński, W. and Hanczakowska, E. (2017). Chemical composition of leguminous seeds: part I—content of basic nutrients, amino acids, phytochemical compounds, and antioxidant activity. *European Food Research and Technology*, 234(8): 1385–1395.

Gupta, D. K., Tripathi, R. D., Rai, U. N., Dwivedi, S., Mishra, S., Srivastava, S. and Inouhe, M. (2006). Changes in amino acid profile and metal content in seeds of *Cicer arietinum* L. (chickpea) grown under various fly-ash amendments. *Chemsphere*, 65: 939-945.

Gupta, N., Ajay, K. G. and Sameer, S. B. (2018a). Biochemical characterisation of lectin from wild chickpea (*Cicer reticulatum* L.) with potential inhibitory action against human cancer cells. *Journal of Food Biochemistry*, 43(2):1-10.

Gupta, N., Prakash, S. B. and Sameer, S. B. (2018b). Chickpea lectin inhibits human breast cancer cell proliferation and induces apoptosis through cell cycle arrest. *Protein & Peptide Letters*, 25: 1-8.

Gupta, R. K., Gupta, K., Akanksha, S., Mukul, D., Irfan, A. A. and Premendra D. D. (2016). Look insight: Health risks and benefits of chickpea (*Cicer arietinum*) consumption. *Journal of Agricultural and Food Chemistry*, 65(1): 6-22.

Guzmán-Uriarte, L., Sánchez-Magaña, L. M., Angulo-Meza, G. Y., Cuevas-Rodríguez, E. O., Gutiérrez-Dorado, R., Mora-Rochín, S., Milán-Carrillo, J., Valdez-Ortiz, A. and Reyes-Moreno, C. (2013). Solid state bioconversion for producing common bean (*Phaseolus vulgaris* L.) functional flour with high antioxidant activity and antihypertensive potential. *Food and Nutrition Sciences*, 4: 480-490.

HaiRong, M., HuaBo, W., Zhen, C., Yi, Y., ZhengHua, W., Madina, H., Xu, C. and Akber, A. H. (2013). The estrogenic activity of isoflavones extracted from chickpea *Cicer arietinum* L sprouts *in vitro*. *Phytotherapy Research*, 27(8): 1237-42.

Hall, C., Hillen, C. and Robinson, J. G. (2016). Composition, nutritional value and health benefits of pulses. *Cereal Chemistry*, 94(1): 11-31.

Hamid, A. and Kalsoom, S. (2017). Comparative analysis of nutritional composition and effect of dietary fiber extracts of chickpea and Bengal gram on blood glucose and cholesterol levels of male induced diabetic and hypercholesterolemic rats. *Pakistan Journal of Zoology*, 49(2): 487-492.

Han, H. and Baik, B. K. (2008). Antioxidant activity and phenolic content of lentils (*Lens culinaris*), chickpeas (*Cicer arietinum* L.) peas (*Pisum sativum* L) and soybeans (Glycine max) and their quantitative changes during processing. *International Journal of Food Science and Technology*, 43: 1971-1978.

Han, I. H. and Baik, B. K. (2006). Oligosaccharide content and composition of legumes and their reduction by soaking, cooking, ultrasound and high hydrostatic pressure. *Cereal Chemistry*, 83: 428–433.

Hefnawy, T. H. (2011). Effect of processing methods on nutritional composition and anti-nutritional factors in lentils (*Lens culinaris*). *Annals of Agricultural Science*, 56(2): 57–61.

Heiras-Palazuelos, M., Ochoa-Lugo, M. I., Gutiérrez-Dorado, R., López-Valenzuela, J. A., Mora-Rochín S., Milán-Carrillo, J., Garzón-Tiznado, J. A. and Reyes-Moreno, C. (2012). Technological properties, antioxidant activity and total phenolic and flavonoid content of pigmented chickpea (*Cicer arietinum* L.) cultivars. *International Journal of Food Sciences and Nutrition*, 64(1): 69-76.

Hermsdorff, H. H., Zulet, M. Á., Abete, I. and Martínez J. A. (2011). A legume-based hypocaloric diet reduces proinflammatory status and improves metabolic features in overweight/obese subjects. *European Journal of Nutrition*, 50(1): 61-69.

Hosseinpour-Niazi, S., Mirmiran, P. Hedayati, M. and Azizi, F. (2015). Substitution of red meat with legumes in the therapeutic lifestyle change diet based on dietary advice improves cardiometabolic risk factors in overweight type 2 diabetes patients: a cross-over randomized clinical trial. *European Journal of Clinical Nutrition*, 69: 592–597.

Howard, M. (2002). The role of hormone replacement therapy in the prevention of alzheimer disease. *Archives of Internal Medicine*, 162(17): 1934-1942.

Huskisson, E., Maggini, S. and Ruf, M. (2007). The role of vitamins and minerals in energy metabolism and well-being. *The Journal of International Medical Research*, 35: 277 – 289.

Jandacek, R. J. (2017). Linoleic acid: A nutritional quandary. *Healthcare*, 25 (5): 1-8.

Jason, R. C. and Timothy J. G. (2011). The role of environmental exposures in neurodegeneration and neurodegenerative diseases. *Toxicological Sciences*, 124(2): 225–250.

Jenkins, D. J., Kendall, C. W., Augustin, L. S., Mitchell, S., Sahye-Pudaruth, S., Blanco Mejia, S., Chiavaroli, L., Mirrahimi, A., Ireland, C., Bashyam, B., Vidgen, E., de Souza, R. J., Sievenpiper, J. L., Coveney, J., Leiter, L. A. and Josse, R. G. (2012). Effect of legumes as part of a low glycemic index diet on glycemic control and cardiovascular risk factors in type 2 Diabetes mellitus. *Archives of Internal Medicine*, 172 (21): 1653-1660.

Jukanti, A. K., Gaur P. M., Gowda, C. L. L. and Chibbar R. N. (2012). Nutritional quality and health benefits of chickpea (*Cicer arietinum* L.): a review. *British Journal of Nutrition*, 108: S11–S26.

Kalogeropoulos, N., Chiou, A., Ioannou, M., Karathanos, V. T., Hassapidou, M. and Andrikopoulos, N. K. (2010). Nutritional evaluation and bioactive microconstituents (phytosterols, tocopherols, polyphenols, triterpenic acids) in cooked dry legumes usually consumed in the Mediterranean countries. *Food Chemistry*, 121: 682–690.

Khandelwal, S., Udipi, S. A., and Ghugre, P. (2007). Polyphenols and tannins in Indian pulses: effect of soaking, germination and pressure cooking. *Food Research International*, 43: 526-530.

Koratkar, R. and Rao, A. V. (1997). Effect of soya bean saponins on azoxymethane-induced preneoplastic lesions in the colon of mice. *Nutrition and Cancer*, 27: 206–209.

Kou, X., H., Gao, J., Xue, Z., Zhang, Z., Wang, H. and Wang, X. (2013). Purification and identification of antioxidant peptides from chickpea (*Cicer arietinum*, L.) albumin hydrolysates. *LWT - Food Science and Technology*, 50(2): 591-598.

Kumar, S., Kapoor, V., Gill, K., Singh, K., Xess, I, Das, S. N. and Dey, S. (2014). Antifungal and antiproliferative protein from *Cicer arietinum*: A bioactive compound against emerging pathogens. *BioMed Research International*, doi:10.1155/2014/387203.

Luo, L., Shen, L., Sun, F. and Ma, Z. (2013). Immunoprecipitation coupled with HPLC–MS/MS to discover the aromatase ligands from Glycyrrhiza uralensis. *Food Chemistry*, 138: 315–320.

Ma, H. R., Wei, H. B., Chen, Z., Yang, Y., Wang, Z. H., Habasi, M., Cao, X. and Aisa H. A. (2013). The estrogenic activity of isoflavones extracted from chickpea *Cicer arietinum* L sprouts *in vitro. Phytotherapy Research*, 27: 1237–1242.

Maesen, L. J. and Vander, G. (1972). *Cicer*, a monograph on the genus with special reference to the chickpea (*Cicer arietnum*), its ecology and cultivation. Wageningen, The Netherlands.

Magee, P. J., Owusu-Apenten, R., McCann, M. J., Gill, C. I. and Rowland, I. R. (2013). Chickpea (*Cicer arietinum*) and other plant-derived protease inhibitor concentrates inhibit breast and prostate cancer cell proliferation *in vitro*. *Nutrition and Cancer*, 64(5): 741-8.

Mamilla, R. K. and Mishra V. K. (2017). Effect of germination on antioxidant and ACE inhibitory activities of legumes. *LWT - Food Science and Technology*, 75: 51-58.

Martin-Cabrejas, M. A., Sanfiz, B., Vidal., Molla, E., Esteban, and Lopez-Andreu, F. J. (2004). Effect of fermentation and autoclaving on dietary fiber fractions and antinutritional factors of beans (*Phaseolus vulgaris* L.). *Journal of Agricultural and Food Chemistry*, 52: 261-266.

Martinez-Gonzalez, A. I., Díaz-Sánchez, A. G., de la Rosa, L. A., Vargas-Requena, C. L., Bustos-Jaimes, I. and Alvarez-Parrilla, A. E. (2017). Polyphenolic compounds and digestive enzymes: *in vitro* non-covalent interactions. *Molecules*, 22(4): 669.

Martínez-Villaluenga, C., Frías, J. and Vidal-Valverde, C. (2008). Alpha-galactosides: antinutritional factors or functional ingredients? *Critical Reviews in Food Science and Nutrition*, 48: 301–316.

Mateos, L., Persson, T., Kathozi, S., Gil-Bea, F. J. and Cedazo-Minguez, A. (2012). Estrogen protects against amyloid-β toxicity by estrogen receptor α-mediated inhibition of Daxx translocation. *Neuroscience letters*, 506(2): 245-250.

Medina-Godoy, S., Ambriz-Pérez, D. L., Fuentes-Gutiérrez, C. I., Germán-Báez, L. J., Gutiérrez-Dorado, R., Reyes-Moreno, C. and Valdez-Ortiz, A. (2012). Angiotensin-converting enzyme inhibitory and antioxidative activities and functional characterization of protein hydrolysates of hard-to-cook chickpeas. *Journal of the Science of Food and Agriculture*, 92(9): 1974–81.

Mekky, R. H., Fayed, M. R., El-Gindi, M. R., Abdel-Monem, A. R., Contreras, M. M., Segura-Carretero, A. and Abdel-Sattar E. (2016). Hepatoprotective effect and chemical assessment of a selected Egyptian chickpea cultivar. *Frontiers in Pharmacology*, 344 (7): 1-9.

Melcion, J. P. and van der Poel, A. F. B. (1993). Process technology and antinutritional factors: principles, adequacy and process optimization. In: *Recent Advances of Research in Antinutritional Factors in Legume Seeds,* edited by: van der Poel, A. F. B., Huisman, J. and Saini, H. S., 419-434, Wageningen Pers: Wageningen.

Milán-Carrillo, J., Alarcón-Valdéz, C., Gutiérrez-Dorado, R., Cárdenas-Valenzuela, O. G., Mora-Escobedo, R., Garzón-Tiznado, J. A. and Reyes-Moreno, C. (2007). Nutritional properties of quality protein maize and chickpea extruded based weaning food. *Plant Foods for Human Nutrition*, 62: 31-33.

Milán-Carrillo, J., Reyes-Moreno, C., Armienta-Rodelo, E., Cárabez-Trejo, A and Mora-Escobedo, R. (2000). Physicochemical and nutritional characteristics of extruded flours from fresh and hardened chickpeas (*Cicer arietinum* L). *LWT- Food Science and Technology*, 33: 117-123.

Milán-Noris, A. K. (2017). "Nutraceutical Properties of Isolated Starch, Phytochemical Compounds and Bioactive Peptides from Pigmented Chickpea Cultivars Influenced by Cooking or Germination Process.". PhD diss., Instituto Tecnológico y de Estudios Superiores de Monterrey.

Milán-Noris, A. K., Gutiérrez-Uribe, J. A., Santacruz, A., Serna-Saldívar, S. O. and Martínez-Villaluenga, C. (2018). Peptides and isoflavones in gastrointestinal digests contribute to the anti-inflammatory potential of cooked or germinated desi and kabuli chickpea (*Cicer arietinum* L.). *Food Chemistry*, 268(1): 66-76.

Minatel, I. O., Borges, C. V., Ferreira, M. I., Gomez Gomez, H. A., Chen, C. O. and Pereira Lima, G. P. (2016). Phenolic Compounds: Functional Properties, Impact of Processing and Bioavailability. In: *Phenolic Compounds*, edited by: Soto-Hernández, M., doi: 10.5772/66368.

Mitsuoka, T. (1996). Intestinal flora and human health. *Asia Pacific Journal of Clinical Nutrition*, 5: 2–9.

Mohan, R. V., Tresina, P. S. and Daffodil, E. D. (2016). Antinutritional Factors in Legume Seeds: Characteristics and Determination. In: *Encyclopedia of Food and Health,* edited by: Caballero, B., Finglas, P. and Toldra, F., 211-220, Oxford: Academic Press, Chidambaram College, Tuticorin, India.

Murillo, G., Choi, J. K., Pan, O., Constantinou, A. I. and Mehta, R. G. (2004). Efficacy of Garbanzo and Soybean Flour in Suppression of Aberrant Crypt Foci in the Colons of CF-1 Mice. *Anticancer Research*, 24: 3049-3056.

Muzaffar, M., Bakhsh, R., Khan, S. and Khan, S. (1989). Effect of different dietary brans on lipid metabolism in male albino rats. *Nahrung*, 33(2):167-70.

Muzquiz, M. (2000a). Antinutritional factors in proteic sources. In: *International Conference of Food Proteins,* edited by: Vioque, J., Clemente, A., Bautista, J. and Millan. F. Sevilla, Spain.

Muzquiz, M. and Wood, J. (2007). Antinutritional factors. In: *Chickpea Breeding and Management,* edited by Yadav, S. S., Redden, R. J., Chen, W. and Cabi, S.B, 143-166. UK: CAB International.

Muzquiz, M., Burbano, C., Cuadrado, C., and Martin, M. (2000). Analytical methods for determination of compounds with no nutritive value. In: *Handbook on Common Bean Related Laboratory Methods,* 11-26, Spain: Galicia.

Myint, H., Kishi, H., Koike, S. and Kobayashi, Y. (2017). Effect of chickpea husk dietary supplementation on blood and cecal parameters in rats. *Animal Science Journal*, 88(2): 372-378.

Nakagawa, M., Kawashima, Y. and Ushiyama, M. (1975). Hypolipidemic Principles of *Cicer arietinum*: Biochanin-A and Formononetin. *Lipids*, 11(3): 243-246.

Nestel, P., Cehun, M. and Chronopoulos, A. (2004). Effects of long-term consumption and single meals of chickpeas on plasma glucose, insulin, and triacylglycerol concentrations. *American Journal of Clinical Nutrition*, 79(3): 390-5.

Oakenful, D. and Sidhu, G. S. (1990). Could saponins be a useful treatment of hypercholesterolemia? *European Journal of Nutrition*, 44: 79–88.

Omima, E. F., Abdullahi, H. E., and Babiker, E. E. (2010). Effect of Fermentation on Biochemical Characteristics of Sorghum Flour Supplemented with Chickpea Flour. *Journal of Applied Sciences Research*, 6(7): 860–865.

Patterson, C. A., Curran, J. and Der, T. (2017). Effect of Processing on Antinutrient Compounds in Pulses. *Cereal Chemistry*, 94(1): 2-10.

Pedroche, J., Yust, M. M., Girón-Calle, J., Alaiz, M., Millán, F. and Vioque, J. (2002). Utilisation of chickpea protein isolates for production of peptides with angiotensin I-converting enzyme (ACE)-inhibitory activity. *Journal of the Science of Food and Agriculture*, 82: 960–965.

Pihlanto-Leppälä, A. (2000). Bioactive peptides derived from bovine whey proteins: opioid and ACE-inhibitory peptides. *Trends in Food Science & Technology*, 11: 347-356.

Pittaway, J. K., Ahuja, K. D., Cehun, M., Chronopoulos, A., Robertson, I. K., Nestel, P. J. and Ball, M. J. (2006). Dietary Supplementation with Chickpeas for at Least 5 Weeks Results in Small but Significant Reductions in Serum Total and Low-Density Lipoprotein Cholesterols in Adult Women and Men. *Annals of Nutrition and Metabolism*, 50(6): 512-8.

Pittaway, J. K., Robertson, I. K. and Ball, M. J. (2008). Chickpeas May Influence Fatty Acid and Fiber Intake in an Ad Libitum Diet, Leading to Small Improvements in Serum Lipid Profile and Glycemic Control. *Journal of the American Dietetic Association*, 108: 1009-1013.

Prathapan, A., Fahad, K., Thomas, B. K., Philip, R. M. and Raghu, K. G. (2011). Effect of sprouting on antioxidant and inhibitory potential of two varieties of Bengal gram

(*Cicer arietinum* L.) against key enzymes linked to type-2 diabetes. *International Journal of Food Sciences and Nutrition*, 62(3): 234–238.

Preuss, H. G. (2009). Bean amylase inhibitor and other carbohydrate absorption blockers: effects on diabesity and general health. *Journal of the American College of Nutrition*, 28(3), 266-276.

Quintero-Soto, M. F., Saracho-Peña, A. G., Chavez-Ontiveros, J., Garzon-Tiznado, J. A., Pineda-Hidalgo, K. V., Delgado-Vargas, F. and Lopez-Valenzuela, J. A. (2018). Phenolic profiles and their contribution to the antioxidant activity of selected chickpea genotypes from Mexico and ICRISAT collections. *Plant Foods for Human Nutrition*, 73(2): 122-129.

Rachwa-Rosiak, D., Nebesny, E., Budryn, G. (2013): Chickpeas - composition, nutritional value, health benefits, application to bread and snacks: A review. *Critical Reviews in Food Science and Nutrition*, 55(8): 1137-1145.

Raicht, R. F., Cohen, B. I., Fazzini, E. P., Sarwal A. N. and Takahashi, M. (1980). Protective effect of plant sterols against chemically induced colon tumors in rats. *Cancer Research*, 40(2): 403–405.

Ratnavathi, C. V. (2019). Grain Structure, Quality, and Nutrition. In: *Breeding Sorghum for Diverse End Uses,* edited by: Aruna, C., Visarada, K. B. R. S., Venkatesh Bhat, B. and Tonapi V. A., 193-207, Woodhead Publishing Series in Food Science, Technology and Nutrition.

Reyes-Bastidas, M., Reyes-Fernández, E. Z., López-Cervantes, J., Milán-Carrillo, J., Loarca-Piña, G. F. and Reyes-Moreno, C. (2010). Physicochemical, nutritional and antioxidant properties of tempeh flour from common bean (*Phaseolus vulgaris* L.). *Food Science and Technology International*, 16: 426-432.

Reyes-Moreno, C., Cuevas-Rodríguez, E. O., Milán-Carrillo, J., Cárdenas-Valenzuela, O. G. and Barrón-Hoyos, J. (2004). Solid state fermentation process for producing chickpea (*Cicer arietinum* L) tempeh flour. Physicochemical and nutritional characteristics of the product. *Journal of the Science of Food and Agriculture*, 84: 271–278.

Reyes-Moreno, C., Milán-Carrillo, J., Rouzaud-Sandez, O., Garzón-Tiznado, J. A. and Mora-Escobedo, R. (2002). Dehulling/Softening/Extrusion (DSE): Technological Alternative to Improve Nutritional Quality of Chickpea (*Cicer arietinum* L.). *Agrociencia*, 36: 181-189.

Reyes-Moreno, C., Romero-Urías, C., Milán-Carrillo, J., Valdéz-Torres, B. and Zárate-Márquez, E. (2000). Optimization of the solid state fermentation process to obtain tempeh from hardened chickpeas (*Cicer arietinum* L.). *Plant Foods for Human Nutrition*, 55: 219–228.

Reyes-Moreno, C., Rouzaud-Sandez, O., Milán-Carrillo, J., Garzón-Tiznado, J. A. and Camacho-Hernández, L. (2001). Hard-to-cook tendency of chickpea (*Cicer arietinum* L) varieties. *Journal of the Science of Food and Agriculture*, 81(10): 1008-1012.

Rice-Evans, C. A., Miller, N. J. and Paganga, G. (1997). Antioxidant properties of phenolic compounds. *Trends in Plant Science*, 2: 152–159.

Rochín-Medina, J. J., Gutiérrez-Dorado, R., Sánchez-Magaña, L. M., Milán-Carrillo, J., Cuevas-Rodríguez, E. O., Mora-Rochín, S., Valdez-Ortiz, A. and Reyes-Moreno, C. (2015). Enhancement of nutritional properties, and antioxidant and antihypertensive potential of black common bean seeds by optimizing the solid state bioconversion process. *International Journal of Food Sciences and Nutrition*, 66(5): 498–504.

Ruiz, R. G., Price, K., Rose, M., Rhodes, M. and Fenwick, R. (1996). A preliminary study on the effect of germination on saponin content and composition of lentils and chick peas. *Zeitschrift fur Lebensmittel-Untersuchung und -Forschung*, 203(4): 366–369.

Sales-Campos, H., Souza, P. R., Peghini, B. C., da Silva, J. S. and Cardoso, C. R. (2013). An overview of the modulatory effects of oleic acid in health and disease. *Mini-Reviews in Medicinal Chemistry*, 13(2): 201-10.

Salgado, P., Lallésa, J. P., Toulleca, R., Mouratob, M., Cabralb, F. and Freire, J. P. B. (2001). Nutrient digestibility of chickpea (*Cicer arietinum* L.) seeds and effects on the small intestine of weaned piglets. *Animal Feed Science and Technology*, 91: 197–212.

Sánchez-Chino, X. M., Martínez, C. J., León-Espinosa, E. B., Leticia Garduño-Siciliano, Álvarez-González, I., Madrigal-Bujaidar, E., Vásquez-Garzón, V. R., Baltiérrez-Hoyos, R. and Dávila-Ortiz, G. (2018). Protective effect of chickpea protein hydrolysates on colon carcinogenesis associated with a hypercaloric diet. *Journal of the American College of Nutrition*, 38(2): 162-170.

Sánchez-Chino, X., Jiménez-Martínez, C., Dávila-Ortiz, G., Álvarez-González, I. and Madrigal-Bujaidar, E. (2015). Nutrient and Nonnutrient Components of Legumes, and Its Chemopreventive Activity: A Review. *Nutrition and Cancer*, 0(0), 1–10.

Sánchez-Chino, X. M., Martínez, C. J., Garzón, V. R. V., González, I. Á., Treviño, S. V., Bujaidar, E. M., Ortiz, G. D. and Hoyos, R. B. (2017). Cooked Chickpea Consumption Inhibits Colon Carcinogenesis in Mice Induced with Azoxymethane and Dextran Sulfate Sodium. *Journal of the American College of Nutrition*, 36(5):391-398.

Sánchez-Magaña, L. M., Cuevas-Rodríguez, E. O., Gutiérrez-Dorado, R., Ayala-Rodríguez, A. E., Valdez-Ortiz, A., Milán-Carrillo, J. and Reyes-Moreno, C. (2014). Solid-state Bioconversion of chickpea (*Cicer arietinum* L.) by Rhizopus oligosporus to improve total phenolic content, antioxidant activity and hypoglycemic functionality. *International Journal of Food Science and Nutrition*, 65(5): 558-564.

Sánchez-Mata, M. C., Peñuela-Teruel, M. J., Cámara-Hurtado, M., Diéz-Marqués, C. and Torija-Isasa, E. M. (1998). Determination of mono-, di-, and oligosaccharides in legumes by high-performance liquid chromatography using an amino-bonded silica column. *Journal of Agricultural and Food Chemistry*, 46: 3648–3652.

Sardari, S., Amiri, M., Rahimi, H., Kamalinejad, M., Narenjkar, J. and Sayyah, M. (2015). Anticonvulsant effect of *Cicer arietinum* seed in animal models of epilepsy: introduction of an active molecule with novel chemical structure. *Iranian Biomedical Journal*, 19 (1): 45-50.

Segev, A., Badani, H., Kapulnik, Y., Shomer, I., Oren-Shamir, M. and Galili, S. (2010). Determination of polyphenols, flavonoids, and antioxidant capacity in coloured chickpea (*Cicer arietinum* L.). *Journal of Food Science*, 75: S115–S119.

Shahidi, F. (1997). Beneficial Health Effects and Drawbacks of Antinutrients and Phytochemicals in Foods. In: *Antinutrients and Phytochemicals in Food*, edited by: Shahidi, F., 1-9, Memorial University of Newfoundland.

Sharma, A. and Sehgal, S. (1992a). Effect of processing and cooking on the antinutritional factors of faba bean (*Vicia faba*). *Food Chemistry*, 43:383-385.

Sharma, A. and Sehgal, S. (1992b). Effect of domestic processing, cooking and germination on the trypsin inhibitor activity and tannin content of faba bean (*Vicia faba*). *Plant Foods for Human Nutrition*, 42: 127-133.

Sharma, R. D. (1979). Isoflavones and Hypercholesterolemia in Rats. *Lipids*, 14(6): 535-9.

Sharma, R. D. (1980). Effect of hydroxy acids on hypocholesterolemia in rats. *Atherosclerosis*, 37: 463–468.

Sharma, R. D. (1984). Hypocholesterolemic effect of hydroxy acid components of Bengal gram. *Nutrition Reports International*, 29: 1315–1322.

Sharma, S., Yadav, N., Singh, A. and Kumar, R. (2013). Nutritional and antinutritional profile of newly developed chickpea (*Cicer arietinum* L) varieties. *International Food Research Journal*, 20(2): 805-810.

Shen, P., Liu, M. H., Ng, T. Y., Chan, Y. H. and Yong, E. L. (2006). Differential effects of isoflavones, from astragalus membranaceus and pueraria thomsonii, on the activation of PPARα, PPARγ, and adipocyte differentiation *in vitro*. *Journal of Nutrition*, 136, 899-905.

Shi, L., Arntfield, S. D. and Nickerson, M. (2018). Changes in levels of phytic acid, lectins and oxalates during soaking and cooking of Canadian pulses. *Food Research International*, 107: 660-668.

Shi, W., Hou, T., Liu, W., Guo, D. and He, H. (2019). The hypolipidemic effects of peptides prepared from *Cicer arietinum* in ovariectomized rats and HepG2 cells. *Journal of the Science of Food and Agriculture*, 99(2): 576-586.

Shimelis, E. A. and Rakshit, S. K. (2007). Effect of processing on antinutrients and *in vitro* digestibility of kidney bean (*Phaseolus vulgaris* L) varieties grown in East Africa. *Food Chemistry*, 103: 161-172.

Shinohara, S., Gu, Y., Yang, Y., Furuta, Y., Tanaka, M., Yue, X., Wang, W., Kitano, M. and Kimura, H. (2016). Ethanol extracts of chickpeas alter the total lipid content and expression levels of genes related to fatty acid metabolism in mouse 3T3-L1 adipocytes. *International Journal of Molecular Medicine*, 38(2): 574-84

Singh, U. and Jamburathan, R. (1982). Distribution of Seed protein fractions and Amino Acid in Different Anatomical Parts of Chickpea (*Cicer arietinum* L.) and Pigeon pea (*Cajanus cajan* L.). *Plant Foods for Human Nutrition*, 32: 347-54.

Singh, U., Kherdekar, M. S. and Jambunathan, R. (1982). Studies on desi and kabuli chickpea (*Cicer arietinum* L.) cultivars. The levels of amylase inhibitors, levels of oligosaccharides and in vitro starch digestibility. *Journal of Food Science*, 47: 510–512.

Slavin, J. (2013). Fiber and prebiotics: mechanisms and health benefits. *Nutrients*, 5: 1417–1435.

Smith, I. E. and Dowsett M. (2003). Aromatase inhibitors in breast cancer. *New England Journal of Medicine*, 348: 2431–2442.

Sotelo, A., Flores, F and Hernandez, M. (1987). Chemical composition and nutritional value of Mexican varieties of chickpeas (*Cicer arietinum*). *Plant Foods for Human Nutrition*, 37: 299-306.

Sreerama, Y. N., Vadakkoot, B. S. and Vishwas, M. P. (2012). Phenolic compounds in cowpea and horse gram flours in comparison to chickpea flour: Evaluation of their antioxidant and enzyme inhibitory properties associated with hyperglycemia and hypertension. *Food Chemistry*, 133: 156–162.

Thompson, L. U. (1993). Potential health benefits and problems associated with antinutrients in foods. *Food Research International*, 26: 131–149.

Tormo, M. A., Gil-Exojo, I., Romero de Tejada, A. and Campillo, J. E. (2004). Hypoglycaemic and anorexigenic activities of an α-amylase inhibitor from white kidney beans (*Phaseolus vulgaris*) in Wistar rats. *British Journal of Nutrition*, 92: 785-90.

Torres-Fuentes, C., Contreras, M. M., Recio, I., Alaiz, M. and Vioque, J. (2015). Identification and characterization of antioxidant peptides from chickpea protein hydrolysates. *Food Chemistry*, 180: 194-202.

Trinidad, T. P., Mallillin, A. C., Loyola, A. S., Sagum, R. S. and Encabo, R. R. (2010). The potential health benefits of legumes as a good source of dietary fibre. *British Journal of Nutrition*, 103: 569–574.

Tsao, R. (2010). Chemistry and biochemistry of dietary polyphenols. *Nutrients*, 2: 1231–46.

Udahogora, M. (2012). Health benefits and bioactive compounds in field peas, faba beans, and chickpeas. In: *Cereals and Pulses Nutraceutical Properties and Health Benefits,* edited by Shahidi, F. Liangli Yu, L. and Tsao, R., 199-216. Wiley-Blackwell.

Urbano, G., López-Jurado, M., Aranda, P., Vidal-Valverde, C., Tenorio, E. J. and Porres, E. (2000). The role of phytic acid in legumes: antinutrient or beneficial function? *Journal of Physiology and Biochemistry*, 56: 283–294.

Urias-Orona, V., Huerta-Oros, J., Carvajal-Millán, E., Lizardi-Mendoza, J., Rascón-Chu, A. and Gardea, A. A. (2010). Component Analysis and Free Radicals Scavenging Activity of *Cicer arietinum* L. Husk Pectin. *Molecules*, 15(10): 6948-6955.

Veena, A., Urooj, A. and Shashikala, P. (1995). Effect of processing on the composition of dietary fibre and starch in some legumes. *Die Nahrung*, 39(2): 132-138.

Velickovic, T. D. C. and Stanic-Vucinic, D. J. (2018). The role of dietary phenolic compounds in protein digestion and processing technologies to improve their antinutritive properties. *Comprehensive Reviews in Food Science and Food Safety*, 17: 82-103.

Venkidasamy, B., Selvaraj, D. Nile, A. S., Ramalingam, S., Kaia, G. and Nilea, S. H. (2019). Indian pulses: A review on nutritional, functional and biochemical properties with future perspectives. *Trends in Food Science & Technology*, 88: 228-242.

Vulevic, J., Drakoularakou, A., Yaqoob, P., Tzortzis, G. and Gibson, G. R. (2008). Modulation of the fecal microflora profile and immune function by a novel trans-galactooligosaccharide mixture (B-GOS) in healthy elderly volunteers. *The American Journal of Clinical Nutrition*, 88(5): 1438–1446.

Vuolo, M. M., Lima, V. S. and Maróstica, M. R. (2019). Phenolic Compounds: Structure, Classification, and Antioxidant Power. In: *Bioactive Compounds: Health Benefits and Potential Applications,* edited by: Segura-Campos, M. R., 33-55, Woodhead Publishing.

Wahby, M. M., Mohammed, D. S., Newairy, A. A., Abdou, H. M. and Zaky, A. (2017). Aluminum-induced molecular neurodegeneration: The protective role of genistein and chickpea extract. *Food and Chemical Toxicology*, 107(A): 57-67.

Wang, J., He, C., Wu, W. Y., Chen, F., Wu, Y. Y., Li, W. Z., Chen, H. Q. and Yin, Y. Y. (2015). Biochanin A protects dopaminergic neurons against lipopolysaccharide-induced damage and oxidative stress in a rat model of Parkinson's disease. *Pharmacology, Biochemistry and Behavior*, 138: 96–103.

Wang, Y., Wang, Z., Fu, L., Chen, Y. and Fang, J. (2013). Legume consumption and colorectal adenoma risk: a meta-analysis of observational studies. *PLoS One*, 8(6): e67335.

Wood, J. A. and Grusak, M. A. (2007). Nutritional value of chickpea. In: *Chickpea Breeding and Management,* edited by: Yadav, S. S., Redden, R. J., Chen, W. and Cabi, S.B, 101–142. UK: CAB International.

Wood, J. A. Knights, J. E. and Harden, S. (2008). Milling performance in desi-type chickpea (*Cicer arietinum* L.): effects of genotype, environment and seed size. *Journal of the Science of Food and Agriculture*, 88: 108–115.

Wu, Z., Song, L., Feng, S., Liu, Y., He, G., Yioe, Y., Liu, S. Q. and Huang, D. (2012). Germination dramatically increases isoflavonoid content and diversity in chickpea (*Cicer arietinum* L.) seeds. *Journal of Agricultural and Food Chemistry*, 60: 8606-8615.

Xu, B. and Chang, S. K. C. (2009). Phytochemical profiles and health-promoting effects of cool-season food legumes as influenced by thermal processing. *Journal of Agricultural and Food Chemistry*, 57: 10718–10731.

Xue, Z., Gao, J., Zhang, Z., Yu, W., Wang, H. and Kou, X. (2012). Antihyperlipidemic and antitumor effects of chickpea albumin hydrolysate. *Plant Foods for Human Nutrition*, 67(4): 393-400.

Xue, Z., Hou, X., Yu, W., Wen, H., Zhang, Q., Li, D. and Kou, X. (2018). Lipid metabolism potential and mechanism of CPe-III from chickpea (*Cicer arietinum* L.). *Food Research International*, 104: 126-133.

Xue, Z., Wen, H., Zhai, L., Yu, Y., Li, Y., Yu, W., Cheng, A., Wang, C. and Kou X. (2015). Antioxidant activity and anti-proliferative effect of a bioactive peptide from chickpea (*Cicer arietinum* L). *Food Research International*, 77(2): 75-81.

Yan, X., Ye, R. and Chen, Y. (2015). Blasting extrusion processing: the increase of soluble dietary fiber content and extraction of soluble-fiber polysaccharides from wheat bran. *Food Chemistry*, 180, 106-115.

Yanagihara, K., Ito, A., Toge, T. and Numoto, M. (1993). Antiproliferative effects of isoflavones on human cancer cell lines established from the gastrointestinal tract. *Cancer Research*, 53: 5815–5821.

Yang, Y., Zhou, L., Gu, Y., Zhang, Y., Tang, J., Li, F., Shang, W., Jiang, B., Yue, X. and Chen, M. (2007). Dietary chickpeas reverse visceral adiposity, dyslipidaemia and insulin resistance in rats induced by a chronic high-fat diet. *British Journal of Nutrition*, 98(4): 720-6.

Ye, Z., Wang, W., Yuan, Q., Ye, H., Sun, Y., Zhang, H., Zeng, X. (2016). Box–Behnken design for extraction optimization, characterization and *in vitro* antioxidant activity of *Cicer arietinum* L. hull polysaccharides. *Carbohydrate polymers*, 147: 354-364.

Yilmazer-Musa, M., Griffith, A. M., Michels, A. J., Schneider E.2 and Frei B. (2015). Inhibition of α-amylase and α-glucosidase activity by tea and grape seed extracts and their constituent catechins. *Journal of Agricultural and Food Chemistry,* 60(36): 8924–8929.

Yust, M. M., Pedroche, J., Giron-Calle, J., Alaiz, M., Millan, F. and Vioque, J. (2003). Production of ace inhibitory peptides by digestion of chickpea legumin with alcalase. *Journal of Food Chemistry*, 81: 363–369.

Zafar, T. A. and Kabir, Y. (2017). Chickpeas suppress postprandial blood glucose concentration, and appetite and reduce energy intake at the next meal. *Journal of Food Science and Technology*, 54(4): 987–994.

Zhang, T., Li, Y., Miao, M. and Jiang, B. (2011). Purification and characterisation of a new antioxidant peptide from chickpea (*Cicer arietium* L.) protein hydrolysates. *Food Chemistry*, 128: 28-33.

Zhang, Y., He, Y., Liu, C., Liu, C. and Li, S. (2018). *In vitro* screening and isolation of human aromatase inhibitors from *Cicer arietinum* by a novel continuous online method combining chromatographic techniques. *Journal of Separation Science*, 41(2): 483-492.

Zhang, Y., Su, D., He, J., Dai, Z., Asad, R., Ou, S., Zeng, X. (2017). Effects of ciceritol from chickpeas on human colonic microflora and the production of short chain fatty acids by *in vitro* fermentation. *LWT - Food Science and Technology*, 79: 294-299.

Zia-ul-Haq, M., Iqbal, S., Ahmad, S., Iqbal Bhanger, M., Wiczkowski, W. and Amarowicz, R. (2008). Antioxidant potential of desi chickpea varieties commonly consumed in Pakistan. *Journal of Food Lipids*, 15: 326–342.

In: Handbook of Chickpeas
Editors: Albert T. Lund and Noah D. Schultz
ISBN: 978-1-53616-374-2

Chapter 2

NUTRITIONAL VALUES, HEALTH BENEFITS AND MULTIPLE USES OF KABULI CHICKPEAS

Muhammad Aasim[1,*], Zeliha Ustun Argon[2], Ashfaq Ahmed Khan[3] and Sibel Day[4]

[1]Department of Biotechnology, Faculty of Science, Necmettin Erbakan University, Konya, Turkey

[2]Department of Biosystem Engineering, regli Faculty of Engineering and Natural Sciences, Necmettin Erbakan University, Konya, Turkey

[3]Department of Geography, Faculty of Literature, Karabuk University, Karabuk, Turkey

[4]Department of Field Crops, Faculty of Agriculture, Ankara University, Ankara, Turkey

ABSTRACT

Chickpea (*Cicer arietinum* L.) is a key dietary component due to its cheap and inexpensive protein, dietary fibres, vitamins and minerals along with other significant functional properties. Chickpea is one of the oldest cultivated edible legumes with high demand in the modern World. The consumption of chickpea is increasing steadily due to its demand especially in the developing and under-developed countries. Chickpeas provide wide taste and texture variety to cereal-based diet and ensures a balanced diet for both a vegetarian and non-vegetarian population. Chickpea is an herbaceous, annual legume and can be classified into three main groups of desi, kabuli, and pea on the basis of seed structure and consumption. However, the major cultivated chickpea types are Kabuli, and desi types. Kabuli varieties are preferred for consumption as whole seeds in different forms, whereas desi types are typically processed into flour and splitted form.

* Corresponding Author's Email: mshazim@gmail.com.

Turkey is the major producer of both types of kabuli and desi chickpea. Turkey is also a major contributor of kabuli type chickpea and is ranked 5th after India, Australia, Myammar and Ethiopia and is ranked 1st in the Mediterranean region in terms of production. In Turkey, chickpea is consumed in different ways from raw to cooked or roasted. The production and consumption of desi chickpea is limited and confined to a specific area of Turkey. The nutritional value and their effects on human health varies for both types and affected by number of factors. This study highlights the nutritional values of domesticated Turkish chickpeas used in different forms and also highlights the impact of different factors which regulate the nutritional values.

Keywords: chickpea, Desi, Kabuli, health, nutrition, Turkey

INTRODUCTION

The chickpea (*Cicer arietinum* L.) is an herbaceous, annual legume with diverse uses such as being a human food and an animal feed. The chickpea is an important dietary component in most of countries and is considered to be an inexpensive protein source along with a wide range of functional properties and applications. The chickpea also has a deep history of cultivation and culture of consumption (Knights and Hobson 2016; Megias et al. 2016). On a global scale, chickpeas are produced mainly in India, Turkey and Pakistan and they are consumed all over the world with different purposes, these purposes including the creation of a balanced diet with high-quality proteins, carbohydrates, minerals and fiber (Thavarajah and Thavarajah 2012; Aydemir and Yemenicioglu 2013; Dadon et al. 2017). This thus improves the nutritional status of both a vegetarian and non-vegetarian population and the enrichment of the variety, taste, and texture of the cereal-based diet (Thavarajah and Thavarajah 2012; Aydemir and Yemenicioglu 2013; Dadon et al. 2017).

The chickpea can be classified into three main groups of Desi, Kabuli, and pea on the basis of seed size and shape (Knights and Hobson 2016). Kabuli varieties are preferred for consumption as whole seeds, in many meals, salads and humus or as an ingredient in gluten-free products. Kabuli chickpea seeds are more rounded and thinner in shape compared to the Desi type and have a white to cream-beige, and smooth seed coat. The size of Kabuli chickpeas is generally larger than the Desi varieties, weighing 100–750 mg. The cotyledon fraction is about 92% and the seed coat fraction is 6.5%, which is lower than the Desi type. The general nutrition composition of Kabuli chickpeas is given as 17.8–31% protein, 56.7–63% carbohydrate, 3.1–6.8 % fat, 45% starch, 3.8–8.0% fibre, and 1.9–3.6% ash (Boye, Zare and Pletch 2010; Arntfield and Maskus 2011; Aydemir and Yemenicioglu 2013, Knights and Hobson 2016, Sharma et al. 2016; Çetiner and Bilek Ersus 2018), minerals and vitamins (Zia-Ul-Haq et al. 2007a). The main proteins existing in chickpeas are albumins and globulins with small traces of glutelins and prolamines (Saharan and Khetarpul 1994).

The chickpea is considered to be a cheap and nutritious food source in many developing countries of the world including Turkey. The most common variety in Turkey is the light seeded Kabuli type with a small amount of Desi chickpea cultivation in the South Eastern Anatolian region of Turkey (Akın 2018). Turkey is one of the major producers of chickpeas in the world with production of 0.47 million tons of chickpeas in the year 2017 (FAO 2019). According to the Turkish Bureau of Statistics, 50 % of the total cultivated area of legumes constituted chickpeas with a share of 40.3% in total production in the country in 2017 (TUİK, 2018). However, a significant decrease in the production and cultivated area of chickpeas has been observed for several years. The cultivated area of chickpeas increased from 89,000 ha in 1961 to 880,000 ha in 1990, however, it decreased to 395,000 ha in the year 2016 (TUİK, 2018). The production volume of chickpeas in Turkey in 2018 reached up to 630.000 mt from 0.5144 mil. ha and the consumption average was 5-6 kg per person/year (Gülümser 2016; TUİK, 2018). The main chickpea planting areas are reported to be in Central Anatolia, the Aegean and Mediterranean Regions, having 60.6% of the total production (Anonymous 2018). The Kırşehir and Ankara provinces of Central Anatolia rank first and second with 34,000 and 30,800 hectares of chickpea cultivated area respectively while the Uşak province of the Aegean Region ranks third with 28,600 ha (TUIK 2018)

Again, the Central Anatolia, Aegean and Mediterranean regions lead in terms of chickpea production in the country (Figure 2). According to the statistics, Kırşehir province stands first with production of 46,000 tons of chickpeas in the year 2017, followed by Ankara and Konya with each producing 35,000 tons. Although the production of chickpeas is increased by 3.3% in the year 2017 as compared to the previous year, there is an overall decrease in the production in the long run. The decrease in cultivated area, crop input price fluctuations, changes in climatic patterns, effects of diseases and lack of fertility are among the important causes of decreases in the production. The decrease in cultivated area of chickpeas is associated with effect of plant diseases and a higher crop input price (TUIK 2018).

There are many cultivated chickpea landraces due to the cultivation practices and diversity in consuming types. The main strains of Kabuli chickpeas in Turkey are namely Akcin-91, Gokce-97, Er-99, Uzunlu-99, Kusmen-99, Dikbas, Inci, Hasanbey, Seckin, Azkan, Yaşa-05, and Aziziye-94 (Bayrak and Onder 2017; Macar, Macar and Mart 2017). The Desi chickpea is also cultivated in some parts of Turkey with limited information about its cultivation, nutritive values and consumption. This study presents an insight about the nutritional, health values and multiple forms of using cultivated Kabuli chickpeas with emphasis on work related to Turkish chickpeas.

1. Nutritive Values of Kabuli Chickpeas

1.1. Protein Contents of Kabuli Chickpeas

The protein content of Kabuli chickpeas averages 17.8-29% and *in vitro* protein digestibility is 52.4-69% for the whole seed and 72.7–79.1% for *dhal* which is higher than the Desi types. The amount of protein in Turkish chickpeas is found to be 27.1, 24.1, 21.9 mg/ml for Inci, Hasanbey and Seckin varieties respectively (Macar, Macar and Mart 2017). A study by Özer et al. (2010) on 91 Kabuli chickpea landraces revealed that the protein varies between a range of 17.55-23.31%. The wide protein range was attributed to genetic variation, different soil and environment requirements or interactions of different factors. The protein content can be affected by practices during plantation such as irrigation. Kırnak et al. (2017) showed that irrigation of 50% during the blooming, and 50% during bean maturation periods will help to increase the protein content of chickpeas (Khan, Akmal and Kissana 1985; Alajaji and El-Adawy 2006; Boye, Zare and Pletch 2010).

Due to their high protein content, pulses have been a point of interest for protein isolates with their functional properties in recent years. The isolates found have high fat and water absorption rates, which is suitable for inclusion within cheese, bakery and meat products. Due to the chickpea protein isolates, protein content being higher than those of chickpea flours, isolates are more suitable to be used in various high protein products as an ingredient which can provide additional protein. The studies conducted with different chickpea varieties found that Kabuli and Desi chickpeas' protein isolates differ significantly from each other in terms of oil and water absorption, foaming properties, gelation capacity, enthalpy of denaturation and denaturation temperature. Kabuli chickpea protein isolates and flour ash, fat and protein contents were 1.14%, 0.94%, 94.4% and 2.91%, 1.21%, 26.7% respectively. The protein isolates' solubility was lowest between a pH range of 4.0–5.0. Water absorption capacity of Kabuli chickpea isolates was 2.34 g/g which is lower than Desi chickpeas. While, the oil absorption capacity (OAC) which reflects that the emulsifying capacity was higher than Desi types with a value of 3.96 g/g, probably because of higher amount of non-polar amino acids in the Kabuli chickpeas. The least gelation concentration, which indicates gelling ability of proteins, foaming capacity, stability and thermal properties of Kabuli chickpeas isolates was also in an acceptable level for the intended applications. From the point of amino acids and digestibility isolates, the Kabuli variety was found to be better than Desi varieties but both varieties have better properties than soybean isolates and can be accepted as an excellent protein source for human nutrition (Kaur and Singh 2007; Boye, Zare and Pletch 2010; Arntfield and Maskus, 2011).

The purpose of isolation of proteins is to increase the protein content and to reduce the amount of anti-nutritional factors. Protein isolates processing occurs in various steps such as dehulling, air classification to separate the protein and starch fractions. Alkaline extraction/isoelectric precipitation, acid extraction, water extractions, salt extraction (micellization) and ultrafiltration are different extraction techniques, which serve the intended purpose and also depend on the variety of the chickpea. In the last step, proteins are purified by pulse protein extraction and extracted protein is recovered. Pulses proteins are utilized in different novel applications, such as in beverages, gluten-free products, meat products, desserts and dairy products, cereal based products. Functional properties of concentrates, protein flours and its isolates are solubility, water-fat binding capacity, emulsifying properties, gelation (least gelling concentration) which affect the later applications of proteins (Boye, Zare and Pletch 2010; Arntfield and Maskus, 2011).

1.2. Oil and Fatty Acids of Kabuli Chickpeas

Chickpea oil is rich in unsaturated fatty acids like Omega-6 (linoleic acid) and omega-9 (oleic acid) which accounts 46%-62% and 21.23%-30.52% of total fatty acids in chickpeas respectively. The fatty acid composition of Kabuli chickpeas (from Canada was found as myristic acid (0.66-0.31%), palmitic acid (10.45-10.51%), stearic acid (2.62-2.46%), oleic acid (19.37-18.91%), linoleic acid (58.77-60.08%) and linolenic acid (2.98-2.74%). Total saturated fatty acids were 15.01-14.49%, total unsaturated fatty acids were 83.52-84.06%, total monounsaturated fatty acids were 21.86-21.56% and total polyunsaturated fatty acids were 61.91-62.92%. Oil yields were also found to be at 8.23-8.39 %. Özer et al. (2010) evaluated the 91 Turkish chickpea landraces and found them to have 4.45-6.11% fat contents. Although the lipid content of chickpeas is generally low, regular and preferential additions to the daily diet of chickpeas could make a small, but significant contribution to a balanced and recommended dietary fat level, which are associated with lower cholesterol levels, decrease the risk of cardiovascular diseases, digestive disorders, inflammation, arthritis, cancer and some other chronic diseases (Boye, Zare and Pletch 2010; Knights and Hobson, 2016; Megias et al. 2016; Dadon et al. 2017; Padhi et al. 2017; Parca, Koca and Unay 2018).

1.3. Carbohydrates

Carbohydrate is the main constituent mainly starch in seed and accounts an average of 55-63.5% of the total seed. The starch amount is between 30-60% and *in vitro* starch digestibility of the Kabuli variety is 37–60%, which is higher than the Desi variety. The starch contents in Turkish landraces revealed the range of 3.486-4.18% (Özer et al. 2010) which are reported to be higher than Sicilian (Italy) landraces (Patane 2006) and lower than Spanish genotypes (Gil et al. 1996). Digestibility of pulses is affected by different physical and chemical factors such as composition, structure and size of the starch granules and it can be increased by applying cooking, roasting or autoclaving processes. Studies show that starch has a negative correlation with amylose and crude protein. 30–40% of total starch is amylose and therefore, the chickpea is accepted as a low glycemic index pulse which can support heart disease, diabetes and obesity treatments (Sharma et al. 2013; Wang et al. 2017). Approximately 10% of the starch is as soluble sugars such as raffinose and stachyose, which are oligosaccharides. Human digestive system (enzymes) is unable to break down these components and bacterial degradation produces gases and causes intestinal discomfort (Knights and Hobson, 2016). Chickpea grains are a good source of ciceritol (50%) and stachyose (35%) which are the main a-galactooligo-saccharide (a-GOS). Another a-GOS in chickpea is raffinose, which can support human intestinal health by regulating the intestinal microbial composition. Raffinose family oligosaccharides are not only important as a selective stimulator of beneficial bacteria, but also act as important immuno-stimulators in humans and animals. Therefore, the oligosaccharides, main carbohydrate components of chickpea, can be accepted as prebiotics and significant health promoters due to their immunostimulatory mechanisms and antioxidant effects in humans (Dadon et al. 2017).

1.4. Mineral Contents

Kabuli chickpea is a rich source of calcium (Ca) (93.4–197.4 mg/100 g), copper (Cu) (0.5–1.1 mg/100 g), iron (Fe) (3.7–10.5 mg/100 g), magnesium (Mg) (84.4–158.7 mg/100 g), phosphorus (P) (196.7-370.5 mg/100g), potassium (K) (731.4–998.0 mg/100 g), selenium (Se) (15.3–56.3 μg/100 g), zinc (Zn) (1.4–7.4 mg/100 g), sulfur (S) (141.7-222.4 mg/100g), sodium (Na) (40.3-102.7 mg/100g), boron (B) (0.7-1.3 mg/100g) and manganese (Mn) (1.1- 2.3 mg/100g). Therefore, an important amount of the recommended daily allowance (RDA) of numerous minerals can be provided with a single chickpea serving. Calcium is a mineral, which is important especially for pre-school children, adolescents, pregnant and lactating women to build bone-density. The amount of calcium, which is generally limited in low-income sectors' diets, is contributed by chickpeas at 3.7% of adults' and 7% of children's (1-3 years) daily recommended intake. Calcium deficiency can cause osteoporosis in adults and rickets in children. Daily iron and zinc intake also can be met with chickpeas. Potassium is important to maintain of energy metabolism and cell membrane. Phosphorous is key element in the human metabolism, phosphoproteins, phospholipids, ATP, nucleic acids hydroxyapetite which is an important bone component. Considering the mineral composition of chickpeas, cultivating appropriate varieties may decrease micronutrient malnutrition (Thavarajah and Thavarajah 2012; Sharma et al. 2013; Dadon et al. 2017, Erbas Kose and Mut 2019). Comparing the treatments in terms of mineral losses with microwave cooking and autoclaving resulted in the greatest retention of all minerals respectively. Cooking chickpeas in boiling water causes minerals to leach from the chickpea seeds into the cooking water and thus causes loss of Fe (8%), Cu (15%) and K (24%) (Alajaji and El-Adawy 2006; Wang et al. 2010a).

1.5. Vitamins

Chickpeas contain essential vitamins including niacin (B3) (0.2 mg/100g), riboflavin (B2) (0.9 mg/100 g), β-carotene (a precursor for vitamin A), folic acid (172mg/100g), thiamin (B1) (0.1 mg/100 g), Vitamin A (4.4 mg/100g), vitamin C (2.1 mg/100g), vitamin E (0.6 mg/100g), pyridoxine (B6) (28.2 mg/100g), vitamin K (6.6 mg/100g), and cobalamin (B12) (0.5 mg/100g). Some studies showed that the risk of cardiovascular diseases is inversely correlated with the amount of antioxidants β-carotene and vitamin E. Vision, cell division/differentiation processes, bone growth and other key developmental processes are regulated by vitamin A. Folate also helps to decrease the risk of neural tube defects (NTDs) in newborn babies. Pantothenic acid (B5) (0.286mg/100 gr), found in chickpeas is involved in the metabolism and breakdown of fats, carbohydrates and proteins. Pyridoxine plays an important role in the neurotransmitters and enzymes synthesis, immune system function, metabolism of red blood cells and blood glucose control. Folic acid is important in DNA replication, RNA synthesis, growth and maintenance of new cells. Vitamin C is essential for fatty acid metabolism, neurotransmitters' synthesis, and for connective tissues (Dadon et al. 2017; Venkidasamy et al. 2019). The cooking process can cause losses in B vitamins starting from pyridoxine to riboflavin, thiamin and niacin. Losses can be prevented or decreased by changing the process to autoclaving or microwaving (Alajaji and El-Adawy 2006).

1.6. Secondary Metabolites of Kabuli Chickpeas

Chickpea protein concentrates are considered to be a "hidden source" of health-promoting isoflavones. Biochanin A (8-31 µg/g), formononetin (1-4 µg/g), daidzein (6 µg/g), and genistein, Biochanin A glucoside, flavonol kaempferol were detected in dry chickpeas or in its flour with six different studies. The isoflavone profile can be affected with cooking, soaking and industrial dehydration depending on the chickpea variety. Germination also rapidly increases the level of major isoflavone biochanin A in chickpeas and also effects its distribution. Using an alkaline extraction process and precipitation at the isoelectric pH during the preparation of chickpea protein concentrates can release the isoflavones, mostly biochanin A, in its aglycone form and the concentrate can contain four times more isoflavone than flour (Megias et al. 2016). Another study also showed double anti-inflammatory activity of germinated chickpeas (GC) compared to cooked chickpeas (CC) digests. This result was associated with greater biochanin A, peptides and formononetin contents in the GC as the protein content is higher in GC than in the raw chickpeas. Therefore, chickpeas may contribute to the prevention of bowel inflammatory diseases and gut health (Milán-Noris et al. 2018; Kilincer and Demir, 2019). Also Biochanin A was found with the highest oestrogenic activity and followed by daidzein and formononetin. Daidzeins is responsible for regulating breast and prostate cancers, heart disease, menopause, plaque and osteoporosis symptoms. Also applying a proteins and isoflavones combined diet can enhance the hypocholesterolaemic activity of isoflavones (Peksen and Artik 2005; Venkidasamy et al. 2019).

Studies on chickpeas showed that phenolic compounds were in different distributions and concentrations in cotyledon, embryo axis, and seed shell sections. The maximum total phenolics and tannins contents are concentrated in the seed peel and can easily be removed by peeling. The present flavonoids are mainly quercetin, kaempferol, myricetin, daidzein and genistein. These flavonoids are mostly concentrated on the embryonic axis. Chlorogenic, gallic, β-coumaric acid and altered aldehydes are the predominant phenolic acids in raw and cooked chickpeas compared to other compounds. The studies with the Kabuli variety in Turkey showed the total phenolic content as 119.2-176.8 mg gallic acid per gram of extract (GAE)/100 g. Another study showed the difference of total phenolics between chickpeas with a seed coat and coatless chickpea as 62 mg GAE/100g and 82 g GAE/100 g respectively. However, the amount of phenolic in foods is affected by extraction conditions and analytical methods used. In addition, climatic conditions, agronomic applications, harvesting, ripening phase, storage, industrial and local processes and genetic factors affect the structure and properties of phenolics (Macar, Macar and Mart 2017; Sayar and Karataş 2017; Sarioglu and Velioglu 2018; Kilincer and Demir, 2019). A study by Milan Noris et al. (2018) showed that phenolic fraction's anti-inflammatory activity in digests was 7 fold greater than the protein fraction ($p < 0.05$). Total polyphenol concentrations for different chickpea varieties are reported from 0.5 to 6.8 mg/g. The study compared the cooked and germinated chickpeas and found that germinated chickpea digest (IC50=93 μg/mL) contained 24 peptides which are based on legumin and vicilin. It was also showed that concentrated proteins of germinated chickpea proteins may support the treatments of bowel inflammatory diseases by preventing inflammation in the lower gut.

In addition, chickpeas are rich in carotenoids which can boost mineral absorption by human digestive system and is the most efficient precursor of vitamin A. Carotenoids in chickpeas found in different forms such as beta-carotene (166–431 μg/100 g), xanthophyll (9.0–19.7 mg/100 g) and canthoxanthine (21.0–67.9 mg/100 g). Apart from β-carotene, lutein and zeaxanthin (6-17 %) are also present in chickpea. These components protect the eye from macular degeneration while acting as potent antioxidants. Quercetin and ferulic acid present in chcikpea are beneficial for the human body along with low phytic acid (PA; 5.8–13.6 mg/g) (Thavarajah and Thavarajah 2012; Dadon et al. 2017; Padhi et al. 2017).

Pulse seeds and extracts have the potential to prevent or treat hypertension due to Angiotensin-converting enzyme (ACE) inhibitors. Another component are lectins in the chickpea which can help to enhance the immune system by showing natural killer activity, antibody synthesis, lymphocyte proliferation, and cytokine regulation as active immunomodulatory agents. Therefore, these components are highly significant for the activation of certain immune system activities, inhibiting certain cancer types and controlling or preventing obesity (Roy, Boye and Simpson 2010). The α-tocopherol content of chickpeas is generally more than the other edible pulses such as green peas, mungbeans, red kidney beans and lentils. The d-tocopherol has an anti-oxidant property and, along with the a-tocopherol, makes chickpea oil stable against oxidation and contributes to longer shelf life.

2. Raw and Cooked Kabuli Chickpea

A survey done in USA with 19 years old adult participants indicates that chickpea consumers had higher dietary fiber (24.4 ± 0.7 g/d), vitamin A (787 ± 42 RAE mcg/d), vitamin E (10.1±0.5 mcg/d), vitamin C (119 ± 8 mg/d), folate (627 ± 16 mcg/d); polyunsaturated fatty acids (19.5 ± 0.4 g/d), potassium (3103 ± 59 mg/d); magnesium (385 ± 13 mg/d); and iron (17.4 ± 0.5 mg/d) intake. Whereas, saturated fatty acids (SFA) (22.4 ± 0.7 g/d), total fat (76.4 ± 4.5 g/d) and cholesterol (227 ± 8 mg/d) were present in low level. The chickpea consumers also showed higher health index (HEI) results (62.2 ± 1.3) compared to non-consumers. In the light of the results, chickpea consumption was related to valuable nutrient intake, diet quality and consumption should be supported (O'Neil, Nicklas and Fulfoni 2014). The comparison of nutritional values of raw and cooked chickpeas is given in Table 1. The analysis revealed the lower nutritional values of cooked chickpeas compared to raw chickpeas. That is also one of the possible reasons of using raw chickpeas more frequently in Turkey and other countries like Pakistan and India

3. Roasted Chickpea (Leblebi)

Roasted chickpea *(Leblebi)* is an authentic and traditional product obtained by roasting chickpeas at different temperatures. *Leblebi* is a famous snack in Turkey and is prepared after many heat treatments, roasting and recession processes of chickpeas. There are different types of *leblebi* produced and consumed in Turkey like dehulled *leblebi*

(*Sarı leblebi* and *Girit leblebi)* and non-dehulled *leblebi (Beyaz leblebi* and *Sakız leblebi)*. Their production processes also differ from each other (Coşkuner and Karababa 2004).

Table 1. Nutritional profile of raw and cooked chickpea

Nutrient	Unit	Value per 100 g[1]	
		Chickpeas, Dry (16056) [2]	Chickpeas, Cooked (16057) [2]
Macronutrients			
Energy	Kcal	378	164
Protein	g	20.47	8.86
Fat	g	6.04	2.59
Carbohydrate	g	62.95	27.42
Fiber	g	12.2	7.6
Sugar	g	10.7	4.8
Minerals			
Calcium	mg	57	49
Iron	mg	4.31	2.89
Magnesium	mg	79	48
Phosphorus	mg	252	168
Potassium	mg	718	291
Sodium	mg	24	7
Zinc	mg	2.76	1.53
Copper	mg	0.656	0.352
Manganese	mg	21.306	1.030
Selenium	g	0	3.7
Vitamins			
Vitamin C	mg	4.0	1.3
Thiamin	mg	0.477	0.116
Riboflavin	mg	0.212	0.063
Niacin	mg	1.541	0.526
Pantothenic acid	mg	1.588	0.286
Vitamin B6	mg	0.535	0.139
Folate	g	557	172
Choline	mg	99.3	42.8
Vitamin B12	g	0	0
Vitamin A	IU	67	27
Vitamin D	g	0	0
Vitamin K	g	9.0	4.0
Vitamin E	mg	0.82	0.35
Total Fatty acids	g	6.04	2.59
Saturated	g	0.60	0.27
Monounsaturated	g	1.38	0.583
Polyunsaturated	g	2.73	1.156

[1] Data obtained from the USDA National Nutrient Database for Standard Reference.

[2] Nutrient Database Number (NDB No.) in the USDA Food Composition Databases.

Coating of *leblebi* with honey, red pepper, sesame, sugar, chocolate and salt enriched products make product variation for consumers. *Leblebi* can be stored for a long time about six to twelve months due to their low moisture content (Coşkuner and Karababa 2004).

The chickpeas in the market are double-roasted chickpeas. During the double-roasting process, the chickpeas are subjected to a second heat treatment by means of roasting systems, thereby imparting the peculiar taste, odor and physical properties that enhance the appeal. In one study by Saglam and Seydim (2017), no important changes in ash, fiber and crude protein contents were found. In contrast, a significant reduction in moisture and total starch content was observed. During roasting and browning of chickpea samples, the color changed to brown, which may be related with the Maillard reaction of reducing sugars with amino compounds such as amino acids or peptides, or by caramelization of sugars with the effect of temperature. The reason for the decrease in hardness with roasting is that there is a significant amount of air gap in the chickpea samples. In the second roasting process, the water in the chickpeas changes from liquid form to a gas form and the vapor pressure of the water increases. As the roasting process continues, steam loses water from the starch structure and this leads to the formation of a porous structure. Thus, it is stated that the chickpeas reach a consumable hardness. According to the sensory analysis results, it was observed that the spotting value of chickpeas increased with increasing temperature and duration of roasting whereas non-spotting color decreased. This increase has been instrumental in increasing the attractiveness of *leblebi*. It was also observed that raw smell and raw taste properties decreased and burnt smell and burnt taste properties increased with increasing roasting temperature and duration. According to the results of sensory analysis, it was stated that chickpeas grilled for 8 minutes at 180 ± 5°C, and 8 minutes at 150 ± 5°C and 6 minutes at 150 ± 5°C were the most popular ones (Saglam and Seydim 2017).

With the process of different *leblebi* production stages, chickpeas are exposed to heat treatment in each stage. Water is added after heat treatment to increase moisture. During the processes of *leblebi* production chickpea seeds come through with structural changes (Bilgir 1976; Köksal 1998). During this process, proteins and carbohydrates are modified with additional caramelization of polysaccharides on the seed surface at last roasting stage. After final processing stage, increased volume and with decreased density and kernel weight of *leblebi* occurs (Tekeli 1965; Gülümser 1988; Köksel et al. 1998; Aydın 2000). Heating during the production process might lead to a retained, modified or totally changed flavor of food materials.

4. Health Benefits of Kabuli Chickpeas

A positive correlation has been found between a pulses-rich diet and a reduced risk of various health problems. Nutraceutical compounds in pulses contribute to maintain wellness with the dietary fiber, antimicrobial peptides, phenolics and flavonoids and phytochemicals in their contents. Dietary fibers facilitate growth of Bifidobacteria and Lactic acid bacteria. Fermented products reduce the risk of colon cancer and induce hypocholesterolemic and hypoglycaemic effects. Thus, the basic metabolic rate and glycemic index in the body decrease and the risk of cardiovascular diseases and Type II diabetes also is reduced. Other factors also support the body to reduce the cholesterol level, inflammation, ulcerative diseases and acts as anticarcinogens. As mentioned earlier, with the components in them such as carotenoids, antioxidants, isoflavone, biochanin A, fatty acids, vitamins and minerals, chickpeas are effective against some types of cancers such as prostate, lung, bowel, breast, and stomach cancer. Chickpeas also show anti-diabetic activity, induce hypoglycaemic and hypocholesterolaemic effects. Similarly, in addition to the animal experiments, if the chickpeas are added into human diets, total cholesterol and fasting insulin levels can decrease. Thus chickpeas can be preferred in obesity treatments and weight loss programs. The polyunsaturated fatty acids also have positive effects on insulin, serum lipids and haemostatic factors. Chickpeas also have a role in prostaglandins synthesis, as result blood pressure can decrease and the constriction process of smooth muscles can be supported. Sterols, tocopherols and tocotrienols are other important component of chickpea oil. These elements can display anti-inflammatory, antimicrobial, anti-tumour, anti-ulcerative and cholesterol lowering effects. The age related diseases such as macular degenaration or senile treatments can also be supported with chickpea components lutein and zeaxanthin. Chickpeas are also used in many different ways in traditional medicine such as tonics, aphrodisiacs, appetizers, as a suppressor of thirst and heartburn, as stimulants and as treatments for parasitic worms. From the view of microbiology, chickpeas' different parts (fruit skin, seed extracts, aerial parts) show antibacterial effects against Gram-positives; *Bacillus subtilis, Staphylococcus aureus, Enterococcus faecalis*, gram-negatives *Pseudomonas aeruginosa, , Escherchia coli* and *Klebsiella pneumoniae* and antifungal activity against *Candida albicans* (Peksen and Artik 2005; Roy, Boye and Simpson 2010; Brummer et al. 2015; Dadon et al. 2017; Çetiner and Bilek Ersus 2018; Sarioglu and Velioglu 2018; Venkidasamy et al. 2019).

Chickpea consumption is helpful for reducing the risk of chronic diseases and health optimization with some physiological effects. Diets, which contain a high amount of fiber-rich chickpeas, can reduce total plasma levels in obese subjects. Another study showed that the main fatty acid of chickpea, linoleic acid has a positive effect on insulin sensitivity and serum lipids level (Dadon et al. 2017). The content of anti-nutritional factors of chickpeas such as protease and amylase inhibitors and protein-binding phenolics are generally low. The activity of these components can be reduced by cooking and processing (Arntfield and Maskus 2011; Knights and Hobson 2016). Pea, chickpeas, beans and lentils are not included in the main allergens. Although proteins of these legumes are classified as allergic, these reactions are limited to Europe, Asia and the Mediterranean, where legumes are highly consumed. Severe anaphylaxis is rarely seen with these plants (Boye, Zare and Pletch 2010. 2010). The proteins and peptides with their high content in chickpeas can act as an anti-carcinogen in colon cancer with antioxidants (Megias et al. 2016).

5. Effects of Different Applications on the Nutritional Value of Chickpeas

There are traditional and technological methods, which can reduce the antinutrient compounds in chickpeas. The type of chickpea, processing conditions, tannins and trypsin content of the product are the main factors affecting the nutiritional value and the quality of the product. The applications and methods, which can affect the characteristic of components are hot water, hot-dry, germination and fermentation and other treatments or applications. Hot water applications include (i) inactivation of enzyme inhibitors and lectins, (ii) reducing tannins and other anti-nutrients, (iii) increasing digestibility and quality of protein and carbohydrates, (iv) long-term cooking, which reduces protein quality and vitamin B value, and (v) high temperature lysine and sulfurous amino acid, which reduces bioavailibility. Controlled cooking of chickpeas is highly significant for maintaining its quality. If controlled cooking is not performed, the added salt and soaking in water may reduce protein quality. Besides that, short-term cooking at high temperature causes similar effects to boiling in water and short-term roasting at high temperature results in reductions in nutrition value and this is classified as hot- dry applications. Germination and fermentation of chickpeas may exert significant impact on chickpeas' quality and nutritional values like (i) increases vitamin content, especially vitamin C, (ii) reduces the factors that cause stomach gas, (iii) reduces the amount of tannin, (iv) increases protein digestibility and quality, and (v) reduces the dietary fiber and minerals. Whereas, grinding and milling of chickpeas may results in shorter cooking time and changes the structural properties of carbohydrates. Fertilization, irrigation, bacterial vaccination techniques, physical, chemical and the biological structure of soil also have positive/negative effects on the nutritional value of pulses like chickpeas (Peksen and Artik 2005; Torres et al 2016; Ertop and Bektaş 2018).

CONCLUSION

Chickpeas are rich in phenolic compounds and are a good source of mineral micronutrients. This pulse has moderate levels of carotenes and is low in antinutrients (e.g., PA). These favorable mineral bioavailability food matrix factors suggest that the chickpea is a good source of bioavailable minerals (e.g., Fe, Zn, Se) and serves as a whole food solution to micronutrient malnutrition in developing countries and as an ultimate source of protein with low fat and cholesterol but is rich in fiber, for the increasing lacto-ovo/vegan population. Kabuli chcikpeas are more preferred in Turkey and other parts of the World in different forms but flour made from Kabuli chickpeas is commonly available in the markets in Turkey due to their high demand during the last decade, mostly due to increased number of foreigners in Turkey. Raw chickpea is also consumed during June-July in Turkey by the public in the parks and during walking on the roads. Although, consumption of chickpea in Turkey is relatively high, more extensive research work about nutritional values of different forms and cooking style of Kabuli chickpeas will be needed.

REFERENCES

Akın, Fatima. "Effects of Salt Stress on Germination and Growth of Black and White Chickpea (*Cicer arietinum* L.) Genotypes Under *in vitro* Conditions. *In vitro koşullarda siyah ve beyaz nohut (cicer arietinum l.) genotiplerinde tuz stresinin çimlenme ve büyüme üzerine etkileri.*" Graduate School of Natural and Applied Sciences, Necmettin Erbakan University, Konya, Turkey (2018).

Alajaji, Saleh A., and El-Adawy Tarek A. "Nutritional Composition of Chickpea (*Cicer arietinum* L.) as Affected by Microwave Cooking and Other Traditional Cooking Methods." *Journal of Food Composition and Analysis*19, no. 8 (2006): 806-12. doi:10.1016/j.jfca.2006.03.015.

Anonymous, "*Nohut Ürün Raporu*, 2018." Ministry of Food Agriculture and Livestock (2019).

Arntfield, S. D., and Maskus H. D. "Peas and Other Legume Proteins." In *Handbook of Food Proteins*, 233-66. Philadelphia, PA: Woodhead Publishing Limited, 2011.

Aydemir, Levent Yurdaer, and Ahmet Yemenicioğlu. "Potential of Turkish Kabuli Type Chickpea and Green and Red Lentil Cultivars as Source of Soy and Animal Origin Functional Protein Alternatives." *LWT - Food Science and Technology* 50, no. 2 (2013): 686-94. doi:10.1016/j.lwt.2012.07.023.

Aydın, F. Nohudun Kullanımı Ve Leblebi üretimi. *Proceedings of Hububat Ürünleri Teknolojisi Kongre Ve Sergisi* [*Proceedings of Grain Products Technology Congress and Exhibition*], Turkey, Gaziantep. 2002. 503-513.

Bayrak, Hakan, and Mustafa Önder. "Konya Ekolojisi'nde Tarımı Yapılan Yerel Nohut Popülasyonları Ve Çeşitlerinin (*Cicer arietinum* L.) Tarımsal, Teknolojik Ve Besinsel Karakterlerinin Belirlenmesi (Determination of Agricultural, Technological and Nutritional Characteristics of Local Populations and Varieties of Chickpea (*Cicer arietinum* L.) Cultivated in Konya Ecological Region)." *Tarla Bitkileri Merkez Araştırma Enstitüsü Dergisi* 26 (2017): 52-61. doi:10.21566/tarbitderg.359119.

Boye, Joyce, Zare, Fatemeh and Pletch, Alison. "Pulse Proteins: Processing, Characterization, Functional Properties and Applications In Food and Feed." *Food Research International* 43 (2010): 414-431. doi:10.1016/j.foodres.2009.09.003.

Brummer, Yolanda, Kaviani, Mina, and Tosh, Susan M. "Structural and Functional Characteristics of Dietary Fibre in Beans, Lentils, Peas and Chickpeas." *Food Research International* 67 (2015): 117-25. doi:10.1016/j.foodres.2014.11.009.

Çetiner, Miray and Bilek, Seda Ersus. "Bitkisel Protein Kaynakları (Plant Protein Sources)." *Çukurova Tarım ve Gıda Bilimleri Dergisi* 33, no. 2 (2018): 111–126.

Coşkuner, Yalçın, and Karababa, Erşan. "Leblebi: A Roasted Chickpea Product as a Traditional Turkish Snack Food." *Food Reviews International* 20, no. 3 (2004): 257-74. doi:10.1081/fri-200029424.

Dadon, Shimrit Bar-El, Abbo, Shahal and Reifen, Ram. "Leveraging Traditional Crops for Better Nutrition and Health - The Case of Chickpea." *Trends in Food Science & Technology* 64 (2017): 39-47. doi:10.1016/j.tifs.2017.04.002.

Ertop, Hendek Müge, and Bektas, Müberra. "Enhancement of Bioavailable Micronutrients and Reduction of Antinutrients in Foods wih Some Processes," *Food and Health* 4, no. 3 (2018):159-165.

Ertop, Hendek Müge, and Bektas, Müberra. "Enhancement of Bioavailable Micronutrients and Reduction of Antinutrients in Foods wih Some Processes," *Food and Health* 4, no. 3 (2018):159-165.

FAO. "*Chickpeas.*" Food and Agriculture Organization. (2019). http://www.fao.org/home/en/

Gil, Juan, Salvador Nadal, Diego Luna, M. Teresa Moreno, Antonio De Haro. "Variability of some physico-chemical characters in desi and kabuli chickpea types." *Journal of The Science of Food and Agriculture* 71 (1996):179–184.

Gülümser, A. "Nohutun hasattan sonra değerlendirilmesi ve leblebi yapımı." *Ondokuz Mayıs Üniversitesi, Ziraat Fakültesi Dergisi* 3 (1988): 251-260.

Gülümser, Ali. "Dünyada Ve Türkiye'De Yemeklik Dane Baklagillerin Durumu (Situation of Pulse in Turkey and World)." *Tarla Bitkileri Merkez Araştırma Enstitüsü Dergisi* 25 (ÖZEL SAYI-1) (2016) 292-292. doi:10.21566/tarbitderg.280511.

Kaur, Maninder, and Singh, Narpinder. "Characterization of Protein Isolates from Different Indian Chickpea (*Cicer arietinum* L.) Cultivars." *Food Chemistry* 102, no. 1 (2007): 366-74. doi:10.1016/j.foodchem.2006.05.029.

Khan, M. Akmal, and Abdus Sattar Kissana. "Nutritional Evaluation of Some Commercial Baby Foods Consumed in Pakistan." *Journal of the Science of Food and Agriculture*36, no. 12 (1985): 1271-274. doi:10.1002/jsfa.2740361211.

Kilinçer, Fatma Nur and Demir, Mustafa Kürşat. "Çimlendirilmiş Bazi Tahil Ve Baklagillerin Fiziksel Ve Kimyasal Özellikleri (Antinutritional Factors and Nutritive Values of Food Grain Legumes)" *GIDA (The Journal of Food)* 44, no. 3 (2019): 419–429.

Kırnak, H., Varol, İ. S., Irik, H. A., and Ozaktan, H., "Nohut Bitkisinde Farklı Gelişim Dönemlerinde Uygulanan Ek Sulamanın Ham Protein, Ham Yağ ve Ham Kül İçeriğine Etkisi (Effects of Supplementary Irrigation at Different Periods on Crude Protein, Oil and Ash Content of Chickpea)." *Gaziosmanpaşa Bilimsel Araştirma Dergisi* (GBAD) 6, (2017): 168–176.

Knights, E. J., and K. B. Hobson. *Encyclopedia of Food Grains*. 2nd ed. UK: Academic Press, Elsevier. 2016.

Köksel, H., D. Sivri, M. G. Scanlon, and W. Bushuk. "Comparison of Physical Properties of Raw and Roasted Chickpeas (leblebi)." *Food Research International* 31, no. 9 (1998): 659-65. doi:10.1016/s0963-9969(99)00042-3.

Kose, Ozge Doğan Erbas and Mut, Zeki. "Türkiye' de yetiştirilen nohut çeşit ve hatlarının mineral madde içeriklerinin belirlenmesi," *Derim* 36, no. 1 (2019): 73–78.

Macar, Tuğçe Kalefetoğlu, Macar, Oksal, and Mart, Dürdane Inci. "Variability in Some Biochemical and Nutritional Characteristics in Desi and Turkish Kabuli Chickpea (*Cicer arietinum* L.) Types." *Celal Bayar Üniversitesi Fen Bilimleri Dergisi*, 2017, 677-80. doi:10.18466/cbayarfbe.339330.

Megías, Cristina, Cortés-Giraldo, Isabel, Alaiz, Manuel, Vioque, Javier, and Girón-Calle, Julio. "Isoflavones in Chickpea (*Cicer arietinum*) Protein Concentrates." *Journal of Functional Foods* 21, (2016): 186-92. doi:10.1016/j.jff.2015.12.012.

Milán-Noris, Ada Keila, Gutiérrez-Uribe, Janet Alejandra Santacruz, Arlette, Serna-Saldívar, Sergio O., and Martínez-Villaluenga, Cristina. "Peptides and Isoflavones in Gastrointestinal Digests Contribute To The Anti-Inflammatory Potential of Cooked Or Germinated Desi and Kabuli Chickpea (*Cicer arietinum* L.)." *Food Chemistry* 268, (2018): 66-76. doi:10.1016/j.foodchem.2018.06.068.

O'Neil, C. E., Nicklas, T. A., Fulgoni III, V.L "Chickpeas and Hummus Are Associated with Better Nutrient Intake, Diet Quality, and Levels of Some Cardiovascular Risk Factors: National Health and Nutrition Examination Survey 2003-2010." *Journal of Nutrition & Food Sciences* 04, no. 01 (2014). doi:10.4172/2155-9600.1000254.

Özer, Sertaç, Karaköy, Tolga, Toklu, Faruk, Baloch, Faheem S., Kilian, Benjamin, and Özkan, Hakan. "Nutritional and Physicochemical Variation in Turkish Kabuli

Chickpea (*Cicer arietinum* L.) Landraces." *Euphytica* 175, no. 2 (2010): 237-49. doi:10.1007/s10681-010-0174-3.

Padhi, Emily M. T., Liu, Ronghua, Hernandez, Marta, Tsao, Rong, and Ramdath, D. Dan. "Total Polyphenol Content, Carotenoid, Tocopherol and Fatty Acid Composition of Commonly Consumed Canadian Pulses and Their Contribution To Antioxidant Activity." *Journal of Functional Foods* 38, (2017): 602-611. doi:10.1016/j.jff. 2016.11.006.

Parca, Filiz, Koca, Yakup Onur, and Unay, Aydın. "Nutritional and Antinutritional Factors of Some Pulses Seed and Their Effects on Human Health." *International Journal of Secondary Metabolite* 5, (2018): 331-342. doi:10.21448/ijsm.488651.

Patane, C. "Variation and relationship among some nutritional traits in sicilian genotypes of chickpea (*Cicer arietinum* L.)." *Journal of Food Quality* 29, (2006):282–293.

Peksen, Erkut and Artik, Cengiz. "Antibesinsel Maddeler ve Yemeklik Tane Baklagillerin Besleyici Değerleri (Antinutritional Factors and Nutritive Values of Food Grain Legumes)." Anadolu Tarım Bilimleri Dergisi 20, (2005): 110-120.

Roy, F., Boye, J. I. and Simpson, B. K. "Bioactive Proteins and Peptides in pulse Crops: Pea, Chickpea and Lentil." *Food Research International* 43, no 2 (2010): 432-442. doi:10.1016/j.foodres.2009.09.002.

Sağlam, Hidayet, and Aseydim, Atıf Can. 2017. "Leblebi Üretiminde İkinci Kavurma Koşullarının Leblebi'nin Fizikokimyasal Özellikleri Ve Duyusal Kalitesi Üzerine Etkisi (The Effect of Double Roasting Parameters on the Physicochemical Properties and Sensory Quality of Leblebi (A Roasted Chickpea Snack)." *Harran Tarım Ve Gıda Bilimleri Dergisi* 21 (3): 279-292. doi:10.29050/harranziraat.339334.

Saharan, K., and N. Khetarpaul. "Protein Quality Traits of Vegetable and Field Peas: Varietal Differences." *Plant Foods for Human Nutrition* 45, no. 1 (1994): 11-22. doi:10.1007/bf01091225.

Sarıoğlu, Gül, and Velioğlu, Y. Sedat. "Baklagillerin Bileşimi (Composition of Pulses)." *Akademik Gıda 16*, (2018): 483-496. doi:10.24323/akademik-gida.505547

Sayar, Sedat, and Karataş, Selen Çalışkantürk. "Nohutta Tane (Tohum) Kabuğunun Tüm Tanenin Fiziksel, Kimyasal Ve Beslenme Özellikleri Üzerine Etkisi (Effect of Seed Coat On The Physical, Chemical and Nutritional Properties of Whole Chickpea)." *Gida/The Journal of Food* 42, no. 4 (2017): 468-76. doi:10.15237/gida.gd16092.

Sharma, S., Yadav, N., Singh A., and Kumar R., "Nutritional and antinutritional profile of newly developed chickpea (*Cicer arietinum* L) varieties," *International Food Research Journal* 20, no. 2 (2013): 805–810.

Tekeli, Sait Tahsin. "Ziraat sanatları." *Ankara Üniversitesi Ziraat Fakültesi Yayınları,* Ankara (1965).

Thavarajah, Dil, and Thavarajah, Pushparajah. "Evaluation of Chickpea (*Cicer arietinum* L.) Micronutrient Composition: Biofortification Opportunities To Combat Global

Micronutrient Malnutrition." *Food Research International* 49, (2012): 99-104. doi:10.1016/j.foodres.2012.08.007.

Torres, J., Rutherfurd, S. M., Muñoz, L. S., Peters M., and Montoya C. A. 2016. "The Impact of Heating and Soaking on the *In vitro* Enzymatic Hydrolysis of Protein Varies in Different Species of Tropical Legumes," *Food Chemistry* 194 (2016): 377-382.

Torres, J., Rutherfurd, S., Muñoz, M. L. S., Peters, M. and Montoya C. A.. 2016. "The Impact of Heating and Soaking on the *In vitro* Enzymatic Hydrolysis of Protein Varies in Different Species of Tropical Legumes," *Food Chemistry* 194 (2016): 377-382.

TUIK (2018), "Dry pulses, 1988-2018," [Online]. Available: http://tuik.gov.tr/PreTablo.do?alt_id=1001. [Accessed: 11-Jun-2019].

Venkidasamy, Baskar, Selvaraj, Dhivya, Nile, Arti Shivraj , Ramalingam, Sathishkumar, Guoyin Kai, and Shivraj Hariram Nile. "Indian Pulses: A Review on Nutritional, Functional and Biochemical Properties With Future Perspectives." *Trends in Food Science & Technology* 88, (2019): 228-242. doi:10.1016/j.tifs.2019.03.012.

Wang, N., Hatcher D. W., Tyler R. T., Toews R., and Gawalko E. J.. "Effect of Cooking on the Composition of Beans (Phaseolus Vulgaris L.) and Chickpeas (*Cicer arietinum* L.)." *Food Research International* 43, no. 2 (2010a): 589-94. doi:10.1016/j.foodres.2009.07.012.

Wang, Runfeng, Gangola, Manu P., Jaiswal, Sarita, Gaur, Pooran M., Båga, Monica, and Chibbar, Ravindra N. "Genotype, Environment and Their Interaction Influence Seed Quality Traits in Chickpea (*Cicer arietinum* L.)." *Journal of Food Composition and Analysis* 63, (2017): 21-27. doi:10.1016/j.jfca.2017.07.025.

Zia-Ul-Haq, Muhammad, Ahmad, Mansoor, Iqbal, Shahid, Ahmad, Shakeel, and Ali, Hakoomat. "Characterization and Compositional Studies of Oil from Seeds of Desi Chickpea (*Cicer arietinum* L.) Cultivars Grown in Pakistan." *Journal of the American Oil Chemists Society* 84, no. 12 (2007a): 1143-148. doi:10.1007/s11746-007-1136-3.

In: Handbook of Chickpeas
Editors: Albert T. Lund and Noah D. Schultz
ISBN: 978-1-53616-374-2

Chapter 3

NUTRITIONAL VALUES, HEALTH BENEFITS AND MULTIPLE USES OF DESI CHICKPEAS

Muhammad Aasim[1,*] and Ashfaq Ahmad Khan[2]
[1]Department of Biotechnology, Faculty of Science,
Necmettin Erbakan University, Konya, Turkey
[2]Department of Geography, Faculty of Literature,
Karabuk University, Karabuk, Turkey

ABSTRACT

Desi type chickpeas (*Cicer arietinum* L.) or microsperma are small sized, dark colored and angular shaped chickpeas with high nutritive and medicinal values. Desi chickpeas are considered as most primitive chickpeas due to its resemblance to the wild ancestor of chickpea. The desi chickpeas are commonly consumed and cultivated in India and Pakistan but also show its presence in other countries like Turkey, Mexico, Iran and some African countries. Desi chickpeas are used in different forms such as cooked, boiled or roasted. Whereas, *dhal* (splitted dehulled chickpeas) and *besin* (chickpea flour) are very popular in India and Pakistan for making different foods like *helva* and *chapatti* etc. Desi chickpeas are highly rich in dietary fibers, proteins, vitamins, minerals, carbohydrates and other compounds which are essential for dietary program. Apart from its nutritional values, desi chickpeas are considered as healthy food for curing diabetes, cardiovascular risks, cancer and skin diseases. However, the knowledge about desi chickpea consumption and its advantages are less known. This study enlightens the nutritional values, health benefits, and its uses in different forms along with its comparison with Kabuli chickpeas.

* Corresponding Author's Email: mshazim@gmail.com.

Keywords: chickpea, Desi, Kabuli, health, nutrition, Turkey

INTRODUCTION

Desi type chickpeas, also known as microsperma, are smaller sized seeds which are angular in shape and possess a yellow to brown colored, rough seed coats (Purushothaman et al. 2014). The leaves of Desi chickpeas are generally smaller as compared to Kabuli (Mohammadi 2015) and contains approximately 5 cm long 9-15 leaflets resembling with the fern-leaf structure. Desi chickpeas possess fewer primary branches, lower cold tolerance and iron deficiency (Hawtin and Singh,1979). The flowers are generally pink colored (Jukanti et al. 2012) and pods contain generally 2-3 ovules per pod with 1-2 seeds per pod being produced. The seeds are generally smaller than the Kabuli type and weigh 0.1 to 0.3g (Moreno and Cubero 1978). Cultivated Desi chickpeas are short duration, early maturing and high yielding. They also show high mechanical and frost resistance and are less prone to insect damage. The plants change their color of leaves and stems from red to purple due to anthocyanins (Jukanti et al. 2012) which are dependent on the plant's position to sunlight, but it has no effect on plant yield or quality. The change in color also occurs due to heat or drought stress and may lead to yield loss or decreased quality. There are some Desi chickpea types which have white flowers and no anthocyanin (Gaur et al. 2010). They are considered as good nitrogen fixer, but adverse environmental conditions may lead to decreased N_2 fixation (Agarwal et al. 2012).

The seed size of Desi types is relatively smaller than Kabulis and expressed by 100 or 1000 seed weight. A study by Rybiński et al. (2019) reported the lowest 100 seed wt value for Ethopian accession CAR 31 (12.3 g) and accession 43 (14.6 g) and highest seed wt of 26.6 g for Turkish accession named CAR 82. They revealed that seed thickness is not associated with geographical origin and Desi types had lower seed thickness compared to Kabulis. Turkish accession CAR 82 showed highest seed thickness that was recorded as 7.16 mm (Rybiński et al. 2019). The seed coat of Desi chickpeas in percentage is relatively higher than Kabulis and possess 80% of total crude fiber along with anti-nutritional polyphenols that play a significant impact on nutritive value, processing quality and cooking (Singh et al. 1980). Desi chickpeas are rich in neutral detergent or crude fiber and total tannins and poor in non-fibrous carbohydrates, soluble sugars and crude protein. It was reported that the nutritive value of Desi types was lesser for ruminants (Khan et al 1995; Maheri-Sis et al 2008). A study by Khalil et al. (2007) revealed that the crude fat contents increased with sprouting. Although, it was relatively less than Kabuli types but statistically similar with each other.

Chickpeas show a narrow genetic base which emerged from domestication (Varshney et al. 2013). The Desi type chickpeas are more primitive than Kabulis and were isolated geographically for many years and considered as the earliest variety on the basis of its resemblance to wild ancestor of chickpea (*C. reticulatum*) (Mohammadi, 2015). Both genotypes show common characteristics despite vast morphological differences and possess 436 common alleles out of 1683 alleles detected with 47% to 54% common alleles frequency (Upadhyaya et al. 2008). Although, mean gene diversity of both germplasms was almost similar, Desis are less diverse than Kabulis but possess more unique and rare alleles compared to Kabulis (Purushothaman et al. 2014).

Desi chickpeas show wide distribution in South and Southeast Asia (Aydemir and Yemenicioğlu 2013) along with cultivation in Ethiopia, Mexico, Turkey and Iran (Mohammadi 2015). However, its consumption is mostly confined to the Middle East and Southeast Asia (Aydemir and Yemenicioğlu 2013; Varshney etal. 2013). It is reported that approximately 80-85% of the total chickpea area is acquired by Desi types in Asia and Africa (Pande et al. 2005). They are cultivated mainly in winter in subtropical and hilly tropical areas (Purushothaman et al. 2014). The best germination occurs at 5°C soil temperature but soil temperature at least 7°C or above is more suitable for seedling vigor (Saskatchewan Pulse Growers 2019).

Desi chickpeas are very popular and commonly used in sub-continent due to its high nutritional value and health benefits. Different forms and products like *dhal* and *besin* are used for making delicious foods and snacks. Desi chickpeas are highly nutritive as Kabuli chickpeas. However, their cultivation is confined to specific areas in some countries of the World like Turkey, where Desi chickpea cultivation and consumption is limited to few areas. In Turkey, *dhal* made from desi chickpea is also available but the cooking quality and taste is not equal to those available in Pakistan. The *dhal* available in Turkey is very hard and difficult to cook possibly due to very low moisture contents. The consumption of desi chickpea is not very high in Turkey and they are mainly available for online purchase while the price is also 3-4 fold more than Kabuli chickpeas.

Desi chickpeas show great resemblance with Kabuli chickpeas in terms of its nutritional value and compounds. However, little information is available about its detailed nutritional values, health benefits and consumption style. This work presents an insight into the potential uses of Desi chickpeas as food and medicinal products for curing certain diseases and disorders as well as its consumption style and nutritive values.

1. Multiple Forms of Using Desi Chickpeas

Chickpeas are widely used in different forms as they are highly valuable for human nutrition and health. The Desi chickpeas are used as (i) whole or (ii) split into two parts after shelling to make *dhal*, or (iii) ground into fine flour called *besan* in the Indian Subcontinent. The main use of Desi chickpeas in the Indian subcontinent is a split cotyledon after milling known as *dhal* or making flour (*besan*) for the preparation of different local snacks (Chavan et al. 1986; Hulse 1991). *Besan* is used pure or mixed with wheat flour to make *roti* or *chapatti* (bread) and for making sweets (helva) (Khan et al. 1995). A comprehensive study by Khan et al. (1995) highlighted the use of Desi chickpeas in different forms as food in Pakistan. These include *dhal* (split Desi chickpeas), *roti* (local hand made bread) from the flour, *pakoray* (a snack made from chickpea flour by frying in oil), *helva* (sweet dish made from chickpea flour) and roasted chickpea (soaked and roasted in sand). They also investigated the chemical compositions of these products and found maximum carbohydrates in roti, protein in dhal and roasted chickpeas, fats in *helva* and crude fiber in *pakora*s and roasted chickpeas. Furthermore, they also highlighted the nutritional and health benefits of Desi chickpea products while *roti* made from Desi chickpea flour is considered to be beneficial for diabetic patients. On the other hand, chickpeas are used for making soups/salads, stews, or consumed after being boiled, roasted, salted or fermented in many areas of Asia, Africa and other continents (Gecit 1991).

The seed coat or testa of Desi chickpeas is thicker with tight adherence to cotyledons making it difficult to split into two cotyledons for making *dhal* (Tikle and Mishra 2019). Shrivastava, Dixit and Kumar (2017) reported the *dhal* recovery of 61.3-82.6% in Desi chickpea types. The husk recovery is also higher in Desi chickpea compared to Kabuli types (Tikle and Mishra 2019) and considered as animal feed in Pakistan and India.

2. Nutritive Values of Desi Chickpeas

2.1. Carbohydrate Contents

Among edible pulses, chickpeas have relatively higher total carbohydrate content (Table 2) in the form of monosaccharides (fructose, galactose, glucose, and ribose,), disaccharides (maltose, sucrose,) or oligosaccharides (ciceritol, raffinose, stachyose, and verbascose) (Han and Baik 2006) with slight variation among Desi and Kabuli chickpeas. Oligosaccharides are generally fermented by colonic bacteria due to no absorption or hydrolyzing by human digestive system and results in the release of gases or flatulence (Hedley 2001) which may cause abdominal discomfort (Singh 1985). The amount of gas production by chickpeas is relatively higher than other pulses due to higher oligosaccharides contents (Jaya, Naik and Venkataraman 1979). The amount of α-Galactosides is next to sucrose in plants (Jones et al. 1999; Han and Baik 2006) and accounts ~62% of total sugar in chickpeas (Sanchez-Mata et al. 1998). A study by Wang and Duan (2004) revealed the α-galactosides fractions in Desi chickpeas as raffinose (0.5) and stachyose (1.6).

2.2. Polysaccharides

Starch is generally a major stored polysaccharide in all pulses (Chibbar, Ambigaipalan and Hoover 2010) including chickpeas which accounts for 41-50% of the total carbohydrates (Dalgetty and Baik 2003; Özer et al. 2010). Desi chickpeas have relatively less soluble sugars than Kabuli type (Jambunathan and Singh 1980). Although, cereals like wheat contain higher starch contents than chickpeas (USDA 2010), amylose contents in chickpeas are higher than wheat (Guillon and Champ 2002). On the other hand, the *in vitro* starch digestibility values (ISDV) of chickpeas are higher than some pulses (Rehman and Shah 2005) but lower than cereals (Madhusudhan and Tharanathan 1996).

2.3. Dietary Fibre

Dietary fibre (DF) is comprised of lignin, poly/oligosaccharides, and other plant-based substances (AACC 2001) which are indigestible by human small intestines. Dietary fibers can be classified as soluble (slow digestion in the colon) and insoluble (metabolically inert and aid in bowel movement) (Tosh and Yada 2010). In raw chickpea seeds, total DF content (DFC) is 18-22 g per 100-g (Aguilera et al. 2009; Tosh and Yada 2010) which is relatively higher or equal to some pulses (Tosh and Yada 2010; Jukanti et al. 2012). The Desi chickpeas have relatively higher total DFC and insoluble DFC as compared to the Kabuli types which is possibly due to relatively more thick hulls and seed coat and it accounts 11.5 % of total seed weight compared to 4.3-4.4 % of Kabuli types (Rincon, Martinez and Inbanez 1998). Contrarily, no significant differences of soluble DFC between both chickpea types due to similar proportion of hemicelluloses contents was reported by Singh (1984).

2.4. Protein Content

Chickpeas have comparatively higher bioavailability of protein (Sánchez-Vioque et al. 1999; del Mar Yust et al. 2003) and protein quality (Kaur, Singh and sodhi 2005) compared to other pulses. However, protein concentration is almost similar compared to certain pulses like lentils, black grams, red and white kidney beans (Rehman and Shah 2005). The protein concentration of chickpeas varies with dehulling process and varied between 17-22% and 25.3-28.9% respectively for before or after the dehulling process (Hulse 1991; Badshah et al. 2003). Both Kabuli and Desi chickpeas possess variable and inconsistent crude protein concentrations (Singh and Jambunathan 1981; Rincon, Martinez and Inbanez 1998) but Kabuli type chickpeas have relatively higher protein digestibility than Desi chickpeas (Paredes-Lopez, Ordorica-Falomir and Olivares-Vázquez 1991; Sánchez-Vioque et al. 1999). A study by Macar, Macar and Mart (2017) on Desi chickpeas revealed the total protein levels range of 18.6-27.1 mg/ml among genotypes with minimum from Desi chickpea.

2.5. Fat Content and Fatty Acid Profile

Total fat content in raw chickpea seeds is variable and low crude fat contents (~ 2.05 g 100-g) in Desi chickpea was reported by Shad et al. (2009). Whereas, 2.90-7.42% fat contents in Desi chickpeas was reported by Wood and Grusak (2007). A study by Rybiński et al. (2019) revealed the highest fat content in Turkish origin Desi chickpeas which were recorded at 13.7% for CAR 16 accession. Earlier, Shah et al. (2013) reported higher fat content for Desi seed type (5.7%) compared to Kabuli seeds (4.5%). Desi chickpeas contain relatively higher linoleic acid and lower oleic acid contents compared to Kabuli chickpeas (Jukanti et al. 2012). A study by Wang and Daun (2004) revealed the presence of different fatty acids in Desi chickpeas (Table 3). The results revealed the presence of higher linoleic acid contents (61.62%) followed by oleic acid (22.31%). Whereas, Lignoceric and Nervonic were not detected (Table 1).

Table 1. Fatty acid profiles of chickpea seeds Fatty Acid

Fatty acids	(%) oil (Wang and Daun 2004)
Arachidic (C20:0)	0.51
Behenic (C22:0)	0.37
Eicosadienoic (C20:2)	0.12
Erucic (C22:1)	0.13
Gadoleic (C20:1)	0.50
Lauric (C12:0)	0.02
Lignoceric (C24:0)	ND
Linoleic (C18:2)	61.62
Linolenic (C18:3)	3.15
Myristic (C14:0)	0.22
Nervonic (C24:1)	ND
Oleic (C18:1)	22.31
Palmitic (C16:0)	9.09
Palmitoleic (C16:1)	0.26
Stearic (C18:0)	1.16

2.6. Oil Characteristics

Chickpea has low oil contents but possess high medicinal and nutritional values because it contains certain secondary metabolites like sterols, tocopherols and tocotrienols (Zia-Ul-Haq et al. 2007a). Zia-Ul-Haq et al (2009) investigated the different sterols and tocopherols contents of Desi chickpea and revealed the presence of different sterols like campesterol (12.06-13.67%), clerosterol (1.94-4.01%), β-sitosterol (73.12-76.10 %), stigmasterol (4.92-5.38%), Δ5-avenasterol (3.12-5.72 %) and Δ7- avenasterol (0.79-1.21 %). Besides, they also reported different tocopherols (mg/100g of oil) like α-tocopherol (32.99-34.82), β- tocopherol (1.67-1.89), γ- tocopherol (185.08-186.02) and δ-tocopherol (7.93-8.88). However, there was no detection of γ-tocotrienols in Desi chickpeas in their study. The physical and chemical characteristics of Desi chickpeas were investigated by Zia-Ul-Haq et al (2007a) and Shad et al (2009). Total oil (%) of Desi chickpea was reported as (5.88-6.87%) with brown and yellow color by Zia-Ul-Haq et al (2009). These characteristics were not reported by Shad et al. (2009). The acid value, calorific value, Iodine value, saponification value and unsaponifiable matter given in both studies were almost similar. However, Zia-Ul-Haq et al (2009) conducted more detailed oil characteristics of Desi chickpeas compared to Shad et al. (2009) (Table 2).

Table 2. Chickpea seed oil: physical and chemical characteristics

Characteristics	Zai-Ul-Haq et al. (2007a)	Shad et al. (2009)
Total oil (%)	5.88-6.87	-
Acid values (mg KOH/g)	2.55-2.67	2.40-2.50
Calorific value (kca/100g sample)	-0.92-3.11	368-373
Color	Brown-Yellow	-
Diacylglycerols (DAG; %)	0.7-1.6	-
Iodine values (Wijs method)	111.87-113.69	112.56-113.87
Monoacylglycerols (MAG; %)	2.2-2.7	-
Oxidation value	13.09-22.34	-
p-Anisidine value	5.39-8.74	-
Peroxide value(mequiv/kg)	3.97-6.37	-
Refractive index (400C)	1.48	-
Relative density (400C; g/cm3)	0.96	-
Saponification values (mg KOH/g)	183.98-185.64	178.90-180.64
Specific gravity	-	0.9339-0.9346
Triacylglycerols (TAG; %)	55.7-63.2	-
Unsaponifiable matter (% w/w)	2.99-3.71	

2.7. Minerals

Chickpeas are good source of minerals and meet the demand of minerals for millions of people in Asia and Africa where daily diet is based mainly on cereals. The mineral analysis of both types of chickpeas (Table 3) presents the identical mineral contents except calcium which is higher in Desi types (Ibanez et al. 1998; Wang and Duan 2004). Other mineral contents were in variable concentration in both studies. The present gene pool of the chickpea and its high nutritive value offers to assess the genetic variability for identifying the potential genotypes for future breeding strategies (Dwivedi et al. 2012) with the main focus on variable nutrients for its biofortification programs (Kahraman, Pandey and Khan 2017). To date, a number of research works on mineral compositions of different Desi chickpeas have been reported. Ibanez et al. (1998) used 16 Desi cultivars for the estimation of genetic variability and reported higher Ca and Mg contents and lower K contents in Desi cultivars compared to Kabuli. Whereas, Cu, Fe, Mn, Na and Zn contents were almost same in both chickpea types. Ibrikci et al. (2003) reported the 1.3-1.8-fold macronutrient and 1.5-2.4-fold micronutrient content in leaves of 19 Kabuli and Desi types chickpeas. Later, Zia-Ul-Haq et al (2007a) reported Zn contents (3.5-6.0 mg per 100 g) and Fe contents (2.4-4.1 mg per 100 g) in seeds using four Desi chickpea cultivars. In another study, desi chickpea cultivars also contained significant amount of Zn, Cu and Ca (Zia-Ul-Haq et al 2007b). Bueckert et al. (2011) reported Lower Zn, higher Mg and Ca contents in Desi types compared to Kabuli types. Whereas, Diapari et al. (2014) investigated the 94 different chickpeas accessions comprised of both types and cultivated on different sites and found significant variability for Fe and Zn concentrations and recommended 2 Desi chickpeas accessions for genetic biofortification as a donor plant. Upadhyaya et al. (2016) investigated the genetic variation for Fe and Zn and achieved an average of 63.3 ppm (Fe) and 46.2 ppm (Zn) for both accessions. They also identified 8 major QTL responsible for regulating Fe and Zn in chickpea.

Table 3. Mineral constituents of chickpea seed (mg 100-g Minerals)

	Ibáñez et al. (1998)	**Wang and Daun (2004)**
Copper	1.25	1.00
Iron	4.51	5.90
Zinc	3.57	3.60
Manganese	1.72	3.40
Calcium	210.0	161.70

2.8. Amino Acid Profile

Pulses are generally low in sulfur-rich amino acids (methionine, cystine) and amino acid profiles of both Kabuli and Desi types are almost similar (Table 4), reported by Wang and Daun (2004) and Wang et al. (2010). Their studies revealed the presence of different amino acids in Desi chickpeas. These studies also revealed the variable amino acid contents in both studies. However, Tryptophan was not detected in study carried out by Wang et al. (2010).

Table 4. Amino acid content in Desi chickpea seeds

Amino acids	Wang & Daun (2004)	Wang et al. (2010)
Alanine	4.10	3.40
Arginine	9.80	8.11
Aspartic acid	12.80	10.59
Cystine	1.40	0.15
Glutamic acid	16.00	16.70
Glycine	3.90	3.12
Histidine	2.20	2.66
Isoleucine	3.60	3.70
Leucine	7.00	6.30
Lysine	5.90	5.55
Methionine	1.50	2.05
Phenylalanine	5.30	5.42
Proline	4.80	3.95
Serine	6.00	4.96
Threonine	4.30	3.23
Tryptophan	0.90	N/D
Tyrosine	2.30	2.55
Valine	4.00	3.60

2.9. Vitamins

Like minerals, pulses are also considered as an excellent vitamin source and chickpeas provide cheap vitamins like folic acid, tocopherols riboflavin (B2), pyridoxine (B6) and panthothenic acid (B5) (Çiftçi et al. 2010). A study by Wang and Duan (2004) revealed the presence of niacin (1.72 mg 100^{-g}), Panthothenic acid (1.09 mg 100^{-g}), Pyridoxine (0.30 mg 100^{-g}), Riboflavin (0.21 mg 100^{-g}), Thiamin (0.29 mg 100^{-g}),

Vitamin C (1.65 mg 100-g), α-tocopherol (1.91 mg 100-g) and γ-tocopherol (9.33 mg 100-g) and folic acid (206.48 μg 100-g).

3. Health Benefits of Desi Chickpeas

Chickpea seeds are an excellent source of food and feed due to their excellent nutritional value. They are also rich in bioactive compound content like phenolic and flavonoids (Magalhães et al. 2017). Desi chickpeas with dark seed coat (Desi) possess relatively higher antioxidant activities due to the phenolic fraction (Segev et al. 2011). Khetarpaul (2018) investigated the antioxidant potential of Desi chickpeas and found increased phenolic contents, DPPH free radical scavenging activity (%) and total flavonoids contents in leaves after 30 and 45 days of sowing. However, there was no difference between 45 and 60 days for both cultivars. Macar, Macar and Mart 2017. (2017) investigated the total phenolic contents and total flavonoid level of Turkish Kabuli and Desi chickpeas and recorded total phenolic content of the Desi type (128.8 mg GAE/100 g seed), which was significantly higher than other Kabuli genotypes. Whereas, the total flavonoid level of ICC 3996 (116.2 mg CE/100 g) was closer to İnci (114.3 mg CE/100 g) and Hasanbey (118.1 mg CE/100 g) cvs of Kabuli origin. Although, Desi chickpeas have lot of health benefits as they are rich in protein, carbohydrates, fats, oils, secondary metabolites, vitamins and nutrients. Their usage as medicinal plants and their health benefits are still not documented properly. There are some websites which highlight the use of Desi chickpeas which are based on traditional knowledge of its use for health rather than scientific based knowledge.

The unique combination of antioxidants, anthocyanins, cyaniding, delphindin, petunidin, ALA and phytonutrients in black chickpea are helpful for preventing cardiovascular risks due to preventing oxidative stresses, keeping blood vessels healthy (Segev et al. 2011; Jukanti et al. 2012). Similarly, folate in Desi chickpeas help to maintain the arteries normal by keeping the homocysteine levels lower (Saba 2019). Soluble fiber is helpful for bile and ferries excretion while, insoluble is helpful for constipation and other digestive disorders (Rinco'n et al. 1998). Other advantages of using dietary fibers include hunger cravings and weight loss; keeping the cholesterol level low by binding the bile acids (Segev et al. 2011; Saba 2019). Stabilization of soluble sugars by controlling the absorption and release of blood sugar due to low glycemic index. Slow digestion of sugar is helpful to prevent type-2 diabetes, prevention of anemia due to high Fe contents (Dreyhan 2019; Saba 2019). Desi chickpeas are an excellent alternative protein source and its use reduce the risk of breast cancer due to presence of saponins. Other medicinal uses of Desi chickpeas are curing leucorrhea and preventing miscarriage in women. Besides, it helps reduce the risk of cancer due to breakdown of fibers in colon by bacteria into short chain fatty acids and act as an energy source for colon cells and keep them healthy, curing vomiting and breakdown of stones in kidneys or bladder. They are also healthy for skin due to absence of sodium and cholesterol and can be used for treating leukoderma, curing ringworm infections, skin facial care, healing properties, promoting hair growth, preventing hair loss and greying and reduced dandruff are other advantages of using Desi chckpea (Koradia 2019; Saba 2019).

CONCLUSION

Desi chickpra is an important edible legume and important constituent of dietary program in India and Pakistan. In Sub-Continent, different foods, snacks and local sweets and salads are made from desi chickpeas or from *dhal* and *besin* (chickpea flour). Besides, flour is also used as face cleaner and also used for curing skin diseases and improving human health as desi chickpeas are effective against diabetes and other chronic diseases. However, research work related to health benefits is limited and need more extensive investigations to explore the detailed benefits on scientific basis. Most of the current health benefits associated with desi chickpeas are without any scientific documentation. Desi chickpeas are highly nutritive and in recent years, nutritive profile and screening of some desi chickpeas have been reported but still require more research. The research work on different aspects of Desi chickpea is also limited in Turkey as compared to Kabuli chickpeas and there is need to do more research work on nutritional aspects and making products like *leblebi*. It is also required to highlight the use of desi chickpeas for medicinal purposes and its consumption as raw.

REFERENCES

Agarwal, Gaurav, Shalu Jhanwar, Pushp Priya, Vikash K. Singh, Maneesha S. Saxena, Swarup K. Parida, Rohini Garg, Akhilesh K. Tyagi, and Mukesh Jain. "Comparative Analysis of Kabuli Chickpea Transcriptome with Desi and Wild Chickpea Provides a Rich Resource for Development of Functional Markers." *PLoS ONE* 7, no. 12 (2012). doi:10.1371/journal.pone.0052443.

Aguilera, Yolanda, María A. Martín-Cabrejas, Vanesa Benítez, Esperanza Mollá, Francisco J. López-Andréu, and Rosa M. Esteban. "Changes in Carbohydrate Fraction during Dehydration Process of Common Legumes." *Journal of Food Composition and Analysis* 22, no. 7-8 (2009): 678-83. doi:10.1016/j.jfca.2009.02.012.

Aydemir, Levent Yurdaer, and Ahmet Yemenicioğlu. "Potential of Turkish Kabuli Type Chickpea and Green and Red Lentil Cultivars as Source of Soy and Animal Origin

Functional Protein Alternatives." *LWT - Food Science and Technology* 50, no. 2 (2013): 686-94. doi:10.1016/j.lwt.2012.07.023.

Badshah, Amal, M. D. Zadid Khan, Nizkat Bibi, Muhammad Zuhaib Khan, Sajjad Ali Ehsan Ali, Muhammad Ashraf Chaudry and Mohammad Saeed Khattak. "Quality studies of newly evolved chickpea cultivars." *Advances in Food Sciences* 25, (2003): 95-99.

Bueckert, Rosalind A., Dil Thavarajah, Pushparajah Thavarajah, and Janet Pritchard. "Phytic Acid and Mineral Micronutrients in Field-grown Chickpea (*Cicer arietinum* L.) Cultivars from Western Canada." *European Food Research and Technology* 233, no. 2 (2011): 203-12. doi:10.1007/s00217-011-1495-8.

Chavan, J. K., S. S. Kadam, D. K. Salunkhe, and Larry R. Beuchat. "Biochemistry and Technology of Chickpea(*Cicer arietinum* L.) Seeds." *C R C Critical Reviews in Food Science and Nutrition* 25, no. 2 (1987): 107-58. doi:10.1080/10408398709527449.

Chibbar, Ravindra N., Priyatharini Ambigaipalan, and Ratnajothi Hoover. "Review: Molecular Diversity in Pulse Seed Starch and Complex Carbohydrates and Its Role in Human Nutrition and Health." *Cereal Chemistry Journal* 87, no. 4 (2010): 342-52. doi:10.1094/cchem-87-4-0342.

Ciftci, Harun, Ahmet Ozkaya, B. Sıktı Cevrimli, Adıl Bakoglu. "Levels of Fat-Soluble Vitamins in Some Foods. *Asian Journal of Chemistry* 22, (2010). 1251-1256.

Dalgetty, David D., and Byung-Kee Baik. "Isolation and Characterization of Cotyledon Fibers from Peas, Lentils, and Chickpeas." *Cereal Chemistry Journal* 80, no. 3 (2003): 310-15. doi:10.1094/cchem.2003.80.3.310.

del Mar Yust, Maria, Justo Pedroche, Julio Girón-Calle, Manuel Alaiz, Francisco Millán, and Javier Vioque. "Production of Ace Inhibitory Peptides by Digestion of Chickpea Legumin with Alcalase." *Food Chemistry* 81, no. 3 (2003): 363-69. doi:10.1016/s0308-8146(02)00431-4.

Diapari, Marwan, Anoop Sindhu, Kirstin Bett, Amit Deokar, Thomas D. Warkentin, and Bunyamin Tar'An. "Genetic Diversity and Association Mapping of Iron and Zinc Concentrations in Chickpea (*Cicer arietinum* L.)." *Genome* 57, no. 8 (2014): 459-68. doi:10.1139/gen-2014-0108.

Dreyhan, Wiran. "Powerful Health Benefits of Kala channa (Black Chickpeas)". (2019). https://drhealthbenefits.com/food-bevarages/grains/health-benefits-of-kala-chana.

Dwivedi, Sangam L., Kanwar L. Sahrawat, Kedar N. Rai, Matthew W. Blair, Meike S. Andersson, and Wolfgang Pfeiffer. "Nutritionally Enhanced Staple Food Crops." *Plant Breeding Reviews*, 2012, 169-291. doi:10.1002/9781118358566.ch3.

Gaur, P. M., S. Tripathi, C. L. L. Gowda, G. V. Ranga Rao, H. C. Sharma, S. Pande and M. Sharma. *Chickpea Seed Production Manual.* International Crops Research Institute for the Semi-Arid Tropics. Patancheru 502 324, Andhra Pradesh, India.

Gecit, Hasan Huseyin. *Chickpea utilization in Turkey*. Proceedings of Uses of Tropical Grain Legumes, ICRISAT INDIA, AP. 1991, 69-74.

Guillon, Fabienne, and Martine M. J. Champ. "Carbohydrate Fractions of Legumes: Uses in Human Nutrition and Potential for Health." *British Journal of Nutrition* 88, no. S3 (2002): 293-306. doi:10.1079/bjn2002720.

Han, In Hwa, and Byung-Kee Baik. "Oligosaccharide Content and Composition of Legumes and Their Reduction by Soaking, Cooking, Ultrasound, and High Hydrostatic Pressure." *Cereal Chemistry Journal* 83, no. 4 (2006): 428-33. doi:10.1094/cc-83-0428.

Hawtin, G. C., and K. B. Singh. *Kabuli-desi Introgression: Problems and Prospects.* Proceedings of International Workshop on Chickpea Improvement, India, Hayderabd AP. 1979. 51-60.

Hulse, J. H. *Nature, Composition and Utilization of Pulses.* Proceedings of Uses of Tropical Grain Legumes, ICRISAT INDIA, AP. 1991. 11-27.

Ibáñez, M.victoria, Francisco Rincón, Manuel Amaro, and Beatriz Martínez. "Intrinsic Variability of Mineral Composition of Chickpea (*Cicer arietinum*, L.)." *Food Chemistry* 63, no. 1 (1998): 55-60. doi:10.1016/s0308-8146(97)00221-5.

Ibrikci, Hayriye, Sharon Jb Knewtson, and Michael A. Grusak. "Chickpea Leaves as a Vegetable Green for Humans: Evaluation of Mineral Composition." *Journal of the Science of Food and Agriculture* 83, no. 9 (2003): 945-50. doi:10.1002/jsfa.1427.

Jambunathan, R., and U. Singh. Studies on Desi and Kabuli Chickpea (*Cicer arietinum* L.) Cultivars. 1. Chemical Composition. Proceedings of the International Workshop on Chickpea Improvemen. India, ICRISAT. 1980. 61-66.

Jaya, T. V., H. S. Naik and L. V. Venkataraman. "Effect of germinated legumes on the rate of in-vitro gas production by clostridium perfringens". *Nutrition Reports International* 20 (1979): 393- 401.

Jones, D. A., M. S. Dupont, M. J. Ambrose, J. Frias, and C. L. Hedley. "The Discovery of Compositional Variation for the Raffinose Family of Oligosaccharides in Pea Seeds." *Seed Science Research* 9, no. 4 (1999): 305-10. doi:10.1017/s0960258599000318.

Jukanti, Arvind K., Pooran M. Gaur, Cholenahalli Laxmipathi Gowda, and Ravindra N. Chibbar. "Nutritional Quality and Health Benefits of Chickpea (*Cicer arietinum* L.): A Review." *British Journal of Nutrition* 108, no. S1 (2012). doi:10.1017/s00071145 12 000797.

Kahraman, Abdullah, Anamika Pandey, and Mohd. Kamran Khan. "Nutritional Diversity Assessment in Chickpea-A Prospect for Nutrient Deprived World." *Harran Tarım Ve Gıda Bilimleri Dergisi* 21, no. 3 (2017): 357-63. doi:10.29050/harranziraat.339496.

Kaur, Maninder, Narpinder Singh, and Navdeep Singh Sodhi. "Physicochemical, Cooking, Textural and Roasting Characteristics of Chickpea (*Cicer arietinum* L.) Cultivars." *Journal of Food Engineering* 69, no. 4 (2005): 511-17. doi:10.1016/j. jfoodeng.2004.09.002.

Khalil, Abdul Wajid, Aurang Zeb, Fazal Mahmood, Saima Tariq, Amal Badshah Khattak, and Hamidullah Shah. "Comparison of Sprout Quality Characteristics of

Desi and Kabuli Type Chickpea Cultivars (*Cicer arietinum* L.)." *LWT - Food Science and Technology* 40, no. 6 (2007): 937-45. doi:10.1016/j.lwt.2006.05.009.

Khan, M. Akmal, Neghmana Akhtar, Ihsan Ullah, and Saeeda Jaffery. "Nutritional Evaluation of Desi and Kabuli Chickpeas and Their Products Commonly Consumed in Pakistan." *International Journal of Food Sciences and Nutrition* 46, no. 3 (1995): 215-23. doi:10.3109/09637489509012551.

Khetarpaul, Sema Neelam. "Antioxidant activity in leaves of chickpea (*Cicer arietinum* L.) varieties desi and kabuli (on dry matter basis) in India". *European Journal of Biotechnology and Bioscience* 6 (2018): 12-14.

Koradia, Hardip. "Immense Health Benefits of Black Chickpeas (Kala channa)". (2019). https://www.ayurhealthtips.com/immense-health-benefits-of-chickpeas/

Macar, Tuğçe Kalefetoğlu, Oksal Macar, and Dürdane Inci Mart. "Variability in Some Biochemical and Nutritional Characteristics in Desi and Turkish Kabuli Chickpea (*Cicer arietinum* L.) Types." *Celal Bayar Üniversitesi Fen Bilimleri Dergisi*, 2017, 677-80. doi:10.18466/cbayarfbe.339330.

Madhusudhan, Basavaraj, and Rudrapatnam N. Tharanathan. "Structural Studies of Linear and Branched Fractions of Chickpea and Finger Millet Starches." *Carbohydrate Research* 284, no. 1 (1996): 101-09. doi:10.1016/0008-6215(95)004 04-1.

Magalhães, Sara C. q., Marcos Taveira, Ana R. j. Cabrita, António J. m. Fonseca, Patrícia Valentão, and Paula B. Andrade. "European Marketable Grain Legume Seeds: Further Insight into Phenolic Compounds Profiles." *Food Chemistry* 215, (2017): 177-84. doi:10.1016/j.foodchem.2016.07.152.

Maheri-Sis, Naser, Mohammad Chamani, Ali Asghar Sadeghi, Ali Mirza-Aghazadeh, Abolfazl Aghajanzadeh-Golshani. "Nutritional evaluation of kabuli and desi type chickpeas (*Cicer arietinum* L.) for ruminants using in vitro gas production technique". *African Journal of Biotechnology* 7, (2008): 2946–2951.

Mohammadi, Khosro. "Nutritional Composition of Iranian Desi and Kabuli Chickpea (*Cicer arietinum* L.) Cultivars in Autumn Sowing". *International Journal of Agricultural and Biosystems Engineering* 9, no 5 (2015): 550 - 553.

Moreno, Maria-Teresa, and J. I. Cubero. "Variation in *Cicer arietinum* L." *Euphytica* 27, no. 2 (1978): 465-85. doi:10.1007/bf00043173.

Özer, Sertaç, Tolga Karaköy, Faruk Toklu, Faheem S. Baloch, Benjamin Kilian, and Hakan Özkan. "Nutritional and Physicochemical Variation in Turkish Kabuli Chickpea (*Cicer arietinum* L.) Landraces." *Euphytica* 175, no. 2 (2010): 237-49. doi:10.1007/s10681-010-0174-3.

Pande, S., K. H. M. Siddique, G. K. Kishore, B. Bayaa, P. M. Gaur, C. L. L. Gowda, T. W. Bretag, and J. H. Crouch. "Ascochyta Blight of Chickpea (*Cicer arietinum* L.): A Review of Biology, Pathogenicity, and Disease Management." *Australian Journal of Agricultural Research* 56, no. 4 (2005): 317. doi:10.1071/ar04143.

Paredes-López, O., C. Ordorica-Falomir, and M. R. Olivares-Vázquez. "Chickpea Protein Isolates: Physicochemical, Functional and Nutritional Characterization." *Journal of Food Science* 56, no. 3 (1991): 726-29. doi:10.1111/j.1365-2621.1991.tb05367.x.

Purushothaman, R., H. D. Upadhyaya, P. M. Gaur, C. L L. Gowda, and L. Krishnamurthy. "Kabuli and Desi Chickpeas Differ in Their Requirement for Reproductive Duration." *Field Crops Research* 163, (2014): 24-31. doi:10.1016/j.fcr.2014.04.006.

Rehman, Zia-Ur, and W. H. Shah. "Thermal Heat Processing Effects on Antinutrients, Protein and Starch Digestibility of Food Legumes." *Food Chemistry* 91, no. 2 (2005): 327-31. doi:10.1016/j.foodchem.2004.06.019.

Rincón, Francisco, Beatríz Martínez, and M. Victoria Ibáñez. "Proximate Composition and Antinutritive Substances in Chickpea (*Cicer arietinum* L) as Affected by the Biotype Factor." *Journal of the Science of Food and Agriculture* 78, no. 3 (1998): 382-88. doi:10.1002/(sici)1097-0010(199811)78:33.0.co;2-j.

Rybiński, Wojciech, Maciej Banda, Jan Bocianowski, Elzbieta Starzycka-Korbas, Michal Starzycki and Kamila Nowosad. "Estimation of the Physicochemical Variation of Chickpea Seeds (*Cicer arietinum* L.)." *International Agrophysics* 33, no. 1 (2019): 67-80. doi:10.31545/intagr/103889.

Saba. "19 Amazing benefits of Black Chickpeas (Kala Channa) for Skin, Hair and Health" (2019). https://www.stylecraze.com/articles/benefits-of-black-chickpeas-for-skin-hair-and-health/#gref

Sánchez-Mata, M. Cortes, M. José Peñuela-Teruel, Montaña Cámara-Hurtado, Carmen Díez-Marqués, and M. Esperanza Torija-Isasa. "Determination of Mono-, Di-, and Oligosaccharides in Legumes by High-Performance Liquid Chromatography Using an Amino-Bonded Silica Column." *Journal of Agricultural and Food Chemistry* 46, no. 9 (1998): 3648-652. doi:10.1021/jf980127w.

Sánchez-Vioque, R., A. Clemente, J. Vioque, J. Bautista, and F. Millán. "Protein Isolates from Chickpea (*Cicer arietinum* L.): Chemical Composition, Functional Properties and Protein Characterization." *Food Chemistry* 64, no. 2 (1999): 237-43. doi:10.1016/s0308-8146(98)00133-2.

Saskatchewan Pulse Growers. "Nourishing the world with Profitable Pulse Production". (2019). https://saskpulse.com/growing-pulses/chickpeas/seeding/

Segev, Aharon, Hana Badani, Liel Galili, Ran Hovav, Yoram Kapulnik, Ilan Shomer, and Shmuel Galili. "Total Phenolic Content and Antioxidant Activity of Chickpea (*Cicer arietinum* L.) as Affected by Soaking and Cooking Conditions." *Food and Nutrition Sciences* 02, no. 07 (2011): 724-30. doi:10.4236/fns.2011.27099.

Shad, Muhammad Aslam, Humayun Pervez, Zafar Iqbal Zafar, Muhammad Zai-Ul-Haq, Haq Nawaz. "Evaluation of biochemical composition and physicochemical parameters of oil from seeds of desi chickpea varieties cultivated in arid zone of Pakistan". *Pakistan Journal of Botany* 41, (2008): 655-662.

Shrivastava, A. K., G. P. Dixit and N. Kumar. "Milling properties of desi and kabuli chickpea (*Cicer arietinum* L.) varieties released in India". *Indian Journal of Agricultural Sciences* 87, no 10 (2017): 1350–1357.

Singh, U. "Nutritional Quality of Chickpea (*Cicer arietinum* L.): Current Status and Future Research Needs." *Qualitas Plantarum Plant Foods for Human Nutrition* 35, no. 4 (1985): 339-51. doi:10.1007/bf01091779.

Singh, U. "Dietary fiber and its constituents in desi and kabuli Chickpea (*Cicer arietinum* L.) cultivars". *Nutrition Reports International* 29, no 2 (1984): 419-426.

Singh, U., and R. Jambunathan. "Studies on Desi and Kabull Chickpea (*Cicer arietinum* L.) Cultivars: Levels of Protease Inhibitors, Levels of Polyphenolic Compounds and in Vitro Protein Digestibility." *Journal of Food Science* 46, no. 5 (1981): 1364-367. doi:10.1111/j.1365-2621.1981.tb04176.x.

Singh, Umaid, Jitendra Kumar, Ramamurthi Jambunathan and John Smithson. "Variability in the seed coat content of desi and kabuli chickpea cultivars." *3. Int. Chickpea Newsl* (1980): 18.

Tikle, Ashwini, and Archana Mishra. "Physical and Milling Properties of Chickpea, *Cicer arietinum* Influenced by Seed Characteristics." *Bioscience Biotechnology Research Communications* 11, no. 1 (2018): 122-27. doi:10.21786/bbrc/11.1/17.

Tosh, Susan M., and Sylvia Yada. "Dietary Fibres in Pulse Seeds and Fractions: Characterization, Functional Attributes, and Applications." *Food Research International* 43, no. 2 (2010): 450-60. doi:10.1016/j.foodres.2009.09.005.

United States Department of Agriculture (USDA). National Nutrient Database for Standard Reference (2010). http://www.nal.usda.gov/fnic/foodcomp/search/

Upadhyaya, Hari D., Deepak Bajaj, Laxmi Narnoliya, Shouvik Das, Vinod Kumar, C. L. L. Gowda, Shivali Sharma, Akhilesh K. Tyagi, and Swarup K. Parida. "Genome-Wide Scans for Delineation of Candidate Genes Regulating Seed-Protein Content in Chickpea." *Frontiers in Plant Science* 7, (2016). doi:10.3389/fpls.2016.00302.

Upadhyaya, Hari D., Sangam L. Dwivedi, Michael Baum, Rajeev K. Varshney, Sripada M. Udupa, Cholenahalli Ll Gowda, David Hoisington, and Sube Singh. "Genetic Structure, Diversity, and Allelic Richness in Composite Collection and Reference Set in Chickpea (*Cicer arietinum* L.)." *BMC Plant Biology* 8, no. 1 (2008): 106. doi:10.1186/1471-2229-8-106.

Varshney, Rajeev K., Chi Song, Rachit K. Saxena, Sarwar Azam, Sheng Yu, Andrew G. Sharpe, Steven Cannon, Jongmin Baek, Benjamin D. Rosen, Bunyamin Taran, Teresa Millan, Xudong Zhang, Larissa D. Ramsay, Aiko Iwata, Ying Wang, William Nelson, Andrew D. Farmer, Pooran M. Gaur, Carol Soderlund, R. Varma Penmetsa, Chunyan Xu, Arvind K. Bharti, Weiming He, Peter Winter, Shancen Zhao, James K. Hane, Noelia Carrasquilla-Garcia, Janet A. Condie, Hari D. Upadhyaya, Ming-Cheng Luo, Mahendar Thudi, C. L L Gowda, Narendra P. Singh, Judith Lichtenzveig, Krishna K. Gali, Josefa Rubio, N. Nadarajan, Jaroslav Dolezel, Kailash C. Bansal,

Xun Xu, David Edwards, Gengyun Zhang, Guenter Kahl, Juan Gil, Karam B. Singh, Swapan K. Datta, Scott A. Jackson, Jun Wang, and Douglas R. Cook. "Draft Genome Sequence of Chickpea (*Cicer arietinum*) Provides a Resource for Trait Improvement." *Nature Biotechnology* 31, no. 3 (2013): 240-46. doi:10.1038/nbt.2491.

Wang, Ning, and James. K. Daun. "The chemical composition and nutritive value of canadian pulses." *Canadian Grain Commission Report* (2004) 19-29.

Wang, Xiansheng, Wenrui Gao, Jusong Zhang, Hua Zhang, Jiangui Li, Xiaoling He, and Hao Ma. "Subunit, Amino Acid Composition and in Vitro Digestibility of Protein Isolates from Chinese Kabuli and Desi Chickpea (*Cicer arietinum* L.) Cultivars." *Food Research International* 43, no. 2 (2010): 567-72. doi:10.1016/j.foodres.2009.07.018.

Wood, J. A., and M. A. Grusak. Nutritional Value of Chickpea. In: Yadav, S. S., Redden, B., Chen, W. and Sharma, B., Eds., *Chickpea Breeding and Management, CAB International,* Wallingford, (2007): 101-142. doi:10.1079/9781845932138.005

Zia-Ul-Haq, Muhammad, Mansoor Ahmad, Shahid Iqbal, Shakeel Ahmad, and Hakoomat Ali. "Characterization and Compositional Studies of Oil from Seeds of Desi Chickpea (*Cicer arietinum* L.) Cultivars Grown in Pakistan." *Journal of the American Oil Chemists Society* 84, no. 12 (2007a): 1143-148. doi:10.1007/s11746-007-1136-3.

Zia-Ul-Haq, Muhammad, Shahid Iqbal, Shakeel Ahmad, Muhammad Imran, Abdul Niaz, and M. I. Bhanger. "Nutritional and Compositional Study of Desi Chickpea (*Cicer arietinum* L.) Cultivars Grown in Punjab, Pakistan." *Food Chemistry* 105, no. 4 (2007b): 1357-363. doi:10.1016/j.foodchem.2007.05.004.

Zia-Ul-Haq, Muhammad, Shakeel Ahmad, Mansoor Ahmad, Shahid Iqbal, Khalid Mahmood Khawar. "Effects of Cultivar and Row Spacing on Tocopherol and Sterol Composition of Chickpea (*Cicer arietinum* L) Seed Oil." *Tarım Bilimleri Dergisi* 15, no. 1 (2009): 25-30. doi:10.1501/tarimbil_0000001068.

In: Handbook of Chickpeas
Editors: Albert T. Lund and Noah D. Schultz
ISBN: 978-1-53616-374-2

Chapter 4

CHICKPEA PROTEINS AND PEPTIDES IN HEALTH

Daniela Guardado-Felix[1,2], PhD, Ada K. Milán-Noris[2], PhD, Sayra N. Serrano-Sandoval[1], PhD, and Sergio O. Serna-Saldivar[1,*]

[1]Tecnologico de Monterrey, Centro de Biotecnología-FEMSA, Escuela de Ingeniería y Ciencias, Monterrey, N.L. México

[2]Laboratorio de Nutracéuticos (18), Facultad de Ciencias Químico-Biológicas, Universidad Autónoma de Sinaloa, Culiacán, Sin, México

ABSTRACT

Chickpea (*Cicer arietinum L.*) is the third most relevant food legume produced worldwide. Its nutritional quality has been considered superior compared to other pulses and its consumption provides several proven health benefits. This legume crop is usually consumed as an alternative to more expensive animal proteins in developing countries. Chickpeas are a significant source of high-quality protein (12-30%) with a good balance of essential amino acids, high protein digestibility and relatively low levels of antinutritional factors. The limiting amino acids are sulphur-containing and therefore can be easily complemented by consumption of cereal-based foods. The amino acid composition, protein quality and functional properties of chickpea proteins are affected by processing methods, such as decortication, soaking, germination, fermentation, boiling, among others, which remove or inactivate antinutritional factors and improve protein digestibility. Chickpea proteins are mainly composed of globulins (56%), glutelins (18%), albumins (12%) and prolamins (3%). Chickpea proteins isolates have been employed as food ingredients to improve both nutritional and functional properties. Besides the nutritional and functional properties of proteins, they may also possess other biological properties. These properties are attributed to active peptides encoded in protein molecules that can be released during gastrointestinal digestion or by

[*] Corresponding Author's Email: sserna@tec.mx.

controlled hydrolyses using exogenous proteases. The enzymatic hydrolysis of chickpea proteins is an important tool for obtaining hydrolysates and bioactive peptides with antioxidant, antifungal, angiotensin I-converting enzyme (ACE) inhibitory (hypotensive), hypocholesterolemic and anticarcinogenic activities. Moreover, the production of selenoproteins during germination has been recently proposed as alternative to diminish oxidative stress and improve health. This is due to the synthesis of selenoproteins, which are used to synthesize Glutathione Peroxide that protects the human system against oxidative stress and cancer.

This chapter describes current information of the following 1) amino acid profile and nutritional protein quality in raw and processed chickpea seeds and flours, 2) health or nutraceutical effects of chickpea protein isolates, hydrolysates or bioactive peptides and 3) generation of selenized proteins during sprouting and their antioxidant and anticancer potential.

Keywords: chickpea, protein, peptides, amino acids, quality, nutraceutical

INTRODUCTION

Approximately 76% of the worldwide population rely on proteins derived from plant sources (FAO, 2014). The most relevant sources are from legumes which are an important component of the diets of those individuals who cannot afford animal proteins or those who are vegetarian by choice.

Unfortunately, protein–energy malnutrition is still observed in infants and young children inhabiting in developing countries. The insufficient protein intake limits growth, tissue repair, decreases the skeletal muscle mass and lymphocyte count (Castaneda et al., 1995). The Recommended Dietary Allowance (RDA) for protein estimated in 0.8 g protein/kg of body weight is the same for all adults, regardless of age or sex. The acceptable macronutrient distribution range (AMDR) specifies that protein should represent 10-35% of an individual's total daily energy intake (Campbell et al., 2008).

Chickpea (*Cicer arietinum L.*) has been reported to have high protein bioavailability and quality with a good balance of essential amino acids, better than some legumes crops such as black gram (*Vigna mungo L.*), green gram (*Vigna radiata L.*) and red gram (*Cajanus cajan L.*) (Milán-Noris et al., 2019).

Chickpea is ranked globally as the third most important legume crop and one of the most consumed in the world, especially in the Mediterranean area as well in the Western world. The global production in 2016 exceeded 12 million tons with an average yield per hectare of 0.96 ton. India is the major producer with an annual production of 7.8 million tons (FAOSTAT 2016). It is estimated that it is grown in 54 countries with nearly 90% of its area planted in developing countries (Shanko et al., 2017). Chickpea seeds are classified as either Kabuli which possesses thin and cream-white seed coat pigmentation or Desi types which presents a thick and pigmented seed coat (Kadam et al., 1989; Jukanti et al., 2012).

Chickpea is used in stews, soups/salads and consumed in roasted, boiled, salted and fermented forms. These different forms of consumption provide consumers with valuable nutrition and potential health benefits (Jukanti et al., 2012).

In recent years, chickpea protein consumption has gained importance due to its high protein quality, amino acid profile and bioactive peptides content, which are known to benefit human health and because it is comparatively less allergenic compared to soybeans (*Glycine max*) or peanuts (*Arachis hypogea*). On the other hand, the selenization of chickpea proteins is an excellent alternative to increase their antioxidant and acticancer potential (Guardado-Félix et al., 2019; Serrano-Sandoval et al., 2019) so these novel products can be incorporated in formulations to improve both the nutritional and nutraceutical values. This chapter reviews investigations regarding the nutritional value of chickpeas proteins and their potential health benefits.

PROTEIN CONTENT

Chickpea *(Cicer arietinum L.)* protein hs received major attention during recent years due to its relative low cost, high digestibility, pleasant bland flavor and other relevant functional properties, which makes it as an important source of protein and energy in most Afro-Asian diets. Its protein content, which ranges from 17 to 22%, varies according to the type (Kabuli or Desi) (Jukanti et al., 2012). Chickpea seed proteins have been fractionated into globulins (56%), albumins (12%), prolamins (2.8%), glutelins (18.1%) and residual proteins (Singh and Jambunathan, 1982). The chickpea proteins are differentially distributed in the anatomical structures of the seed; coat, cotyledons and embryo. The cotyledons constitute 82.9% of the total seed weight whereas the embryo and seed coat constitute a small proportion. The embryo contains 52.1% protein whereas the cotyledon 24.8% and seed coat 4.1% (Table 1) (Singh and Jambunathan, 1982). The embryo is rich in albumins whereas the cotyledons in globulins and seed coat in glutelins and prolamins.

Table 1. Distribution of protein fractions in the different anatomical parts of the chickpea seed (dry weight basis)

Component	Amount (%)	Protein (%)	Protein fractions (%)					
			Albumin	Globulin	Glutelin	Prolamin	Residue	Total
Embryo	1.2	52.1	22.5	50.0	21.4	3.0	1.5	98.4
Cotyledon	82.9	24.8	15.9	62.7	17.5	2.3	1.0	99.4
Seed coat	16.4	4.1	3.5	22.8	33.2	3.4	30.5	93.4
Whole seed	-	21.3	12.6	56.6	18.1	2.8	4.9	95.0

Adapted from Singh and Jambunathan, 1982.

Globulins act as storage proteins representing a source of energy and reduced nitrogen essential for germination and seedling growth. Globulins comprise about 50% of chickpea seed proteins and are composed of two major groups classified according to the sedimentation coefficients, the 11S known as legumin (320-400 kDa) and the 7S named vicilin (145-190 kDa) (Casey et al., 1993).

Chickpea 11S legumin is the major subfraction and electrophretically produce bands between 46.5 to 39.8 and 25.3 to 24.3 kDa related to α and β polypeptide chains. These two subunits are connected by a single disulfide bond (Vairinhos and Murray 1982). On the other hand, chickpea 7S vicilin is also storage protein, which structure and conformation have been studied less extensibly compared to the 11S counterpart. Chang et al. (2012) reported subunit MWs of the chickpea 7S globulin as 70.2, 50.7, 35.0, 33.6 and 18.9 kDa based on SDS-PAGE characterization. Little is known about the allergenicity of chickpeas, however, both vicilin (50kDa) and the basic subunit of legumin (20 kDa) have been previously reported as potential chickpea allergens (Bar-El Dadon et al., 2013). Chickpea is one of the few legumes containing significant amount of glutelins which are considered the major fraction in cereal grains. Glutelins belong to the 11–12S globulin family and are structurally alike to globulins (Takaiwa et al., 1999). However, there is slight information on the molecular characterization of glutelins such as peptide sequences, amino acid sequences and biological and functional properties.

The albumin fraction is more heterogeneous, including most of the proteins needed for cellular metabolism, physiological function, such as the enzymatic activities of lipoxygenases, glycosidases, or proteases. The last are deeply involved in the degradation of storage proteins. Other albumins, such as protease inhibitors or lectins, are implicated in defensive mechanisms (Gueguen, 1991). These have been described as the PA2 albumin and the PA1 or 2S albumin (Casey et al., 1993). Chickpea 2S albumin (~ 20 kDa) is composed of two polypeptides of 10 and 12 kDa. Its native molecular mass determined by gel filtration chromatography is 20 kDa. Amino acid composition shows that it is rich in sulfur amino acids, mainly cysteine with 4.6% of the total amino acid composition (Vioque et al., 1999). The 2S albumin subfraction is a compact globular protein, as suggested by the disulfide structure reported for the 2S albumin of lupin (*Lupinus albus)* (Lilley and Inglis, 1986).

Trypsin Inhibitors in Chickpea

Protease inhibitors are synthesized by plants as defence molecules against pathogens and plant pests. They are divided into four groups depending on the proteolytic anzymes inhibited (serine, cysteine, aspartic and metalloprotease). Two types of serine protease inhibitors have been characterized: Kunitz and Bowman-Birk. Kunitz type inhibitors are single chain polypeptides of ~20 kDa, with two intra-chain disulphide bridges and form

1:1 complex with the target proteinase. The Bowman–Birk inhibitors (BBIs) are also single chain polypeptides, but smaller in size (~8 kDa) and possess seven disulphide bridges and two active domains for trypsin and/or chymotrypsin (Srinivasan et al., 2005). Chickpeas contain relatively low amounts of trypsin inhibitors compared to other legumes such as soybeans, which are capable of binding trypsin, thus inhibiting its activity, and interfering with the digestion of proteins (Birk, 1989). The content of trypsin inhibitors in chickpea has been reported in 2.7 mg/g dw (Savage et al., 1998). Other research found that trypsin inhibitor constitutes the 5.03% of extractable seed protein (Sastry and Murray, 1897). A Kunitz-type trypsin inhibitor protein purified from chickpea seeds was estimated to have a molecular mass of 18 kDa, its IC50 value was determined to be 2.5 mg against trypsin and its inhibitory activity was 114 TIU (trypsin inhibitory units) per milligram of protein (Sharma and Suresh, 2011). The chickpea trypsin inhibitors are distributed in the three characteristic anatomical parts, cotyledon (77.2% to 75.8%), embryonic axis (11.9% to 15.5%), and seedcoat (10.9% to 8.7%) (Sreerama et al., 2010).

The presence of trypsin inhibitors in human and monogastric diets interferes with the intestinal digestion of proteins, causing a negative effect on health and animal growth and efficiency of feed conversion, mainly due to the hyerplasia or enlargement of the pancreas and hypersecretion of digestive enzymes, which chronically causes muscle wasting. This leads to a los of sulphur-rich endogenous proteins, trypsin and chymotrypsin, which would depress growth because legumes seed proteins are generally deficient in sulphur amino acids (Lajolo and Genovese, 2002).

Different food preparation techniques have been successfully used to diminish trypsin inhibitor activity such as soaking, germination, boiling and pressure-cooking. It has been shown that the simple application of water soaking for 12 h on chickpea seeds caused reduction in the levels of trypsin inhibitor by 59.9%. Also, cooking for 90 sec completely inactivated the inhibitor (Singh et al., 2015). Aditionally, the trypsin inhibitor activity was effectively reduced after boiling seeds in water and was totally removed after 300 sec of treatment (Márquez and Alonso, 1999).

Although trypsin inhibitors are characterized by their antinutritional properties they have also been recognized for their anticancer potential. Magee et al. (2012) showed that the viability of MDAMB-231 breast cancer and PC-3 and LNCaP prostate cancer cells were inhibited significantly by the chickpea Bowman-Birk-type protease inhibitor (molecular weight ~8 kDa) at all concentrations tested (25–400 μg/ml).

PROTEIN QUALITY IN RAW AND PROCESSED CHICKPEAS

Table 2 depicts seed protein quality of the raw and processed chickpea.

Table 2. Protein quality of raw and processed Desi and Kabuli chickpeas

Process/Characteristics	Type	TPD (%)	*IVPD* (%)	PER	CPER	PDCAAS	Reference
Raw							
	Kabuli	83.7	72.2	1.59	1.54	0.73	Angulo-Bejarano et al., 2008.
	Kabuli	84.8	72.7	1.47			Milán-Carrillo et al., 2000.
	Kabuli		78.8				Xu et al., 2016.
	Kabuli		76.6				Khalil et al., 2007.
	Kabuli		74.5				Chandra-Hioe et al., 2016.
	Desi		69.7				Khalil et al., 2007.
	Desi		70.5				Chandra-Hioe et al., 2016.
	Desi		76.4				Bai et al., 2018.
Flour cv IAC-Marrocos	NS	83.0	82.5			0.26-0.60	Tavano et al., 2016.
215 cultivars	NS		59.4-76.2				Bhagyawant et al., 2018.
	NS		82.2				Xiao et al., 2015.
Soaking							
Soaked seeds 16 h in 1/10 dilution at 22-24°C; cv Blanco Sinaloa	Kabuli		88.9				Xu et al., 2016.
Hardened							
Hardened seeds (storage 180 days at 33-35°C and 75% HR), cv Blanco Sinaloa	Kabuli	82.5	67.7	1.30			Milán-Carrillo et al., 2000.
Extrusion							
Dehulling/Softening (soaking in 1% salt solution ($NaCl/NaHCO_3$) 12 h 4°C)/Extrusion (Single-screw, 151°C, 189.5 rpm), cv Blanco Sinaloa	Kabuli	91.2	79.5	1.87			Milán-Carrillo et al., 2000.
Hardened seeds (storage 180d at 33-35°C and 75% HR)/Dehulling/Softening (soaking in1% salt solution ($NaCl/NaHCO_3$) 12 h 4°C)/Extrusion (Single-screw, 151°C, 189.5 rpm), cv Blanco Sinaloa	Kabuli	87.3	82.4	1.78			Milán-Carrillo et al., 2000.
Thermal processing							
Boiling flours 20 min, cv Blanco Sinaloa	Kabuli		80.6		2.1-2.2	0.59	Milán-Noris et al., 2019.
Boiling flours 20 min, 9 cultivars	Desi		75.4-82.4		2.3-3.0	0.59-0.82	Milán-Noris et al., 2019.
Flour autoclaving at 121°C/15 min.		83.4	93.2			0.28-0.61	Tavano et al., 2016.
Pressure cooking (20 min)	Kabuli		93.7				Xu et al., 2016.
Microwave cooking (15 min)	Kabuli		92.2				Xu et al., 2016.
Roasting (400°C/30 min)	Kabuli		81.9				Xu et al., 2016.
Infrared Heated 115°C	Desi		73.2			0.63	Bai et al., 2018.
Infrared Heated 135°C	Desi		76.2			0.62	Bai et al., 2018.

Process/Characteristics	Type	TPD (%)	*IVPD* (%)	PER	CPER	PDCAAS	Reference
Tempered 20% moisture / Infrared Heated 115°C	Desi		77.1			0.65	Bai et al., 2018.
Tempered 20% moisture / Infrared Heated 135°	Desi		79.2			0.70	Bai et al., 2018.
Soaking seeds 16 h/ boiling 22 min	Kabuli	85.0		2.32		0.52	Nosworthy et a., 2017.
Soaked seeds 16 h in 1/10 dilution at 22-24°C /Pressure cooking (20 min)	Kabuli		99.7				Xu et al., 2016.
Soaked seeds 16 h in 1/10 dilution at 22-24°C /Microwave cooking (15 min)	Kabuli		98.0				Xu et al., 2016.
Soaked seeds 16 h in 1/10 dilution at 22-24°C /Roasting (400°C/30 min)	Kabuli		86.4				Xu et al., 2016.
Fermentation							
Rhizopus oligosporus, 35°C, 51 h, cv Blanco Sinaloa	Kabuli	88.8	83.2	2.3	2.2	0.92	Angulo-Bejarano et al., 2008.
Cordyceps militaris SN-18, 25°C,7 days	NS		85.5				Xiao et al., 2015.
Natural fermentation, 30°C, 200 rpm, 16h	Desi		72.4				Chandra-Hioe et al., 2016.
Natural fermentation, 30°C, 200 rpm, 16h	Kabuli		75.9				Chandra-Hioe et al., 2016.
Lactobacillus ssp fermentation, 30°C, 200 rpm, 16h	Desi		77.4				Chandra-Hioe et al., 2016.
Lactobacillus ssp fermentation, 30°C, 200 rpm, 16h	Kabuli		76.2				Chandra-Hioe et al., 2016.
Germination							
96 h, room temperature	Desi		93.4				Khalil et al., 2007.
96 h, room temperature	Kabuli		95.0				Khalil et al., 2007.

NS. Not specified. TPD: true protein digestibility, IVPD: in vitro protein digestibility, PER: protein efficiency ratio, CPER: calculated protein efficiency ratio, PDCAAS: protein digestibility corrected amino acid score.

Amino Acid Composition

Amino acid composition plays an important role in the nutritional quality of proteins. In pulses, the limiting amino acids are sulphur-containing, which can be easily complemented by consumption of cereals in the diet. The limiting essential amino acid (EAA) is calculated as the relation among the essential amino acid content and the FAO recommendations for preschool children (FAO/WHO/UNU, 2007).

It is well-known that sulphur-containing amino acids vary slightly according to chickpea cultivar or genotype and agronomical practices like soil fertility. Milán-Noris et al. (2019) assessed the amino acid profiles of 9 Desi and 1 Kabuli cooked chickpea flours and concluded that the profiles slightly varied among cultivars. The content of His, Lys, Leu, Ile, Phe+Tyr and Val were above the FAO recommendations for preschool children (FAO/WHO/UNU, 2007). In 8 of 9 of the Desi chickpea flours, the limiting EAA was Trp which supplied between 73 to 90% of the requirement and in the remaining flour was

Thr which supplied 99% of the two year old infant requirement. In the same study, the Kabuli commercial white colored chickpea supplied 86% Thr, 76% Trp and 95% of Met+Cys. Likewise, Iqbal et al. (2006) identified these three amino acids as the most limiting EAA. Nosworthy et al. (2017) reported in cooked Kabuli chickpeas Trp (61%) and Met+Cys (85%) as the limiting EAA. In a Brazilian chickpea flour the limiting EAA was Met+Cys (75%) followed by Trp (84%) and Val (97%) (Tavano et al., 2016). In another study, raw and infrared cooked Desi chickpeas showed Thr (87%) as the most limiting EAA followed by Trp (95%) (Bai et al., 2018). All these studies clarly show that chickpeas are rich in Lys which is the limiting EAA in all cereal grains; therefore, this pulse is ideal to complement the amino acid profile of cereal-based foods.

Tempeh is a traditional oriental fermented food produced by solid state fermentation (SSF) of soybeans. The product is fermented with Rhizopus ssp, which synthesises hydrolytic enzymes diminish anti-nutritional factors improving the nutritional quality in the final product (Angulo-Bejarano et al., 2008). The SSF of chickpeas improved the amino acid content of Ile, Met+Cys, Phe+ Tyr, Thr and Val. As a result, the chickpea tempeh contained a better EAA profile and shoed that the limiting amino acids were Trp (93%) and Met+Cys (87%). Furthermore, the SSF of chickpeas with C. millitaris SN-18 also improved all EAA in the fermented product (Xiao et al., 2015).

Protein Digestibility

Besides the EAA composition, the seed protein digestibility is an important nutritional parameter affecting the overall protein quality of seeds. In pulses and legumes, this factor is mainly affected by the protein structure and presence of antinutritional factors. The protein digestibility establishes the availability of the amino acids present in food proteins. It is usually measured using rat bioassays (Nosworthy et al., 2017; Tavano et al., 2016). The most appropriate method to quantifying protein digestion is ileal digestibility, because true protein digestibility (TD) does not account for essential amino acids loss by fermentation in the colon. Although TD is more used in part to the practicality compared to ileal-fistulated pigs (Tavano et al., 2016). The use of *in vitro* protein digestibility assays must be considered to assess protein quality since the traditional rat bioassays are expensive, time consuming and questionable from the ethical viewpoint (Tavano et al., 2016; Wolzak et al., 1981). Besides, the chickpea seeds or flours are usually processed in order to reduce antinutritional factors such as trypsin inhibitors, tannins and hemagglutinins and thus improve protein digestibility.

In Vivo

True protein digestibility (TPD) is calculated with the content of nitrogen in food and feces (FAO/WHO/UNU, 2007). Several investogations have reported TPD in raw chickpea samples with values around 83 to 85% (Tavano et al., 2016; Angulo-Bejarano et al., 2008; Milan-Carrillo et al., 2000). The thermal processing of chickpea flours did not improve the TD (Nosworthy et al., 2017; Tavano et al., 2016). The inappropriate storage of pulses at high temperature (≥25°C) and high relative humidity (HR≥65%) enhances the development a the hard to cook (HTC) phenomenon. Chickpea seed hardening decreases the nutritional value, significantly increases cooking time and modifies sensorial and physicochemical characteristics (Milán-Carrillo et al., 2000; Milán-Carrillo et al., 2002). The storage of chickpea at harsh conditions (33-35°C/75% RH) for 160 days decreased TPD value by 2.3% compared to raw flours (84.8%) (Milán-Carrillo et al., 2000). Among the alternatives to improve nutritional value in hardened chickpeas is the combination of dehulling, softening and extrusion cooking. Dehulling mechanically eliminates the seed coat improving protein digestion and reduces tannins and optimum cooking time. The softening in salt solution of chickpea seeds lowered both optimum cooking and anti-nutritional factors. Thermoplastic extrusion or cooking is classified as high temperature/short time process that can generate chickpea protein products with good digestibility and biological values (Milán-Carrillo et al., 2000). The TPD values of chickpeas improved by DS (dehulling/softening) and extrusion to a level of 91.2% in regular chickpeas and to 87.3% when hardened chickpea were processed (Milán-Carrillo et al., 2000). Moreover, the fermentation of chickpeas improved the TPD by 5.1% compared to raw chickpeas (Angulo-Bejarano et al., 2008).

In Vitro

The *in vitro* protein digestibility (IVPD) tries to mimic the effect of the human digestion process using enzyme mixtures or in sequence to proteolysis of protein sample. The Hsu et al. (1977) and Akeson and Stahmann (1964) are the most used methods. The Hsu et al. (1977) method consists of the measurement of the sample pH before and after 10 min of reaction with an enzyme blend of trypsin, chymotrypsin and peptidase. The principle of this technique is that the rate of hydrolysis is related to the H+ ions released by free carboxyl group and broken peptide bonds (Hsu et al., 1977). Other method involves the use of sequential protease addcition (pepsin + pancreatin) to emulate stomack and small intestine digestions. The effectiveness of the hydrolysis is calculated by the quantification of soluble nitrogen in relation to the original sample nitrogen content. The soluble nitrogen is closely related to the free amino groups released during the reaction (Akeson et al., 1964).

Some authors have evaluated IVPD in unprocessed or raw chickpea, the values reported varied from 59 to 82% (Table 2). The Kabuli types commonly depict higher IVPD values compared to the Desi counterparts (Khalil et al., 2007). The seed soaking process is relevant because as explained above reduces cooking time and induces water imbibition in seed which triggers synthesis of giberelins and the physiological stage of germination. The IVPD improved by 10% in soaked chickpeas compared to raw counterparts (Xu et al., 2016). Similarly to TPD, the storage of chickpea reduced 5% IVPD value in hardened chickpea flour compared to raw chickpea flours. Although, the implementation of DS extrusion processing in hardened chickpea flours increased IVPD value by 10.2% compared to raw chickpea flours counterparts and 14.7% compared to hardened chickpea flours (Milán-Carrillo et al., 2000).

Several thermal processes have been employed in chickpea seeds/flours as steaming, boiling, roasting, autoclaving or microwaving. These thermal treatments increases the stability, safety and palatability of chickpeas. During thermal processes, several anti-nutritional factors are significantly diminished mainly due to protein denaturation. Thermally-treated chickpeas had improved protein digestibility values (Sashikala et al., 2015). The IVPD values in chickpea samples after thermal processes varied from 73 to 93%, and the values were higher (77 to 99%) when soaking was combined with the heat treatments (Table 2). The chickpea genotype influenced IVPD values. An investigation which screened ten cooked chickpea flours showed that digestion rates varied from 76 to 82%. Interestingly, a negative correlation with β-glucans and insoluble dietary fiber contents was found indicating that these fiber components restricted the biocatalytic action of peptidases (Milán-Noris et al., 2019; Hansen, 1986).

Fermentation bioprocess are known to improve protein digestibility rates due to the reduction of anti-nutritional factors. The SSF of chickpea with R. oligosporus increased 11% the IVPD values compared with raw chickpea (Angulo-Bejarano et al., 2008). Also, SSF of chickpeas with C. millitaris SN-18 improved IVPD values compared to the raw counterpart (Xiao et al., 2015). Likewise, the natural and Lactobacullus ssp fermentation of Desi chickpea flours increased the IVPD values whereas in Kabuli chickpea these fermentations did not exert significant effects (Chandra-Hioe et al., 2016).

Germination improves the nutritional quality of chickpea by the breaking down certain anti-nutrients factors as phytates and oligosaccharides. The application of this physiological natural process in Desi and Kabuli chickpeas for 96 h improved the IVPD values in both sorts of seeds (Khalil et al., 2007).

Protein Efficiency Ratio

Protein efficiency ratio (PER) is one of the most relevant methods to determine overall protein quality. This bioassay measures the growth of weanling rats fed isonitrogenous diets balanced to contain 1.6% nitrogen or 10% protein but requires special facilities and commonly one month to conduct the test (Milán-Noris et al., 2019; Nosworthy et al., 2017). Fortunately, the *in vitro* calculated PER values have good correlation with the *in vivo* PER assay (Alsmeyer et al., 1974). The *in vitro* value is caluculated based in the *in vitro* protein digestibility and amino acid profile of test proteins.

Milán-Carrillo et al. 2000 evaluated the protein quality of raw, HTC or hardened, and DS extruded chickpea flours. As expected, the storage of seeds slightly decreased PER values compared to raw chickpeas (PER of 1.30 and 1.47). Although the PER values were low, DS extrusion improved PER values in both fresh and hardened chickpeas (Milán-Carrillo et al., 2000). In contrast, the chickpea solid state fermentation improved the PER values to 2.31 compared to the raw chickpea value of 1.59. In the same study the calculated PER values were in close agreement to the *in vivo* PER values (1.54 for raw chickpeas and 2.21for SSF chickpeas) (Angulo-Bejarano et al., 2008).

Nosworthy et al. (2017) evaluated and compared protein quality in different cooked pulse flours. The chickpea cooked flour had a PER of 2.32 compared to the other pulses (beans, lentil and peas) that varied from 0.86 to 1.64. Likewise, recently Milán-Noris et al. (2019) evaluated the calculated PER in ten cooked chickpea flours and predicted values ranged from 2.17 to 2.65. PER values higher than 2 indicate that chickpea proteins are considered as good quality (Milán-Noris et al., 2019).

PDCAAS

The Protein Digestibility Corrected Amino Acid Score or PDCAAS is the most recommended theoretical measurement to determine food protein quality. The method considers the two most critical factors affecting protein quality: indispensable or essential amino acids and protein digestibility. It is calculated based in the limiting amino acid score according to the FAO requirements for preschool infants and protein digestibility (Milán-Noris et al., 2019; Tavano et al., 2016). The highest PDCAAS value is 1, which will provide 100% of the essential amino acids required by a two year old preschool children (FAO/WHO/UNU, 2007). If the protein is completely digested and absorbed, PDCAAS lower than 1 indicates the limitations of amino acids in the evaluated protein. However, most food proteins are not completely digestible (Nosworthy et al., 2017). According to the FAO/WHO (2007), PDCAAS must be above 0.6 to meet amino acids needs.

The chickpea fermentation significantly improved the protein quality, with a 0.92 PDCAAS value, compared to raw chickpeas (0.73 PDCAAS). The PDCAAS in ten cooked chickpea cultivars varied from 0.59 to 0.82 and eight chickpea flours had values above 0.6 (Milá-Noris et al., 2019). Likewise, Bai et al. 2018 evaluated the *in vitro*

PDCAAS in desi chickpea flours and concluded that the untreated flours had a value of 0.62 whereas samples infrared-heated achieved PDCAAS of 0.63 to 0.70 (Bai et al., 2018). Interestingly, the cooked Kabuli chickpea had the lowest PDCAAS value (0.51) (Nosworthy et al., 2017).

Tavano et al. (2016) compared seven *in vitro* protein digestibility methods versus TPD in the calculation of PDCAAS in chickpea samples. *in vitro* assay showed strong correlations to TPD. In this study, PDCAAS values of raw and heated chickpea flours varied from 0.28 to 0.60 and 0.28 to 0.61 correspondingly.

NUTRACEUTICAL HEALTH EFECTS OF CHICKPEA

Nowadays, several beneficial effects of chickpea proteins are known. These proteins can be converted into hydrolysates by the action of enzymes, fermentation, or other processes. Protein hydrolysates are basically defined as a heterogenous mixture of free amino acids, peptides, and oligopeptides that are generated by the action of enzymes or other processes. On the other hand, biopeptides or bioactive peptides are moieties consisting of 3 to 50 amino acid residues in length that have demonstrated positive health benefits beyond basic nutrition (Nasri, 2017; Li-Chan, 2015; Chalamaiah et al., 2018). The biopeptides could be isolated from the complex mixture to elucidate the amino acid sequence and other characteristics. Table 3 depicts main health effects of bioactive peptides obtained after protein hydrolysis of chickpeas.

Protein chickpea hydrolysates are mainly produced by enzymatic hydrolysis of whole protein, but also could be of specific protein fractions. Even though the hydrolysates contain a complex mixture of active and inactive peptides, they have evidenced noted health benefits. The main reported outcome of chickpea hydrolysates is antioxidant. Yust et al. (2012) demonstrated that enzymatic hydrolysis of chickpea protein isolate (CPI) with alcalase followed by flavourzyme yielded hydrolysates rich in antioxidant properties. Chickpea hydrolysates showed significantly higher antioxidant activity determined by the β-carotene bleaching assay, reducing power, and 2,2-diphenyl-1-picrylhydrazyl (DPPH) radical-scavenging effects compared to the non-hydrolyzed protein isolate. In another study, chickpea protein hydrolysates were prepared from fresh and hard-to-cook grains treated with alcalase, papain, and pancreatin. The albumin fraction from fresh grains hydrolyzed with papain or the albumin fraction from HTC grains hydrolyzed with alcalase exhibited the highest antioxidant activity determined with the DPPH assay (46.8 and 46.0 μmol TEAC μg^{-1}, respectively). Wang et al. (2015) showed that chickpea protein hydrolysates obtained by the enzymatic action of alcalase followed by the plastein reaction showed significantly higher antioxidant activity than counterparts obtained from the total protein. The authors concluded that the pH had the greatest influence on the decrease of free amino acids. The modified hydrolysates were enhanced in their hydroxyl radical scavenging activity and reducing power. In other research, different hydrolysate fractions (3.5, 3.5-7.0, and <7.0 kDa) were produced by *in vitro* gastrointestinal digestion from raw or heat-treated chickpea. The hydrolysates (3.5-7.0k Da) from the heat-treated chickpea possessed the highest antiradical activity against $ABTS^+$ (IC_{50}: 41.01 μg/mL). On the other hand, the hydrolysates (3.5-7.0k Da) from the raw chickpea exerted the highest antiradical against DPPH and Fe^{2+} chelating activity (IC_{50}: 20.94 and 52.53 μg mL^1, respectively). The hydrolysate (>3.5kDa) from the heat-treated chickpea demonstrated the highest fibroblast growth-stimulating and Cu^{2+} chelating activities (Karas et al., 2015).

Likewise, some pretreatments can enhance protein hydrolysis and improve the antioxidant capacity of resulting hydrolysates. A CPI was first subjected to high-hydrostatic pressure and then to alcalase hydrolysis. The hydrolysates produced at high pressure (100-300 MPa) possessed higher superoxide anion capturing rate. The 200 MPa for 20 min treatment followed by alcalase treatment produced chickpea hydrolysates with the highest antioxidative activity. This particular combination significantly increased the superoxide radical scavenging activity (39%) and reducing power (0.272) (Zhang et al., 2012). Similarly, to optimize enzymolysis, a chickpea protein was pretreated with ultrasound. The optimum conditions of this pretreatment were 500 W, 28 min and 48ºC. The hydroxyl radical and the reducibility of the hydrolysate were 95.97% and 1.73%. The half inhibitory concentration (IC_{50}) value of hydroxyl radical decreased with the hydrolysates pretreated with ultrasound. It was concluded that the ultrasonic pretreatment enhanced enzymatic hydrolysis and antioxidant activity of the hydrolysates (Du et al., 2013). Lately, other processes like germination has been used to as a pretreatment to produce hydrolysates. Albumin, globulin, and glutelin hydrolysates were produced by the germination process in presence of selenium followed by an *in vitro* gastrointestinal hydrolysis. The hydrolysates from the glutelin fraction <10kDa showed the highest cellular antioxidant capacity in Caco-2 cells (Serrano-Sandoval et al., 2019). Moreover, Kumar-Mamilla et al. (2017) have demonstrated that hydrolysis of chickpea protein occurs during the germination process.

Up to date, food proteins, hydrolysates and peptides have received great attention because of them antihyperlipidemic effect. Studies have demonstrated that chickpea hydrolysates exhibit hypocholesterolemic activity. Hydrolysates obtained from CPI with enzymatic hydrolysis showed significantly higher hypocholesterolemic activity (*in vitro* micellar solubility inhibition, 50%) than the CPI (Yust et al. (2012). Likewise, a chickpea albumin hydrolysate produced by enzymatic process (alcalase and flavourzyme) demonstrated antihyperlipidemic effects. The hydrolysates showed *in vivo* a dose dependent ability to decrease serum lipid levels (serum total cholesterol, triglycerides, low density lipoprotein (LDL)) while increased high density lipoprotein (HDL) cholesterol levels (Xue et al., 2012).

Table 3. Health effect of bioactive peptides found in chickpea

	Health effect	Pretreatment/Hydrolysis	Assay	Main effect	Sequence	Reference
Chickpea hydrolysates	Antioxidant	Alcalase and flavourzyme hydrolysis	*In vitro*, DDPH radical-scavenging effect, β-carotene bleaching, and reducing power	Hydrolysates showed a better antioxidant activity than protein concentrate.		Yust et al., 2012.
	Antioxidant	Hydrostatic high-pressure and alcalase hydrolysis	*In vitro*, superoxide radical scavenging activity and reducing power.	High-pressure treatment at 200 MPa for 20 min with alcalase resulted in products with high antioxidative activity.		Zhang et al., 2012.
	Antioxidant	Ultrasonic pretreatment and alcalase hydrolysis	*In vitro,* hydroxyl radical, reducibility of the antioxidant peptides.	Hydroxyl radical, reducibility of the antioxidant peptides was 95.97% and 1.73, respectively.		Du et al., 2012.
	Antioxidant	Alcalase, and then as a substrate modified by plastein reaction.	*In vitro,* DPPH, ABTS, and reducing power.	Hydrolysates increased the reducing power and their hydroxyl radical scavenging activity.		Wang et al., 2015.
	Hypocholesterolaemic	Alcalase and flavourzyme hydrolysis	*In vitro,* Hypocholesterolaemic activity.	Hydrolysates showed an *in vitro* micellar solubility inhibition of 50%.		Yust et al., 2012.
Chickpea hydrolysates	Antioxidant	Germination in presence of selenium as a pretreatment and pepsin and pancreatin enzymatic hydrolysis.	*In vitro*, cellular antioxidant capacity with Caco2 cells.	Selenized hydrolysates from the glutelin fraction (<10kDa) exhibit the highest cellular antioxidant capacity.		Serrano-Sandoval et al., 2019
	Angiotensin converting enzyme inhibitory (ACE-I) activity	Germination process	*In vitro*, ACE-I activity.	Germinated hydrolysates showed significant ACE inhibition *in vitro*.		Kumar-Mamilla et al., 2017

	Health effect	Pretreatment/Hydrolysis	Assay	Main effect	Sequence	Reference
	Antioxidant	Raw and heat-treated chickpea were hydrolysated by simulating *in vitro* gastrointestinal digestion (saliva, α-amilase, pepsin, and pancreatin).	*In vitro,* DPPH and ABTS radical scavenging activity assays, determination of the ion chelating activity and reducing power, proliferation and fibroblast growth stimulation assays.	Hydrolysates from treated and raw chickpea possessed high antiradical activity. The hydrolysate (>3.5kDa) from the heat-treated chickpea demonstrated the highest fibroblast growth-stimulating activity and the highest Cu^{2+} chelating activity was showed by the heat-treated hydrolysates.		Karas et al., 2015.
	Antihyperlipidemic	Alcalase and flavourzyme hydrolysis	*In vivo,* 10 mice by group (normal diet, high fat diet (HFD), HFD with low, medium and high doses of chickpea hydrolysate for 4 weeks.	The chickpea hydrolysates treatment decreased in a dose manner triglyceride, total cholesterol and LDL-C, also increased fecal fat excretion.		Xue et al., 2012.
Chickpea hydrolysates	Lipid peroxidation inhibition by copper-chelating activity	Pepsin and pancreatin hydrolysis	*In vitro,* the hydrolysates were purified by metal affinity chromatography and β-carotene-bleaching and TBARS assays were evaluated.	Hydrolysates that contain higher histidine content exhibited the highest antioxidant effect.		Torres-Fuentes et al., 2014.
	Iron-chelating activity	Pepsin and pancreatin hydrolysis	*In vitro*, the hydrolysates were purified by metal affinity chromatography and iron-chelating activity was evaluated.	Metal chelation may increase iron solubility and bioavailability and improve iron absorption.		Torres-Fuentes et al., 2012.
	Anticarcinogenic	Alcalase and flavourzyme hydrolysis	*In vivo*, 10 mice by group (normal diet (not inoculated), high fat diet (HFD), HFD with low, medium and high doses of chickpea hydrolysate were inoculated with H-22 tumor cells and fed by 12 days.	The chickpea hydrolysates decreased tumor volume and increased the tumor inhibition rate.		Xue et al., 2012.

Table 3. (Continued)

	Health effect	Pretreatment/Hydrolysis	Assay	Main effect	Sequence	Reference
Chickpea hydrolysates	Anticarcinogenic	Pepsin and pancreatin hydrolysis	*In vivo,* measure of the hypolipidemic and anticancer activity in the animals.	The consumption of hydrolysates by the animals that consumed either diet, the number of aberrant crypts was reduced with the 3 doses tested.		Sánchez-Chino et al., 2019
Chickpea peptides	Antioxidant	Alcalase and flavorzyme hydrolysis	*In vitro*, DPPH, TEAC, and hydroxyl radical scavenging.	Identified hydrolysate exhibited the highest antioxidant activity.	RQSHFANAQP	Kou et al., 2013.
	Antioxidant	Pepsin and pancreatin hydrolysis	*In vitro,* reducing power, FRSA, and CAA.	Chickpea peptides showed great antioxidant activity.	ALEPDHR, TETWNPNHPEL, FVPH and SAEHGSLH	Torres-Fuentes et al., 2015.
	Antioxidant	Synthetic	*In vitro*, DPPH, ABTS, reducing power, and others.	Chickpea peptides at different concentration showed a certain antioxidant capability and reducing activity.	RQSHFANAQP	Xue et al., 2015.
	Antioxidant	Alcalase	*In vitro,* oxidative stress in Caco-2 and HT-29 cell lines in catalase, glutathione reductase and glutathione peroxidase activity.	The peptide treatment elevated the expression of Nrf2 mRNA and relative genes NQO1, HO-1, γ-GCS.	NRYHE	Guo et al., 2014.
Chickpea peptides	Antioxidant	Alcalase	*In vitro,* DPPH, reducing power assay, antioxidant β-carotene bleaching method, metal chelating activity.	The peptides showed high antioxidant activity.	DHG, VGDI	Ghribi et al., 2015.

	Health effect	Pretreatment/Hydrolysis	Assay	Main effect	Sequence	Reference
	Antiinflammatory	Pepsin-Pancreatin hydrolysis	*In vitro, Inhibition of nitric oxide in RAW264.7 macrophaghes as anti-inflammatory effect*	Chickpea peptides from germinated peptides showed higher antiinflammatory effect than peptides from cooked chickpea.	24 peptides for legumin and vicilin.	Milán-Noris et al., 2018.
	Anticarcinogenic	Synthetic	*In vitro*, MCF-7 and MDA-MB-23. 1	Chickpea peptides showed a high inhibition on MCF-7 and MDA-MB-231 cell lines with EC_{50} of 2.38 and 1.50 µmol/mL, respectively.	RQSHFANAQP	Xue et al., 2015.

Recently, the effects of chickpea hydrolysates (DH = 8% (CPH8) and DH = 17% (CPH17) on lipoprotein profile and on oxidant status in cholesterol-fed rats were evaluated. The total serum cholesterol, CPH17 reduced LDL- and HDL-cholesterol levels decreased in both groups fed with hydrolysates and increased the lecithin-cholesterol acyltransferase activity and apolipoprotein A4. The total and esterified hepatic cholesterol levels were 2-fold in both treated groups. Only the treatment with CPH17 decreased VLDL- and HDL- malondialdehyde contents and improved glutathione peroxidase activity (Yahia et al., 2017).

Another promising therapeutic approach against some no-communicable diseases is by the scavenging reactive oxygen species (ROS) and radicals with metal chelators. Metal chelators form soluble and stable complexes that are then passed out through feces or urine (Adjimani and Asare, 2015). Chickpea hydrolysates were purified by metal affinity chromatography and fractionated by gel filtration chromatography. The iron chelating hydrolysates from the fractions showed higher iron-chelating activity than the unpurified counterparts (Torres-Fuentes et al., 2012). Afterwards, the capability of the chickpea hydrolysate to inhibit copper-mediated lipid peroxidation was evaluated. The purified and fractionated hydrolysates were evaluated by the inhibition of β-carotene oxidation, inhibition of copper-mediated fatty acid oxidation, and inhibition of copper-mediated low-density lipoprotein oxidation. The chickpea hydrolysates with the highest histidine amino acid content exerted the highest antioxidant effects. The authors confirmed the high ability of histidine to bind transition metals and act as a donor (Torres-Fuentes et al., 2015).

Besides these health benefits, chickpea hydrolysates have shown anticarcinogenic effect. In vivo studies demonstrated that albumin hydrolysates significantly decreased tumor volume and increased the tumor inhibition rate. Besides, the hydrolysates mice administration led to rise in the superoxide dismutase (SOD) activity and decrease of hepatic malondialdehyde (MDA) (Xue et al., 2015). Sánchez-Chino et al. (2019) evaluated the inhibitory effect of chickpea hydrolyzed protein on azoxymethane (AOM)-induced colorectal cancer in mice fed with a hypercaloric diet. The administration of chickpea protein hydrolysates significantly reduced the cholesterol and the atherogenic index. Besides, the consumption of hydrolysates reduced the number of colon aberrant crypts in a dose related relationship.

Some bioactive peptides have been purified and sequenced in order to assess their health-related effects mainly in connection with their structure and physical characteristics (Nasri, 2017). An antioxidant peptide derived from a chickpea albumin hydrolysate was obtained by alcalase and flavorzyme hydrolysis. The peptide was identified by fractionating the albumin hydrolysate by size exclusion chromatography. According to the DPPH and TEAC assays, the chickpea peptide (CPe)-III exhibited the highest antioxidant activity. The peptide sequence was identified as RQSHFANAQP (1155 Da) by LC-ESI-MS/MS (Kou et al., 2013). A bioactive pentapeptide identified as

NRYHE was evaluated as antioxidant in Caco-2 and HT-29 cell lines. According to the results of the analysis the pentapeptide regulate three important antioxidative enzymes: catalase, glutathione reductase, and glutathione peroxidase. The response of these enzymes was in a dose-dependent manner. Likewise, the pentapeptide treatment modulates the Nrf2 pathway by the activation of key promoters (Guo et al., 2014). In the same manner, chickpea protein was hydrolyzed by enzymatic digestion, purified with affinity chromatography, and fractionated. Antioxidant peptides from chickpea proteins were identified and the reducing power, free radical scavenging (FRSA), and cellular antioxidant activities (CAA) were evaluated. The identified sequences corresponded to legumin protein: ALEPDHR, TETWNPNHPEL, FVPH, and SAEHGSLH. The high antioxidant peptides contained histidine, tryptophan, and phenylalanine residues which are considered as important hydrogen donors (Torres-Fuentes et al., 2015). Xue et al. (2015) synthetized a chickpea peptide (CPe-III) derived from its water-soluble albumin fraction. The reducing power, hydroxyl radical, DPPH radical, ABTS radical scavenging and DCFH-DA assays were evaluated after treatment with the synthetic peptide RQSHFANAQP. The antioxidant effect showed a significant dose-dependent activity and the synthetized peptide inhibited the proliferation of breast cancer cells. In other research, chickpea peptides were obtained by the enzymatic hydrolysis of enzymes followed by the fractioning and purification of the hydrolysates. The sequences (DHG, VGDI) were elucidated using ESI-MS and ESIMS/MS and were identified as the ones that exerted tje highest antioxidant effects. Besides, these peptides did not show haemolytic activity (Ghribi et al., 2015). Moreover, Milán-Noris et al. (2018) evaluated the anti-inflammatory effect of peptides of protein concentrates from cooked and germinated Desi and Kabuli chickpea. The protein concentrates were hydrolyzed by pepsin-pancretin, in a simulated *in vitro* gastrointestinal digestion and absorption. The anti-inflammatory effect of germinated chickpea digests was 2-fold higher than cooked chickpea digests. For the identification of which peptides were the main contributors to the anti-inflammatory effects of germinated chickpea protein digests from Kabuli and Desi cultivars, both samples were fractionated by RP-LC-MS. The resulting separation produced 3 fractions for each cultivar. The F3, from both chickpea type, was further analyzed by LC-MS/MS, in which peptide profile was similar for both cultivars. A total of 24 different peptide sequences were successfully identified (SEGGLIETWNPSNK, NEDEEKGAIVKVK, VKGGLSIITPPEKEPR, GGLSIITPPEKEPR, GGLSIITPPEKEPRQK, LHQNIGSSSS PDIYNPQAGR, LHQNIGSSSSPDIYNPQAGRIK, GRRGSEESEEGDAIVK, GSEE SEEGDAIVK, GSEESEEGDAIVK, SRNPIYSNKFGK, ISREQIEELSKNAK, QQSQE TDVIVK, EQIEELSKNAK ,ISREQIEELSKNAK, VLLEEQEQKPK, EQIEELSKNAK, SPIYSNKFGK, DVGQKTKEVGQDIQAK, IFKISKEDVHGLAPK, HIVDKLQGRDED EEK, SREEETTEWEEEVAK and FSGNRGPLVQPR) as fragments from storage, structural, nutrient reservoir and catabolic proteins. This study showed the potential of germinated chickpea concentrates to exert anti-inflammatory effect in the colon. These findings provide the idea that likewise non-digestible carbohydrates and phenolics, the hydrolysed protein fraction may also contribute to the anti-inflammatory activity of chickpea.

PRODUCTION OF SELENIZED CHICKPEA PROTEINS: ANTIOXIDANT AND ANTICANCER POTENTIAL

Selenium (Se) is an essential micronutrient for humans. It is involved in several biochemical pathways such as protection against oxidative damage and the regulation of immune and reproductive systems (NIH, 2018). This trace mineral acts as a cofactor of selenoenzymes such as glutathione peroxidase (GPx) and thioredoxine reductase (TrxR), which reduce the levels of reactive oxygen species and maintain the cellular redox balance (Hu et al., 2008). For this reason, Se has been classified as chemopreventive due to its ability to reduce the risk, delay the progression and avoid the recurrence of cancer (Brodowska et al., 2016; Guardado-Félix et al., 2019; Kieliszek and Błażejak, 2016). Additionally, it has been proved that Se plays an important role in the function of the brain and prevention of cardiovascular diseases (Benstoem et al., 2015). The recommended dietary Se for humans is 55 μg per day for an adult with average weight of 70 kg (NIH, 2018). It is known that the intake of supranutritional levels of selenium (200 μg/day) reduces cancer risk (Combs et al., 2012); however, daily intakes higher than 350μg are considered toxic in adults (San martin et al., 2012).

Selenium is accumulated in plants mainly in the form of Se-amino acids such as selenomethionine (SeM) and selenocysteine (SeC) (Misra et al., 2015; Fernandes and Gandin, 2015). Due to the similarity with sulfur (S), Se could be able to enter into the S metabolic pathways where would be converted in Se-amino acids. SeM contained in foods can be accumulated in selenoproteins and structural proteins. The half-life of SeM and inorganic selenite in the human system is 252 and 102 days, respectively (Lazo-Vélez et al., 2015). The proper assimilation of Se into foods depends on their biovaility, bioaccessibility, and/or bioactivity of given Se compound. The bioaccessibility of Se in enriched foods crops was found to be highest in the small intestine compared to inorganic forms (Lavu et al., 2016).

Globally, one billion people have Se-deficient diets mainly due to the low concentrations of the mineral in the soil (Lazo-Vélez et al., 2015). Several strategies are known to improve both the Se status and bioavailability in foods. The germination of seeds using selenite or selenate in the soaking water has been explored as an alternative to increase the Se dietary content in foods and therefore supplement the deficit of this mineral worldwide (D´Amato et al., 2018; Garousi et al., 2016; Guardado-Félix et al., 2017; Trolove et al., 2018).

Chickpea seeds are an excellent candidate for selenium enrichment due to their high protein content and proven capacity to accumulate this mineral. Guardado-Félix *et al.* (2017) germinated chickpeas for four days after soaking seeds with 2 mg Na_2SeO_3/100 g and concluded that the Se content increased 115 fold compared to the control sprouts. Also, total isoflavonoids, phenylalanine ammonia lyase (PAL) activity, and antioxidant capacity increased by 83%, 56%, and 33%, respectively (Guardado-Félix et al., 2017). Zhan et al. (2012) showed that Se was mainly incorporated in SeM during germination of chickpeas.

Recently, Serrano Sandoval et al. (2019) showed that Se is accumulated differentially in the protein fractions of chickpeas and that the accumulation of this mineral in proteins, increased antioxidant capacity and digestibility. The authors showed that chickpeas seeds germinated during four days after soaking with sodium selenite (0, 1, or 2mg/100g seeds), accumulated selenium in the order of glutelin > albumin > globulin. Se-enriched protein fractions increased cellular antioxidant activity (CAA) compared to unenriched protein. The highest CAA (51.47%) was shown in ultrafitrated glutelin hydrolysate (<10 kDa) treated with 2 mg Na_2SeO_3. Also, the intensity of vicilin bands (34–37kDa) increased from the second to the fourth day compared with the control samples. Glo digestibility was higher in selenized chickpea sprouts. These results show the potential health benefits of selenization de chickpea proteins. It is worth mentioning that several researchers have demonstrated the comparatively higher antioxidant activity of selenized molecules associated to rice (Liu et al., 2015; Liu et al., 2012), soybeans (Hu et al., 2014), ovalbumin (Can-Peng et al., 2014), and egg white protein (Zhao et al., 2013).

To evaluate the chemopreventive potential of the Se-enriched chickpea sprouts, Guardado- Félix et al. (2019) recently evaluated a diet enriched with selenized chickpea sprouts flour on immune-suppressed mice xenografted with colorectal cancer (HT-29 RFP fluorescent cells). It was found that the high intake of Se (2.29μg/g) with and without isoflavonoids increased glutathione peroxidase (GPx) and thioredoxin reductase (TrxR) activities, cholesterol (CHOL), triglycerides (TRIGL) and low-density lipoprotein cholesterol (LDL-C) levels and decreased significantly the tumor growth by 40%. Thus, the Se accumulated mainly in chickpea sprouts proteins exerted chemopreventive effects mediated mainly by the antioxidant protection of hepatic GPx. These results are supported by a previous research, which showed the chemopreventive and antioxidant potential of seleno-compounds. In previous studies, rats fed Se-enriched diet fortified with selenized mushroom (1μg Se/g diet) increased GPx activity by 1.77-fold compared to the control diet (Maseko et al., 2014). On the other hand, mice fed with a Kaiware radish sprouts-based diet containing 2.0 Se μg/g, increased GPx activity by 127% and decreased the formation of aberrant crypt loci in the colon compared to control animals (Yoshida et al., 2007).

All these investigations indicate that selenized chickpea proteins represent an excellent source of organic dietary selenium with chemopreventive and antioxidant potential and could be potentially used in the prevention and treatment of colorectal cancer.

Conclusion and Perspective

Chickpea is an important source of protein with high digestibility and good content and balance of essential amino acids. Chickpea proteins are staple foods of many cultures, but innovative processing methods are required to further increase their consumption.

Although, the FAO/WHO periodically revise the parameters of protein quality determination in foods, several improvements are needed to achieve its quantification. For instance, a validated method for the determination of protein digestibility with good correlation to *in vivo* assays, especially in terms of ileal protein digestibility. This will improve the calculation of *in vitro* of the protein quality parameters without unnecessary animal assays. Moreover, the protein quality is required to be quantify in processed seeds or prepared foods, main form of consumption. Avoiding the determination in raw or unprocessed pulses, which are known to contain anti-nutritional factors that will interfere with nutritional parameters.

Previous research showed antioxidant, antifungal, hypocholesterolemic and anticarcinogenic activities of peptides and hydrolysates of chickpea proteins. Likewise, anticarcinogenic and antioxidant potential of Se-enriched proteins have been validated *in vitro* and *in vivo* but their stability in processed foods has not always been tested. Chickpea proteins have a wide chemical diversity and potential to be used as bioactive ingredients in nutraceutical formulations. Chickpea proteins are also reservoir of bioactive peptides and as their relase is naturally controlled depending on interaction with microbiota, they may be used to design personalized foods in the future.

References

Angulo-Bejarano P. A., Verdugo-Montoya, N. M., Cuevas-Rodríguez, E. O., Milán-Carrillo, J., Mora-Escobedo, R., Lopez-Valenzuela, J. A., Garzón-Tiznado, J. A. and Reyes-Moreno, C. (2008). Tempeh flour from chickpea (*Cicer arietinum L.*) nutritional and physicochemical properties. *Food Chemistry,* 106(1): 106-112.

Adjimani, J. P., and Asare, P. (2015). Antioxidant and free radical scavenging activity of iron chelators. *Toxicology Reports,* 2:721-728.

Alsmeyer, R. H., Cunningham, A. E., and Happich, M. L. (1974). Equations predict PER from amino acid analysis. *Food Technology*, 28: 34-38.

Akeson, W. R. and M. A. Stahmann. (1964). A Pepsin Pancreatin Digest Index of Protein Quality Evaluation. *The Journal of Nutrition*, 83(3): 257-261.

Bai, T. Nosworthy, M. G., House, J. D. and Nickerson, M. T. (2018). Effect of tempering moisture and infrared heating temperature on the nutritional properties of desi chickpea and hull-less barley flours, and their blends. *Food Research International*, 108: 430-439.

Bar-El Dadon, S., Pascual, C. Y., Eshel, D., Teper-Bamnolker, P., Ibáñez, M. D. P. and Reifen, R. (2013). Vicilin and the basic subunit of legumin are putative chickpea allergens. *Food Chemistry*, 138: 13-18.

Benstoem, C., Goetzenich, A., Kraemer, S., Borosch, S., William, M., Hardy, G. and C. Stoppe. (2015). Selenium and its supplementation in cardiovascular disease-what do we know? *Nutrients*, 3094-3118.

Birk, Y. (1989). Protein protease inhibitors of plant origin and their significance in nutrition J. Huisman, A. F. B. Van der Poel, I. E. Liener (Eds.), *Recent advances of research in antinutritional factors in legume seeds,* Wageningen, The Netherlands: Pudoc, 83-94.

Bhagyawant, S. S., Gautam, A. K., Narvekar, D. T., Gupta, N., Bhadkaria, A., Srivastava, N., Upadhyaya, H. D. (2018). Biochemical diversity evaluation in chickpea accessions employing mini-core collection. *Physiology and Molecular Biology of Plants*, 24(6): 1165-1183.

Brodowska, M. S., Kurzyna-Szklarek, M. and Haliniarz. M. (2016). Selenium in the environment. *Journal of Elementology*, 1173-1185.

Can-Peng, L., Zhikun, H., Xiaoyan, W., Long, Y., Chunyan, Y., Nan, Z., Jie, L. and Hui, Z. (2014). Selenization of ovalbumin by dry-heating in the presence of selenite: Effect on protein structure and antioxidant activity. *Food Chemistry*, 148, 209-217.

Campbell, W. W., Johnson, C. A., McCabe, G. P. and Carnell, N. S. (2008). Dietary protein requirements of younger and older adults. *The American Journal of Clinical Nutrition*. 88(5):1322-9.

Casey, R., Domoney, C., Smith Peas A. M. (1993). Genetics, molecular biology and biotechnology R. Casey, D.R. Davies (Eds.), *Biotechnology in agriculture series,* 10, CAB International, Wallingford, Oxon, UK, 121-164.

Castaneda, C., Charnley, J. M., Evans, W. J. and Crim, M. C. (1995). Elderly women accommodate to a low-protein diet with losses of body cell mass, muscle function, and immune response. *American Journal of Clinical Nutrition,* 62(1):30–39.

Chang, Y. W., Alli, I., Molina, A. T., Konishi, Y., and Boye, J. I. (2012). Isolation and characterization of chickpea (*Cicer arietinum L.*) seed protein fractions. *Food and Bioprocess Technology*, (5)2: 618–625.

Chalamaiah, M., Yu, W., and Wu, J. (2018). Immunomodulatory and anticancer protein hydrolysates (peptides) from food proteins: A review. *Food Chemistry,* 245:205-222.

Chandra-Hioe, M. V., Wong, C. H. M. and Arcot, J. (2016). The potential use of fermented chickpea and faba bean flour as food ingredients. *Plant Foods for Human Nutrition*, 71(1): 90-95.

Combs, G. F., Jackson, M. I., Watts, J. C., Johnson, L. K., Zeng, H., Idso, J.,...Milner, J.A. (2012). Differential responses to selenomethionine supplementation by sex and genotype in healthy adults. *British Journal of Nutrition*, 107: 1514–1525.

D´Amato, R., M. C. Fontanella, B. Falcinelli, G. M. Beone, E. Bravi, O. Marconi, P. Benincasa, and D. Businell. (2018). Selenium biofortification in rice (*Oryza sativa L.*) sprouting: effects on Se yield and nutritional traits with focus on phenolic acid profile. *Journal of Agricultural and Food Chemistry*, 66(16): 4082-4090.

Du, S., Yu, X., and Li, Z. (2013). Ultrasonic pretreatment for preparation of antioxidant peptides from chickpea protein. *Journal of Chinese Institute of Food Science and Technology,* 13(1):80-88.

FAO. 2014. Food Balance Sheets, 1970–2011. Rome, Italy: FAO. http://www.fao.org/faostat/en/#data/FBS/visualize.

FAO/WHO/UNU, (2007) Protein and amino acid requirements in human nutrition. Vol. 935: World Health Organization.

FAOSTAT. Statistical data base. (2016). http://www.fao.org/statistics/es/.

Fernandes, A. P. and Gandin. V. (2015). Selenium compounds as therapeutic agents in cancer. *Biochimica et Biophysica Acta*, 1642-1660.

Garousi, F., Kovács, Domokos-Szabolcsy, B. and Veres, S. (2016). Biological changes of green pea (*Pisum Sativum L.*) by selenium enrichment. *Acta Biologica Hungarica*, 60-72.

Ghribi, A. M., Sila, A., Przybylski, R., Nedjar-Arroume, N., Makhlouf, I., Blecker, C., Attia, H., Dhulster, P., Bougatef, A. and Besbes, S. (2015). Purification and identification of novel antioxidant peptides from enzymatic hydrolysate of chickpea (*Cicer arietinum L.)* protein concentrate. *Journal of Functional Foods,* 12:516-525.

Guardado-Félix, D., Antunes-Ricardo, M., Rocha-Pizaña, M. R. Martínez-Torres, A. C. Gutiérrez-Uribe, J. A. and Serna-Saldivar, S. O. (2019). Chickpea (*Cicer arietinum L.*) sprouts containing supranutritional levels of selenium decrease tumor growth of colon cancer cells xenografted in immune-suppressed mice. *Journal of Functional Foods*. 53: 76-84.

Guardado-Félix, D., Serna-Saldivar, S. O., Cuevas-Rodríguez, E. O., Jacobo-Velázquez, D. A., and Guitiérrez-Uribe, J. A. (2017). Effect of sodium selenite on isoflavonoid contents and antioxidant capacity of chickpea. *Food Chemistry*, 226: 69-74.

Gueguen, J. (1991). Pea and fababean proteins. In *Development in Food Proteins;* Hudson, B. J. F., Ed.; Elsevier: London, U.K., 7: Chapter 2.

Guo, Y., Zhang, T., Jiang, B., Miao, M. and Mu. W. (2014). The effects of an antioxidative pentapeptide derived from chickpea protein hydrolysates on oxidative stress in Caco-2 and HT-29 cell lines. *Journal of Functional Foods,* 7(0): 719-726.

Hansen, W. E. (1986). Effect of dietary fiber on proteolytic pancreatic enzymes *in vitro. International Journal of Pancreatology*, **1**(5): 341-351.

Hsu, H. W., Satterlee, L. D. and Miller, G. A. (1977). A multienzyme technique for estimating protein digestibility. *Journal Food Science.* 42:1269-1273.

Hu, J., Zhao, Q., Cheng, X., Selomuya, C., Bai, C., Zhu, X., Li, X. and Xiong, H. (2014). Antioxidant activities of Se-SPI produced from soybean as accumulation and biotransformation reactor of natural selenium. *Food Chemistry,* 146: 531-537.

Hu, Y., McIntosh, G. H., Leu, R. k. L., Woodman, R. and Young, G. P. (2008). Suppression of colorectal oncogenesis by selenium-enriched milk proteins: apoptosis and K-ras mutations. *Cancer Research.* 68, 4936-4944.

Iqbal, A., Khalil, I. A., Ateeq, N. and Khan, M. S. (2006). Nutritional quality of important food legumes. *Food Chemistry*, 97(2): 331-335.

Jukanti A., Guar, P., Gowda, and Chibbar, R. (2012). Nutritional quality and health benefits of chickpea. *British Journal of Nutrition,* 108: S11-S26.

Kadam, S. S., Desphande, S. S. and Jambhale, N. D. (1989). Seed Structure and Composition. In: *CRC Handbook of World Food Legumes: Nutritional Chemistry, Processing Technology and Utilization* (edited by D.K. Salunkhe & S.S. Kadam). Boca Raton, FL: CRC Press, Inc. 23–50.

Karas, M., Baraniak, B., Rybczynska, K., Gminski, J., Gawel-Beben, K., and Jakubczyk. (2015). The influence of heat treatment of chickpea seeds on antioxidant and fibroblast growth-stimulating activity of peptide fractions obtained from proteins digested under simulated gastrointestinal conditions. *International Journal of Food Science and Technology,* 50:2097-2103.

Khalil, A. W., Zeb, A. Mahmood, F., Tariq, S., Khattak, A. B. and Shah, H. (2007). Comparison of sprout quality characteristics of desi and kabuli type chickpea cultivars *(Cicer arietinum L.). LWT - Food Science and Technology,* 40(6): 937-945.

Kieliszek, M. and S. Błażejak. (2016). Current knowledge on the importance of selenium in food for living organisms: A review. *Molecules*, 21: 609.

Kumar-Mamilla, R., and Kumar-Mishra, V. (2017). Effect of germination on antioxidant and ACE inhibitory activities of legumes. *LWT-Food Science and Technology,* 75:51-58.

Kou, X., Gao, J., Xue, Z., Zhang, Z., and Wang, X. (2013). Purification and identification of antioxidant peptides from chickpea (*Cicer arietinum L.)* albumin hydrolysates. *LWT- Food Science and Technology*, 50(2), 591-598.

Lajolo, F. M. and M. I. Genovese. (2002). Nutritional significance of lectins and enzyme inhibitiors from legumes. *Journal of agricultural and food chemistry*, 50: 6592-6598.

Lavu R. V. S., Van De Wiele, T., Pratti, V. L., Tack, F. and Laing, G. D. (2016). Selenium bioaccessibility in stomach, small intestine and colon: Comparison between pure Se compounds, Se-enriched food crops and food supplements. *Food Chemistry,* 197: 382-387.

Lazo-Vélez, M. A., Chávez-Santoscoy, A. and Serna-Saldívar, S. O. (2015). Selenium-enriched breads and their benefits in human nutrition and health as affected by agronomic, milling, and baking factors. *Cereal Chemistry*, 92(2): 134-144.

Li-Chan, E. C. Y. (2015). Bioactive peptides and protein hydrolysates: research trends and challenges for application as nutraceuticals and functional food ingredients. *Current Opinion in Food Science,* 1:28-37.

Lilley, G. G. and Inglis, A. S. (1986). Amino acid sequence of conglutins, a sulfur-rich seed protein of *Lupinus angustifolium L. FEBS Letters*, 195: 235-241.

Liu, K., Zhao, Y., Chen, F. and Fang. Y. (2015). Purification and identification of Se-containing antioxidative peptides from enzymatic hydrolysates of Se-enriched brown rice protein. *Food Chemistry*, 187: 424-430.

Liu, K., Zhao, Y., Chen, F., Gu, Z. and Bu. G. (2012). Purification, identification, and in vitro antioxidant activities of selenium-containing proteins from selenium-enriched brown rice. *European Food Research Technology*, 234: 61-68.

Magee, P. J., Owusu-Apenten, R., McCann, M. J., Gill, C. I. and Rowland, I. R. (2012). Chickpea (*Cicer arietinum*) and other plant-derived protease inhibitor concentrates inhibit breast and prostate cancer cell proliferation *in vitro*. *Nutrition and Cancer*, 64: 741-748.

Márquez M. C. and Alonso, R. (1999). Inactivation of Trypsin Inhibitor in Chickpea. *Journal of Food Composition and Analysis*, 12: 211-217.

Maseko, T., Howell, K., Dunshea, F. R. and Ng, K. (2014).Selenium-enriched *Agaricus bisporus* increases expression and activity of glutathione peroxidase-1 and expression of glutathione peroxidase-2 in rat colon. *Food Chemistry*, 146: 327-333.

Milán-Noris, A. K., De la Rosa-Millan, J. and Serna-Saldivar, S. O. (2019). Comparative analysis of techno-functional properties, starch digestion and protein quality of pigmented chickpea flours. I*nternational Journal of Food Science & Technology*, 0(0).

Milán-Carrillo, J., Reyes-Moreno, C., Armienta-Rodelo, E. Carábez-Trejo, A. and Mora-Escobedo, R. (2000). Physicochemical and Nutritional Characteristics of Extruded Flours from Fresh and Hardened Chickpeas *(Cicer arietinum L). LWT - Food Science and Technology,* 33(2):117-123.

Milán-Carrillo, J., Reyes-Moreno, C., Camacho-Hernández, C. L. and Rouzaud-Sandez, O. (2002). Optimisation of extrusion process to transform hardened chickpeas *(Cicer arietinum L)* into a useful product. *Journal of the Science of Food and Agriculture,* 82(14):1718-1728.

Milán-Noris, A. K., Gutiérrez-Uribe, J. A., Santacruz, A., Serna-Saldívar, S. O. and Martínez-Villaluenga, C. (2018). "Peptides and isoflavones in gastrointestinal digests contribute to the anti-inflammatory potential of cooked or germinated desi and kabuli chickpea (*Cicer arietinum* L.)." *Food Chemistry* 268:66-76. doi: https://doi.org/10.1016/j.foodchem.2018.06.068.

Misra, S., Boylan, M., Selvam, A., Spallholz, J. E. and Björnsted, M. (2015). Redox-active selenium compounds-from toxicity and cell death to cancer treatment. *Nutrients*, 7, 3536-3556.

National Institutes of Health (NIH) (2018). Selenium. (online) https://ods.od.nih.gov/factsheets/SeleniumHealthProfessional/#h2.

Nasri, M. (2017). Chapter Four- Protein hydrolysates and biopeptides: production, biological activities, and applications in foods and health benefits. A review. *Advances in Food and Nutrition Research,* 81: 109-159.

Nosworthy, M. G., Neufeld, J., Frohlich, P., Young, G., Malcolmson, L. and House, J. D. (2017). Determination of the protein quality of cooked Canadian pulses. *Food Science & Nutrition,* 5(4): 896-903.

Sanmartín, C., Plano, D., Sharma, A. K., and Palop, J. A. (2012). Selenium compounds, apoptosis and other types of cell death: An overview for cancer therapy. *International Journal of Molecular Sciences*, 13: 9649–9672.

Sánchez-Chino, X. M., Jiménez-Martínez, C., León-Espinosa, E. B., Garduño-Siliciano, L., Álvarez-González, I., Madrigal-Bujaidar, E., Vásquez-Garzón, V.R., Baltiérrez-Hoyos, R. *et al.* (2019). Protective effect of chickpea protein hydrolysates on colon carcinogenesis associated with a hypercaloric diet. *Journal of the American College of Nutrition,* 38(2):162-170.

Sashikala, V. B., Sreerama, Y. N., Pratape, V. M., Narasimha, H. V. (2015). Effect of thermal processing on protein solubility of green gram *(Phaseolus aureus)* legume cultivars. *Journal of food science and technology*, 52(3):1552-1560.

Sastry, M. C. S. and Murray, D. R. (1897). The contribution of trypsin inhibitors to the nutritional value of chick pea seed protein. *Journal of the Science of Food and Agriculture*, 40: 253-261.

Savage, G. P., Knowler, K. and Goulden, D. (1998). Trypsin inhibitor and crude protein content of some New Zealand selections of chickpeas, in Recent Advances of Research in Antinutritional Factors in Legume Seeds and Rapeseeds. Wageningen Academic Publishers, Wageningen, 51–54.

Serrano-Sandoval, S. N., Guardado-Félix, D., and Gutiérrez-Uribe, J.A. (2019). Changes in digestibility of proteins from chickpeas (*Cicer arietinum L.)* germinated in presence of selenium and antioxidant capacity of hydrolysates. *Food Chemistry,* 285:290-295.

Shanko, D., Jateni, G., and Debela, A. (2017). Effects of Salinity on Chickpea (*Cicer arietinum L.*) Landraces During Germination Stage. *Journal of biochemistry and molecular biology*. 3: 2-9.

Sharma U. and C. G. Suresh. (2011). Purification, crystallization and x-ray characterization of a kunitz-type trypsin inhibitor protein from the seeds of chickpea (*Cicer arietinum*). *Acta Crystallographica Section F*. 67:714-717.

Singh P. K., Shrivastava, N., Sharma, B., Bhagyawant, S. S. (2015). Effect of domestic processes on chickpea seeds for antinutritional contents and their divergence. *American Journal of Food Science and Technology*, 3(4): 111-117.

Singh, U. and Jambunathan, R. (1982). Distribution of seed protein fractions and amino acids in different anatomical parts of chickpea (*Cicer arietinum L.*) and pigeonpea (*Cajanus cajan L.*) Plant Foods for Human Nutrition 31: 347.

Sreerama, Y. N., Neelam, D. A., Sashikala, V. B. and Pratape, V. M. (2010). Distribution of nutrients andantinutrients in milled fractions of chickpea and horse gram: seed coat phenolics and theirdistinct modes of enzyme inhibition. *Journal of Agricultural and Food Chemistry,* 58(7): 4322–30.

Srinivasan A., Giri1, P. A., Harsulkar, A. M., Gatehouse, J. A. and Gupta, V. S. (2005). A Kunitz trypsin inhibitor from chickpea (*Cicer arietinum L.*) that exerts anti-metabolic effect on podborer (*Helicoverpa armigera*) larvae. *Plant Molecular Biology,* 57: 359–374.

Takaiwa, F., Ogawa, M. and Okita, T. W. (1999). Rice glutelin. In P. R. Shewry, and R. Casey (Eds.), Seed proteins. Dordrecht; Boston: Kluwer Academic. 401-425.

Tavano, O. L., Neves, V. A. and da Silva Júnior, S. I. (2016). In vitro versus in vivo protein digestibility techniques for calculating PDCAAS (protein digestibility-corrected amino acid score) applied to chickpea fractions. *Food Research International,* 89: 756-763.

Torres-Fuentes, C., Alaiz, M., and Vioque, J. (2012). Iron-chelating activity of chickpea protein hydrolysate peptides. *Food chemistry,* 134(3):1585-1588.

Torres-Fuentes, C., Alaiz, M., and Vioque, J. (2014). Chickpea chelating peptides inhibit copper-mediated lipid peroxidation. *Journal Science of Food and Agriculture,* 94(15):3181-3188.

Torres-Fuentes, C., Contreras, M. del M., Recio, I., Alaiz, M., and Vioque, J. (2015). Identification and characterization of antioxidant peptides from chickpea protein hydrolysates. *Food Chemistry,* 180:194-202.

Trolove, S. N., Tan, Y., Morrison, S. C., Feng, L. and J. Eason. (2018). Development of a method for producing selenium-enriched radish sprouts. *LWT- Food Science and Technology*, 95:187-192.

Vairinhos, F., and Murray, D. R. (1982). Changes in polypeptides composition during seed development in chickpea (*Cicer arietinum L.*). *Zeitschrift für Pflanzenphysiologie*, 106(5): 447-452.

Vioque, J., Sánchez-Vioque, R., Clemente, A., Pedroche, J., Bautista, J. and Millán, F. (1999). Purification and partial characterization of chickpea 2S albumin. *Journal of Agricultural and Food Chemistry*, 47: 1405-1409.

Wang, H., Du, S., Zhao, Y., and Qiao, L. (2015). Modification of chickpea protein hydrolysates by alcalase-catalyzed plastein reaction and improvement in antioxidant activity. *Journal of Chinese Institute of Food Science and Technology*, 15(1):34-40.

Wolzak, A., Bressani, R. and Gomez Brenes, R. (1981). A comparison of in vivo and in vitro estimates of protein digestibility of native and thermally processed vegetable proteins. *Plant Foods for Human Nutrition*, 31(1):31-43.

Xiao, Y., Xing, G., Rui, X., Li, W., Chen, X. and Jiang Xiao, M. (2015). Effect of solid-state fermentation with Cordyceps militaris SN-18 on physicochemical and functional properties of chickpea *(Cicer arietinum L.)* flour. *LWT - Food Science and Technology*, 63(2):1317-1324.

Xue, Z., Gao, J., Zhang, Z., Yu, W., Wang, H. and Kou. X. (2012). Antihyperlipidemic and Antitumor Effects of Chickpea Albumin Hydrolysate. *Plant Foods for Human Nutrition*, 67(4): 393-400.

Xue, Z., Wen, H., Zhai, L., Yu, Y., Li, Y., Yu, W., Cheng, A., Wang, C. and Kou. X. (2015). Antioxidant activity and anti-proliferative effect of a bioactive peptide from chickpea (*Cicer arietinum L.*). *Food Research International*, 77, 2: 75-81.

Xu, Y., Cartier, A., Obielodan, M., Jordan, K., Hairston, T., Shannon, A. and Sismour, E. (2016). Nutritional and anti-nutritional composition, and in vitro protein digestibility of Kabuli chickpea *(Cicer arietinum L.)* as affected by differential processing methods. *Journal of Food Measurement and Characterization*, 10(3): 625-633.

Yahia, S., Benomar, S., Dehiba, F., Allaoui, A., Guillen, N., Rodriguez-Yoldi, M. J., Osada, J., and Boualga, A. (2017). Hypocholesterolaemic and antioxidant efficiency of chickpea (*Cicer arietinum*) protein hydrolysates depend on its degree of hydrolysis in cholesterol-fed rat. *Nutrition and Food Science,* 47(2):254-269.

Yoshida, M., Okada, T., Namikawa, Y., Matsuzaki, Y., Nishiyama, T., and Fukunaga, K., (2007). Evaluation of nutritional availability and anti-tumor activity of selenium contained in selenium-enriched *Kaiware radishs* prouts. *Bioscience, Biotechnology, and Biochemistry*, 71: 2198–2205.

Yust, M. del Mar., Millan-Linares, M. del C., Alcaide-Hidalgo, J. M., Millan, F. and Pedroche. J. (2012). Hypocholesterolaemic and antioxidant activities of chickpea (*Cicer arietinum L.*) protein hydrolysates. *Journal Science Food Agricultural*, 92(9): 1994-2001.

Zhang, L., Li, Q., Yang, X., and Xia. Z. (2012). Effects of sodium selenite and germination on the sprouting of chickpeas (*Cicer arietinum L.*) and its content of selenium, formononetin and biochanin A in the sprouts. *Biological Trace Element Research*, 146(3): 376-380.

Zhao, J., Zhao, H., Wang, X., Huang, R., Enomoto, H., He, Z. and Can-Peng, L. (2013). Characteristics and enhanced antioxidant activity of egg white protein selenized by dry-heating in the presence of selenite. *Journal of Agricultural Food and Chemistry*, 61: 3131-3139.

Zhang, T., Jiang, B., Miao, M., Mu, W., and Li, Y. (2012). Combined effects of high-pressure and enzymatic treatments on the hydrolysis of chickpea protein isolates and antioxidant activity of the hydrolysates. *Food Chemistry,* 135:904-912.

Biographical Sketches

Daniela Guardado Félix

Affiliation:

Tecnológico de Monterrey, Centro de Biotecnología-FEMSA, Escuela de Ingeniería y Ciencias, Monterrey, N.L. México.

Facultad de Ciencias Químico-Biológicas (FCQB), Universidad Autónoma de Sinaloa (UAS), Culiacán, Sin, México.

Education:

PhD in Science in Food biotechnology. UAS, Culiacán, Sinaloa, México. 2013-2016.

M.Sc in Food Science and Technology. UAS, Culiacán, Sinaloa, México. 2007-2010.

Biochemical Engineer. UAS. Culiacán, Sinaloa, México. 2001-2006.

Business Address:

Av. Eugenio Garza Sada 2501 Sur, C.P. 64849 Monterrey, NL, México.

FCQB-UAS, AP 1354, CP 80000 Culiacán, Sinaloa, México.

Research and Professional Experience:

2005-2006. Research thesis "Biochemical and functional characterization of proteolytic enzymes of Aguama (*Bromelia pinguin*)" (FCQB, UAS, Sinaloa, México).

2007-2011. Research thesis "Analysis of the regulatory activity of the amaranthin gene in transgenic tobacco" (FCQB, UAS, Sinaloa, México).

2013-2014. Research" Genetic transformation of sugar cane." (FCQB, UAS, Sinaloa, México).

2015-2019. "Enriched of selenium in legumes and cereal; development of foods with high anticancer and antioxidant potential evaluated *in vivo* and *in vitro.*" (Tecnológico de Monterrey, México).

Professional Appointments:

Professor in the Universidad Autónoma de Sinaloa. México.

Postdoctoral researcher in the Tecnológico de Monterrey, México.

Honors:

Honorable Mention of Excellence by the Universidad Autónoma de Sinaloa in the PhD in Biotechnology program. 2017.

Publications from the Last 3 Years:

Guardado-Félix, D., Serna-Saldivar, S. O., Cuevas-Rodríguez, E. O., Jacobo-Velázquez, D. A., and Guitiérrez-Uribe, J. A. (2017). Effect of sodium selenite on isoflavonoid contents and antioxidant capacity of chickpea. *Food Chemistry*, 226: 69-74.

Lazo-Vélez, M. A., Guardado-Félix, D., Avilés-González, J., Romo-López, I. and Serna-Saldívar, S. O. (2018). Effect of germination with sodium selenite on the isoflavones and cellular antioxidant activity of soybean (*Glycine max*). *LWT - Food Science and Technology,* 93: 64-70.

Guardado-Félix, D., Antunes-Ricardo, M., Rocha-Pizaña, M. R. Martínez-Torres, A. C. Gutiérrez-Uribe, J. A. and Serna-Saldivar, S. O. (2019). Chickpea (*Cicer arietinum L.*) sprouts containing supranutritional levels of selenium decrease tumor growth of colon cancer cells xenografted in immune-suppressed mice. *Journal of Functional Foods*. 53: 76-84.

Serrano-Sandoval, S. N., Guardado-Félix, D., and Gutiérrez-Uribe, J.A. (2019). Changes in digestibility of proteins from chickpeas (*Cicer arietinum L.)* germinated in presence of selenium and antioxidant capacity of hydrolysates. *Food Chemistry,* 285:290-295.

Guardado-Félix, D., Serna-Saldivar, S. O., Gutiérrez-Uribe, J. A., and 3 and Chuck-Hernández, C. (2019). Selenium in Germinated Chickpea (*Cicer arietinum L.*) Increases the Stability of Its Oil Fraction. *Plants*, 8, 113; doi:10.3390/plants8050113.

Guardado Félix, D. and Gutiérrez Uribe, J. A. Legumes and Oxidative Stress. Food Chemistry, Function and Analysis No. 8. Legumes: Nutritional Quality, Processing and Potential Health Benefits. Edited by María Ángeles Martín-Cabrejas. The Royal Society of Chemistry, 2019. Chapter 12. pp. 261 - 281.

Guardado-Félix, D., Lazo-Velez, M.A. and Serna-Saldivar, S.O. Protein-Selenized Enriched Breads. Flour and Breads and their Fortification in Health and Disease Prevention (Second Edition). Edited by Victor R. Preedy, Ronald Ross Watson. Academic Press. 2019. Chapter 24. pp. 307-317. doi.org/10.1016/B978-0-12-814639-2.00024-1.

Ada Keila Milán-Noris

Affiliation: Universidad Autónoma de Sinaloa

Education:

PhD in Food Biotechnology (2017), Tec de Monterrey.

M.Sc. in Biotechnology (2011), Tec de Monterrey.

B.S. in Biochemical Engineering (2006), Universidad Autónoma de Sinaloa.

Business Address: Laboratorio de Nutracéuticos (18), Facultad de Ciencias Químico Biológicas, Universidad Autónoma de Sinaloa, Ciudad Universitaria, Blv. de las Américas y Josefa Ortiz de Domínguez, S/N, Culiacán, Sinaloa, México.

Research and Professional Experience:

Effect of cereal and pulses processing on nutritional, nutraceuticals and technological characteristics. Also, elucidation of mechanism effect of bioactive compounds.

Teaching experience (undergraduate): Food Processing, Nutraceuticals and Pharma and Food Development in Biochemical Engineering; and Advanced Chemical Analysis in Genomic Biotechnology.

Professional Appointments:

2018-present. Assistant Professor, Universidad Autónoma de Sinaloa.
2012-2013. Research assistant, Universidad Autónoma de Sinaloa.
2011-2012. Research assistant, Tec de Monterrey.

Honors:

Recognized by the National Researchers System of Mexico (SNI):
Candidate to National Researcher (January 2019-December 2021)

Publications from the Last 3 Years:

Milán-Noris, Ada Keila, Julián De la Rosa-Millan, and Sergio Othón Serna-Saldivar. 2019. "Comparative analysis of techno-functional properties, starch digestion and protein quality of pigmented chickpea flours." *International Journal of Food Science & Technology* 0 (0). doi: 10.1111/ijfs.14144.

Mora-Rochin, Saraid., Milán-Noris, Ada Keila., Milán-Carrillo, Jorge. 2019. "Maize: Processing, Product Development, and Nutritional Aspects." In *Whole Grains*, edited by Annamalai Manickavasagan Shabir Ahmad Mir, Manzoor Ahmad Shah. CRC Press.

Milán-Noris, Ada Keila, Janet Alejandra Gutiérrez-Uribe, Arlette Santacruz, Sergio O. Serna-Saldívar, and Cristina Martínez-Villaluenga. 2018. "Peptides and isoflavones in gastrointestinal digests contribute to the anti-inflammatory potential of cooked or germinated desi and kabuli chickpea (Cicer arietinum L.)." *Food Chemistry* 268:66-76. doi: https://doi.org/10.1016/j.foodchem.2018.06.068.

Milán-Noris, Ada Keila, Julián de la Rosa-Millán, Cuauhutémoc Reyes-Moreno, and Sergio O. Serna-Saldivar. 2017. "Physicochemical, functional properties, and digestion of isolated starches from pigmented chickpea (Cicer arietinum L.) cultivars." *Starch - Stärke* 69 (5-6). doi: doi:10.1002/star.201600152.

Sayra N. Serrano Sandoval

Affiliation: Tecnológico de Monterrey

Education:

2010-2015
B.S. in Pharmaceutical Sciences
Universidad de las Américas Puebla

2016-2017
M.Sc in Biotechnology
Tecnológico de Monterrey

2018-2021
Ph.D of Sciences in Biotechnology
Tecnológico de Monterrey

Business Address: Av. Eugenio Garza Sada 2501 Sur, Tecnológico, 64849, Monterrey, N.L.

Research and Professional Experience:

2014. Research "Anti-E-selectin SAINT-O-Somes for delivery of siRNA into activated endothelial cells" (Rijksuniversiteit Groningen; Groningen, Netherlands)

2015. Research thesis "Variability of multidrug resistance protein-1 activity in bioavailability study" (Laboratorios Clínicos de Puebla; Puebla, Mexico).

2015. Development and validation of bio-analytic methods for the quantification of generic drugs by LC-MS/MS to determine bioequivalence. (Biophade Clinical Research; Cd. De Mexico, Mexico).

2016-2017. Research thesis "Effect of ultrasound treatment on bioactive compounds' composition of soybean and red prickly pear used to develop a functional beverage" (Tecnológico de Monterrey; Nuevo León, Mexico).

2018-2021. Research thesis "Validation of the beneficial effects on health of selenized chickpea proteins" (Tecnológico de Monterrey; Nuevo León, Mexico).

Professional Appointments: N/A

Honors:

- 2015. *Magna Cum Laude* (98/100) by the Universidad de las Américas Puebla in the B.S. in Pharmaceutical Sciences program.
- 2017. *Honorable Mention of Excellence* (97/100) by the Tecnológico de Monterrey in the M.Sc in Biotechnology program.
- 2010-2015. Full scholarship to study the bachelor's degree at Universidad de las Américas Puebla.
- 2014. Full research scholarship to study abroad by University of Groningen.
- 2016-2017. Full scholarship to study the master´s degree at Tecnológico de Monterrey.
- 2018-2021. Full scholarship to study the Ph.D degree at Tecnológico de Monterrey.
- 2018. Full scholarship to attend at the Summer School "FSC 2018- Planetary health and Food Security" in Costa Rica by the Food Security Center, the University of Hohenheim, the Federal Ministry for Economic Cooperation and Development, the Excellence Centers for Exchange and Development (exceed), and the German Academic Exchange Service (DAAD).
- 2016-2021. Maintenance grant by CONACyT for post-graduate studies.

Publications from the Last 3 Years:

- Castro-Pastrana, L.I., Guzmán-Ruiz, O., Serrano-Sandoval, S.N., Pérez-Romano, B., Estrada-Marín, L., Cedillo-Carvallo, B., and Ruiz-Argüelles, A. (2016). Selection of Healthy Volunteers According to their Multidrug Resistance Protein-1 Activity Does Not Affect Bioavailability of Quetiapine. *Current Psychopharmacology* 5(1), 61-68.
- Serrano-Sandoval, S.N., Guardado-Felix, D., Gutierrez-Uribe, J.A. (2019. Changes in digestibility of proteins from chickpeas (*Cicer arietinum L.)* germinated in presence of selenium and antioxidant capacity of hydrolysates. *Food Chemistry* 285, 290-295.

Sergio O. Serna Saldivar

Affiliation: Centro de Biotecnología FEMSA, Escuela de Ingeniería y Ciencias, Tecnologico de Monterrey.

Education:

PhD. Food Science and Technology, Texas A&M University, 1984.
MSc. Scientific Nutrition, Texas A&M University, 1982.
BSc. Agricultural Engineering/Animal Science, 1979.

Business Address:

Instituto Tecnológico y de Estudios Superiores de Monterrey, CIDPRO, Escuela de Ingeniería y Ciencias, Ave Ave. Eugenio Garza Sada 2501 Sur Monterrey N.L. CP 64849 México. Telephone: (81) 83284322. E-mail: sserna@tec.mx

Research and Professional Experience:

- Head of CIDPRO (Protein Research and Development Center (01/2013 – Date). Department Head (10/2010-12/2012 and 06/1994-06/2005) and professor (01/1992-Date). Centro de Biotecnología FEMSA, Escuela de Ingeniería y Ciencias. Instituto Tecnológico y de Estudios Superiores de Monterrey. Campus Monterrey.
- International Consultant-Scientist (08/1991-12/1991, 07/1992-08/1992) EMBRAPA-CTAA (Centro Nacional de Pesquisa de Tecnología Agroindustrial de Alimentos), Río de Janeiro, Brasil.
- Research Associate/Scientist (01/1987-09/1991). Texas A&M University, Soil & Crop Sci. Dept, Cereal Quality Lab.
- Associate Professor/Researcher (01/1985-01/1987). Universidad de Sonora. Graduate Program in Storage and Processing of Grains. Hermosillo, Sonora, México.
- Research Graduate Assistant (08/1982-12/1984). Texas A&M University, Dept. Soil & Crop Sci., Cereal Quality Laboratory.
- Visiting Scientist (05/82-08/82). Universidad Nacional Autónoma de México. Graduate Program at Cuautitlán Izcalli.

Professional Appointments:

- Sistema Nacional de Investigadores (Mexican National Research System)
- National Researcher -
    - Level III 01/05-Date
    - Level II 08/95 – 12/04
    - Level I. 08/92-07/95.
- Candidate National Researcher. 1985 - 1987.
- Mexican Academy of Sciences
    - Professional member 01/94-Date.
- USA Wheat Quality Council, 1987-1990.

- American Association of Cereal Chemists, 1982-Date.
- Institute of Food Technologists, 1986-Date.

Honors:

- Honorific Diploma to outstanding undergraduate students, Instituto Tecnológico y de Estudios Superiores de Monterrey, 1979.
- Member Gamma Sigma Delta, 1982.
- Member Phi Sigma, 1983.
- Coauthor of best poster given by the Corn Refiners Association, American Association of Cereal Chemists, Dallas, Texas, 1990.
- Teaching and Research Excellence Award, ITESM-Campus Monterrey. May 1993.
- Member of the Scientific Diplomatic Mission to China, CONACyT, 1994.
- Teaching and Research Excellence Award, ITESM-Campus Monterrey. May 1996.
- Romulo Garza Award. First Place in Research, ITESM System. January 1998.
- Romulo Garza Award for best book, ITESM System, January 1998.
- "Visiting Adjunct Professor" Texas A&M University.
- Member of Evaluation Committee of "Repatriación y Retención de Científicos" CONACyT (1998-1999)
- Member of Evaluation Committee of Area VI of the National Research System (1999-2001).
- Certification in "Teaching Redesign." ITESM-Campus Monterrey. January 2000.
- Associate Editor, Cereal Chemistry 2002-2003.
- Member of the National Network of Food Sanitation CONACyT (2000 -)
- Third Place "Rómulo Garza Award" ITESM. January 2001.
- Honorific Mention. Redesigned Teaching Courses. ITESM-Campus Monterrey. May 2001.
- Member of the committee "Premio Nacional en Ciencia y Tecnología de Alimentos" sponsored by Coca Cola. 2001-2003.
- Second Place, TIA Research Award. "American Association of Cereal Chemists," Charlotte, NC, October, 2001.
- Teaching and Research Excellence Award for full professors, ITESM-Campus Monterrey, May, 2002.
- Main advisor of first prize student award of the IFT International Division (2003 George F. Stewart IFT International Research Award) "Institute of Food Technologists," Chicago Illinois, July, 2003.
- National Award for Community Service 2003 given by the Federal Secretary of Social Development. September 25, 2003.

- National Scientific and Technological Award "Luis Elizondo" category of Agroindustries and Food Technology, ITESM-System. November, 2003.
- "2004 Excellence Teaching Award" of the "American Association of Cereal Chemists," San Diego, CA. September, 2004.
- Nuevo Leon State Government Award, CORPES, INVITE e IC^2 for Innovation and Technology Development. December, 2004.
- First Place "Premio Rómulo Garza" Research Award. ITESM-System. January 2005.
- Coauthor of best poster presentation of the "36 Congreso de Investigación del Sistema ITESM." January 2006.
- Teaching and Research Excellence Award for full professors, ITESM-Campus Monterrey. May, 2006.
- Member of the Editorial Board of "Journal of Cereal Science," 2007- date.
- Board of Directors American Association of Cereal Chemists (AACC-International), October, 2007-2010.
- "Yum Kax" national award for Research related to Nixtamalization. Second International Congress and Third National Congresses "Del Maíz a la Tortilla." 20-22 october, 2008. Querétaro, Qro., Mexico.
- Teaching and Research Excellence Award for full professors of the Biotechnology and Food Engineering Department, ITESM-Campus Monterrey. Agust, 2009.
- Associate editor, "Cereal Chemistry," 2010-2012.
- Third Place "Romulo Garza Research Award. Tecnológico de Monterrey, January 2010.
- Second Place "Rómulo Garza Book Award." Tecnologico de Monterrey, January 2010.
- First Place "Rómulo Garza Book Award." Tecnologico de Monterrey, January 2011.
- AgroBio México 2012 Award "Research in Biothecnology." October 2012.
- Award for number of scientific publications, best professor for the Food Engineering and best Research project (CIDPRO). School of Biotechnology and Foods, November 2013.
- Member of the Scientific Mission to Kenya sponsored by the Federal Secretary of External Affairs of México, Public Health of Kenya and CIMMyT, November 15-22, 2013.
- Romulo Garza Award to the Emblematic of Tecnológico de Monterrey 2013, Monterrey N.L., Mexico, January 2014.
- Member of the Award Jury Agro Bio 2013.

- Award *"Best Professor for Former Alumni"* Tecnologico de Monterrey. September 4, 2013.
- Coorganization of Fifth International Nixtamalization Congress, Tecnologico de Monterrey, Monterrey, N.L., Mexico 7 to 9 october, 2013.
- Coorganization of the First International Congress of Functional Foods and Nutraceuticals, Tecnologico de Monterrey, Monterrey, N.L., Mexico . June 11 to 13, 2014.
- President of the National "Luis Elizondo 2014 Award" in the scientfic category.
- Member of the Selection Committee of the MIT Award *"Innovators Under 35" Technology Reviews* 2015.
- Vicepresident and president of the National Food Science and Technology Award of Coca Cola -2015 and 2016, respectively.
- Dondé Foundation Award for Trajectory in Research. Latin Food Congress, Cancun Quintana Roo, November 8, 2016.
- First place "Rómulo Garza Award for Publication of Books 2016." ITESM, January, 2017.
- Associate Editor "Journal of Cereal Science" 2018-
- Award for Outstanding Professors selected by the alumni of the Food Engineering Program in the 75 anniversary of Tecnologico de Monterrey, August 2018.
- Doctor Honoris Causa, Universidad de Sonora. June 2019.

Publications from the Last 3 Years:

Books (1)

- Serna Saldivar, S.O. 2018. Corn: Chemistry and Technology. Third Edition, Elsevier-AACCI, 690 p. ISBN 9780128119716

Chapters (13)

- Serna-Saldivar, S. O., Heredia-Olea, E., and Perez-Carrillo, E. 2019. Soybean-Fortified Wheat Flour Tortillas. Chapter 27 (pp. 291-306) In: *Flour and Breads and Their Fortification in Health and Disease Prevention.* V. R. Preedy and R.R. Watson (eds.). Second Edition, Academic Press-Elsevier, London UK.
- Guardado-Felix, D., Lazo-Velez, M., and Serna-Saldivar, S. O. 2019. Protein-Selenized Enriched Breads. Chapter 28 (pp. 307-317) In: *Flour and Breads and Their Fortification in Health and Disease Prevention.* V. R. Preedy and R.R. Watson (eds.). Second Edition, Academic Press-Elsevier, London UK.
- Chuck-Hernandez, C., and Serna-Saldivar, S. O. 2019. Soybean Fortified Corn Tortillas and Related Products. Chapter 29 (pp. 319-332) In: *Flour and Breads*

and Their Fortification in Health and Disease Prevention. V. R. Preedy and R. R. Watson (eds.). Second Edition, Academic Press-Elsevier, London UK.

- Garcia-Lara, S., and Serna-Saldivar, S. O. 2018. Corn History and Culture. Chapter 1 (pp.1-19) In: *Corn: Chemistry and Technology.* S. O. Serna Saldivar, (ed.). Third Edition, Elsevier (Woodhead Publishing), AACC-Intenational. ISBN 9780128119716.
- Garcia-Lara, S., Chuck-Hernandez, C., and Serna-Saldivar, S. O. 2018. Description, Development and Structure of the Corn Kernel. Chapter 6 (pp.147-163) In*: Corn: Chemistry and Technology.* S. O. Serna Saldivar, (ed.). Third Edition, Elsevier (Woodhead Publishing), AACC-Intenational. ISBN 9780128119716.
- Acosta-Estrada, B., Gutierrez-Uribe, J., and Serna-Saldivar, S. O. 2018. Minor Constituents and Phytochemicals of the Kernel. Chapter 14 (pp. 369-403) In: *Corn: Chemistry and Technology.* S. O. Serna Saldivar, (ed.). Third Edition, Elsevier (Woodhead Publishing), AACC-Intenational. ISBN 9780128119716.
- Serna-Saldivar, S. O., and Perez-Carillo, E. 2018. Food Uses of Corn and Dry Milled Fractions. Chapter 16 (pp.435-468) In*: Corn: Chemistry and Technology.* S. O. Serna Saldivar, (ed.). Third Edition, Elsevier (Woodhead Publishing), AACC-Intenational. ISBN 9780128119716.
- Serna-Saldivar, S. O., and Chuck-Hernandez, C. E. 2018. Food Uses of Lime-Cooked Corn with Emphasis in Tortillas and Snacks. Chapter 17 (pp. 469-500), In: *Corn: Chemistry and Technology*. S. O. Serna Saldivar, (ed.). Third Edition, Elsevier (Woodhead Publishing), AACC-Intenational. ISBN 9780128119716.
- Barrera-Arellano, D., Badan-Ribeiro, A. P., and Serna-Saldivar, S. O. 2018. Corn Oil: Composition, Processing and Utilization. Chapter 21 (pp. 593-614) In: *Corn: Chemistry and Technology*. S. O. Serna Saldivar, (ed.). Third Edition, Elsevier (Woodhead Publishing), AACC-Intenational. ISBN 9780128119716.
- Serna-Saldivar, S. O., and Espinosa-Ramirez, J. 2018. Grain Structure and Grain Chemical Composition. Chapter 5 (pp. 85-129) In: *Sorghum & Millets: Chemistry, Technology and Nutritional Attributes,* J. R. N. Taylor and K. G. Duodu (eds.). Second Edition, Elsevier (Woodhead Publishing), AACC-Intenational. ISBN: 978-0-12-811527-5.
- Heredia-Olea, E., Serna-Saldivar, S. O., Perez-Carrillo, E., and Canizo, J. R. 2018. Conversion of High Biomass/Bagasse from Sorghum and Bermuda Grass into Second-Generation Bioethanol. Chapter 5 (pp. 77-90) In: *Biofuels - State of Development,* K. Biernat (ed.), IntechOpen, London, England. http://dx.doi.org/10.5772/intechopen.75064, ISBN 978-1-78923-347-6.
- Ponce-García, N., Ramírez-Wong, B., Escalante-Aburto, A., Torres-Chávez, P. I., and Serna-Saldivar, S. O. 2017. Relationship between wheat physical

properties and quality: a review. Chapter In: *Wheat*. (eds.). Intech. London, England, ISBN 978-953-51-5125-8.

- Serna-Saldivar, S. O., and Rubio-Flores, M. 2016. Role of intrinsic and supplemented enzymes in brewing and beer properties. Chapter 15 pp 273-298 In: *Microbial Enzyme Technology and Food Applications*. R. C. Ray and C. M. Rosell, (eds.). CRC Press Taylor & Francis Group, Boca Raton, FL.

Scientific Referred Research Articles (52)

- Garcia-Amezquita, L. E., Tejada-Ortigoza, V., Perez-Carrillo, E., Serna-Saldívar, S. O., Campanella, O. H., and Welti-Chanes, J. 2019. Functional and compositional changes of orange peel fiber thermally-treated in a twin extruder. *Food and Bioprocess Technology* (In Press).
- Guardado-Félix, D., Serna Saldivar, S. O., Gutiérrez-Uribe, J. A., and Chuck Hernandez, C. 2019. Selenium stress in chickpea (Cicer arietinum L.) increases the stability of its oil fraction and improves the antioxidant capacity in cells. *Food Research International.* (In Press).
- Tejada-Ortigoza, V., Garcia-Amezquita, L. E., Kazem, A. E., Campanella, O. H., Cano, M. P. Serna-Saldívar, S. O., and Welti-Chanes, J. 2019. In vitro fecal fermentation of high pressure treated fruit peels used as dietary fiber sources. *Molecules* 24(4):697.
- Milán-Noris, A. K., De la Rosa-Millan, J., and Serna-Saldivar, S. O. 2019. Comparative analysis of techno-functional properties, starch digestion and protein quality of pigmented chickpea flours. *International Journal of Food Science and Technology* https://doi.org/10.1111/ijfs.14144.
- de la Rosa-Millán. J., Heredia-Olea, E., Perez-Carrillo, E., Guajardo-Flores, D., and Serna-Saldívar. S. O. 2019. Effect of decortication, germination and extrusion on physicochemical and *in vitro* protein and starch digestion characteristics of black beans (*Phaseolus vulgaris* L.). *LWT Food Science and Technology*. 102:330-337.
- Mejía-Manzano, L. A., Barba-Dávila, B. A., Vázquez-Villegas, P., Serna Saldívar, S. O., and González-Valdez, J. 2019. Improved extraction of the natural anticancerigen pristimerin from *Mortonia greggii* root bark using green solvents and Aqueous Two-Phase Systems. *Journal of Separation and Purification Technology* 211: 667-672.
- Ortiz-Islas, S., García-Lara, S., Preciado-Ortiz, R. E., and Serna-Saldivar, S. O. 2019. Fatty acid composition and proximate analysis of improved high oil corn hybrids adapted to subtropical areas. *Cereal Chemistry* 96(2):182-192.
- Guardado-Félix, D., Antunes-Ricardo, M., Rocha-Pizaña, M. R., Martínez-Torres, A. M., Gutiérrez-Uribe, J. A., and Serna Saldivar, S. O. 2019. Chickpea

(*Cicer arietinum* L.) sprouts containing supranutritional levels of selenium decrease tumor growth of colon cancer cells xenografted in immune-suppressed mice. *J. Functional Foods* 56:73-84.

- Hernandez-Reyes, K. E., Montemayor-Mora, G., Garcia Villanueva, C., Pérez-Carrillo, E., Millan de la Rosa, J., and Serna-Saldívar, S. O. 2019. Effect of soybean bagasse addition on texture, sensory properties and protein quality of maize tortillas. *Cereal Chemistry*. 96(2): 283-291.
- Avilés-Gaxiola, S., Chuck-Hernández, C., and Serna-Saldívar, S. O. 2018. Effect of thermal processing and reducing agents on trypsin inhibitor activity and functional properties of soybean and chickpea protein concentrates. *LWT - Food Science and Technology* 98:629-634.
- Montemayor-Mora, G., Hernandez-Reyes, K. E., Heredia-Olea, E., Pérez-Carrillo, E., Chew Guevara, A. A. and Serna-Saldívar, S. O. 2018. Rheology, acceptability and texture of wheat flour tortillas supplemented with soybean residue. *Journal of Food Science and Technology* 55(12):4964-4972.
- Buitimea-Cantúa, N. E., Salazar-García, M. G., Serna-Saldívar, S. O., Buitimea Cantúa, G. V., Magaña-Barajas, E., and Morales-Ortega. A. 2018. Reforumating tortillas with zero trans crystallized vegetable fat produced from palm stearin and high oleic safflower oil blend. *Biotecnia*. XX(3):83-89.
- Mata-Ramírez, D., Serna-Saldívar, S. O., Villela-Castrejón, J., Villaseñor-Durán' M. C. and Buitimea-Cantúa, N. E. 2018. Phytochemical profiles, dietary fiber and baking performance of wheat bread formulations supplemented with roselle (*Hibiscus sabdariffa*). *Journal of Food Measurement and Characterization.* 12(4):2657-2665. https://doi.org/10.1007/s11694-018-9883-4.
- Ponce-García, N., Serna-Saldivar, S. O., and García Lara, S. 2018. Fumonisins and their analogs in contaminated corn and its processed foods. *Food Additives and Contaminants* 35(11).
- Lazo-Vélez M. A., Rosales-Medina M. F., Tinoco-Alvear M., Briones-García, M and Serna-Saldívar S. O. 2108. Nutritional and nutraceutical applications of *Saccharomyces cerevisiae* var. *boulardii* in foods: a review. *Applied Microbiology* 125(4): 943-951, doi.org/10.1111/jam.14037.
- Milán-Noris, A. K., Gutierrez-Uribe, J. A., Santacruz, A., Serna-Saldivar, S. O., and Martinez-Villaluenga, C. 2018. Peptides and isoflavones in gastrointestinal digests contribute to the anti-inflammatory potential of cooked or germinated desi and kabulli chickpea. *Food Chemistry,* 268:66-76.
- Espinosa-Ramírez, J., Garzon, R., Serna-Saldivar, S. O., and Rosell, C. M. 2018. Functional and nutritional replacement of gluten in gluten-free yeast-leavened breads by using -conglycinin concentrate extracted from soybean flour. *Food Hydrocolloids.* 84:353-360.

- Rocha-Villarreal, V., Hoffmann, J. F., Vanier, N. L., Serna-Saldivar, S. O., and García-Lara, S. 2018. Hydrothermal treatment of maize: changes in physical, chemical, and functional properties. *Food Chemistry* 263:225-231.
- Garcia-Amezquita, L. E., Tejada-Ortigoza, V., Serna-Saldívar, S. O., and Welti-Chanes, J. 2018. Dietary fiber concentrates from fruit and vegetable by-products: processing, modification, and application as functional ingredients. *Food and Bioprocess Technology* 11(8):1439-1463.
- Sánchez-Reséndiz, A., Rodríguez-Barrientos, S., Rodríguez-Rodríguez, J., Barba-Dávila, B., Serna-Saldívar, S. O., and Chuck-Hernández, C. 2018. Phosphoesterification of soybean and peanut proteins with sodium trimetaphosphate (STMP): changes in structure to improve functionality for food applications. *Food Chemistry* 260:299-306.
- Buitimea-Cantua, N. E., Torres-Chavez, P. I., Ledesma-Osuna, A. I., Ramirez-Wong, B. Robles-Sanchez, R. M., Serna Guerrero, D., Gutierrez-Uribe, J. A., and Serna-Saldıvar, S. O. 2018. Ferulic, p-coumaric, diferulic and triferulic acids contents of corn tortillas prepared with extruded masa flour and enriched with sorghum (Sorghum bicolor (L) Moench) bran. *International Journal of Food Measurement and Characterization* 12(3): 1633-1640.
- Espinosa-Ramírez, J., Garzon, R., Serna-Saldivar, S. O., and Rosell, C. M. 2018. Mimicking gluten functionality with β-conglycinin concentrate: evaluation in gluten free yeast-leavened breads. *Food Research International* 106: 64-70.
- de la Rosa-Millan, J., Orona-Padilla, J., Flores-Moreno, V., and Serna Saldivar, S. O. 2018. Physicochemical, functional and ATR-FTIR molecular analysis of protein extracts derived from starchy pulses. *International Journal of Food Science and Technology.* 105:371-383 DOI: 10.1111/ijfs.13719.
- Chavez Murillo, C. E., Veyna Torres, J. I., Cavazos Tamez, L. M., de la Rosa Millan, J., and Serna-Saldivar, S. O. 2018. Physicochemical characteristics, ATR-FTIR molecular interactions and in vitro starch and protgein digestion of thermally treated whole pulse flours. *Food Research International* 105:371-383.
- Lazo-Vélez, M. A., Guardado-Félix, D., Avilés-González, J., Romo-Lopez, I., and Serna-Saldívar, S. O. 2018. Effect of germination with sodium selenite on the isoflavones and cellular antioxidant activity of soybean (*Glycine max*). *LWT-Food Science and Technology* 93:64-70.
- Garcia-Amezquita, L. E., Tejada-Ortigoza, V., Heredia-Olea, E., Serna-Saldívar, S. O., and Welti-Chanes, J. 2018. Dietary fiber quantification in fruits and their by-products: Comparison between conventional and integrated AOAC methodologies. *Journal of Food Composition and Analysis.* 67:77-85.

- Rocha-Villarreal, V., Serna-Saldivar, S. O., and García-Lara, S. 2018. Effects of parboiling on the physical, functional and nutritional properties of cereals other than rice. *Cereal Chemistry* 95:79-91.
- Aregueta-Robles, U., Fajardo-Ramírez, O. R., Villela, L., Gutiérrez-Uribe, J. A., Hernández-Hernández, J., López-Sánchez, R. D. C., Scott, S.P., and Serna-Saldívar, S. 2018. Cytotoxic activity of a black bean (Phaseolus vulgaris L.) extract and its flavonoid fraction in both in vitro and invivo models of lymphoma. *Revista de Investigacion Clinica,* 70(1): 32-39.
- Buitimea-Cantúa, N. E., Gutiérrez-Uribe, J. A. and Serna-Saldívar, S. O. 2018. Phenolic-protein interactions that affect food properties and enhance health benefits. *Journal Medicinal Foods,* 21(2):188-198 DOI 10.1089/jmf.2017.0057.
- Zavala-López, M., López-Tavera, E., Figueroa-Cárdenas, J. de D., Serna-Saldívar, S. O. and García-Lara, S. 2018. Screening of major phenolics and antioxidant activities in Teosinte populations and modern maize types. *Journal of Cereal Science* 79:276-285.
- Avilés-Gaxiola, S., Chuck-Hernández, C., and Serna-Saldívar, S. O. 2018. Inactivation methods of trypsin inhibitor in legumes: a review. *Journal of Food Science* 83(1):17-29.
- Tejada-Ortigoza, V., Garcia-Amezquita, L. E., Serna-Saldívar, S. O., Martín-Belloso, O., and Welti-Chanes, J. 2018. High hydrostatic pressure and mild heat treatments for the modification of orange peel dietary fiber: effects on hygroscopic properties and functionality. *Food and Bioprocess Technology,* 11(1):110–121, DOI 10.100/s 11947-017-1998-9.
- Buitimea-Cantúa, N. E., Salazar-García, M. G., Vidal Quintanar, R. L., Serna-Saldivar, S. O., Ortega Ramirez, R., and Buitimea-Cantúa, G.V. 2017. Formulation of zero-trans crystalized fats produced from palm stearin and high oleic safflower oil blends. *Journal of Food Quality* Article ID 1253976, 8 pages https://doi.org/10.1155/2017/1253976.
- Antunes-Ricardo, M., Rodriguez-Rodriguez, C., Gutiérrez-Uribe, J. A., Cepeda-Cañedo, E., and Serna-Saldivar, S. O. 2017. Bioaccessibility, intestinal permeability and plasma stability of isorhamnetin glycosides from Opuntia ficus-indica (L.). *International Journal of Molecular Science.* 18(8);1816-1830, doi:10.3390/ijms1808181.
- Ortiz-Martinez, M., Otero-Pappatheodorou, J. T., Serna-Saldívar, S. O., and García-Lara, S. 2017. Antioxidant activity and characterization of protein fractions and hydrolysates from QPM and normal maize. *Journal of Cereal Science* 76:85-91.
- Carrasco-Gonzalez, J. A., Serna-Saldivar, S. O., and Gutierrez-Uribe, J. A. 2017. Mycochemical changes induced by inorganic selenium enrichment in *P.*

ostreatus fruiting bodies. *Journal of Agriculture and Food Chemistry*, 65(20): 4074–4082.

- Lazo-Velez, M. A., Mata-Ramírez, D., Serna-Saldivar, S. O., and Chuck-Hernandez, C. 2017. Functional effects of soybean concentrates obtained from sprouted seeds enriched in selenium in wheat breadmaking. *Cereal Chemistry* 94(4): 740-745.
- Pérez-Carrillo, E., González-Fernández, A. G., Morales-Garza, S. M. A., Treviño-Garza, E., Guajardo-Flores, S., Serna-Saldívar, S. O., and Sánchez-Hernández, D. 2017. Effect of sodium stearoyl-2-lactylate, carboxymethyl-cellulose and guar-xanthan gums on muffins enriched with soybean milk powder and amaranth flour. *CyTA Journal of Food.* 15:538-543, DOI 10.1080/19476337.2017.1309684.
- Cuellar-Bermúdez, S. P., Barba-Davila, B., Serna-Saldivar, S. O., Parra-Saldivar, R., Rodriguez-Rodriguez, U. J., Morales-Davila, S., Goiris, K., Muylaert, K., and Chuck-Hernández, C. 2017. Deodorization of *Arthrospira platensis* biomass for futher scale up food applications. *Journal of the Science of Food and Agriculture* 97:5123–5130 DOI 10.1002/jsfa.8391.
- Benítez-Azaola, R., García-Varela, R., García-García, R. M., Suárez-Jacobo, A., Cardineau, G. A., Altamirano, J., and Serna-Saldivar, S. O. 2017. *Rhoeo discolor* leaf aqueous extracts as potential natural preservatives in Mexican fresh panela cheese. *Revista Mexicana de Ingenería Química* 16(2): 425-434.
- Tejada-Ortigoza, V., Garcia-Amezquita, L. E., Serna-Saldívar, S. O. and Welti-Chanes, J. 2017. The dietary fiber profile of fruit peels and functionality modifications induced by high hydrostatic pressure treatments. *Food Science and Technology International.* 23(5):396-402.
- Carrasco-Gonzalez, J. A., Serna-Saldivar, S. O., and Gutierrez-Uribe, J. A. 2017. Nutritional and nutraceutical potential of the *Pleurotus* fruiting bodies: strategies to improve pnutraceutical properties and use as food ingredient. *Journal of Food Composition and Analysis.* 58:69-81.
- Guardado-Félix, D., Serna-Saldivar, S. O., Cuevas-Rodríguez, E. O., Jacobo-Velázquez, D. A., and Gutiérrez-Uribe, J. A. 2017. Effect of sodium selenite on isoflavonoid contents and antioxidant capacity of chickpea (*Cicer arietinum* L.) sprouts. *Food Chemistry* 226:69-74.
- Espinosa-Ramírez, J., Garza-Guajardo, I., Pérez-Carrillo, E., and Serna-Saldívar, S. O. 2017. Differences in the functionality and characterization of kafirins extracted from decorticated sorghum flour or gluten meal treated with protease. *Journal of Cereal Science* 73:174-182.
- Pérez-Carrillo, E., Frías-Escobar, A., Gutiérrez-Mendivil, K., Guajardo-Flores, S., and Serna-Saldívar, S. O. 2017. Effects of maize starch addition on

physicochemical and sensory attributes of gluten free cookies produced from nixtamalized masa flour. *Journal of Food Processing*. Vol. 2017, Article ID 6365182, 6 pages, doi:10.1155/2017/6365182.

- Ayala-Soto, F. E., Serna-Saldívar, S. O., and Welti-Chanes, J. 2017. Effect of arabinoxylans and laccase on batter rheology and quality of yeast-leavened gluten-free breads. *Journal of Cereal Science* 73:10-17.
- Heredia-Olea, E., Cortés-Ceballos, E., and Serna-Saldívar, S. O. 2017. Malting sorghum with *Aspergillus oryzae* enhances gluten-free wort yield and extract. *Journal of the American Society of Brewing Chemists* 75(2):116-121.
- Elizondo-Garza, P., Serna Saldivar, S., and Chuck-Hernandez, C. 2017. Protein recovery from skipjack tuna (*Katsuwonus pelamins*) wash water with different pH and temperature combinations. *Revista AMIDIQ* 16 (1): 91-99.
- de la Rosa-Millan, J., Orona-Padilla, J., Flores-Moreno, V., and Serna Saldivar, S. O. 2017. Physicochemical and digestion characteristics of starch and fiber rich subfractions from four pulse bagasses. *Cereal Chemistry* 94(3): 524-531, DOI: 10.1094/CCHEM-08-16-0208-R.
- Buitimea-Cantua, N. E., Torres-Chavez, P. I., Ramirez-Wong, B. Serna-Saldıvar, S. O. Rouzaud-Sández, O., Rosas-Burgos, E. C., Platt-Lucero, L. C., and Salazar-García, M. G. 2017. Phenolic compounds and antioxidant activity of extruded nixtamalized corn flour and tortillas enriched with sorghum bran [*Sorghum bicolor* (L. Moench)]. *Cereal Chemistry* 94(2): 277-283.
- Ponce-García, N., Torres-Chávez, P. I., Figueroa-Cárdenas, J. D., Serna-Saldívar, S. O., Cortez-Rocha, M. O., and Escalante-Aburto, A. 2017. Tempered wheat kernels and their doughs: a study of viscoelasticity for indirect distinctions among wheat classes using small strain test. *Journal of the Science of Food and Agriculture* 97(4):1235-1243 doi:10.1002/jsfa.7855.
- Ochoa-Rivas, A., Chuck-Hernandez, C., and Serna-Salidvar, S. O. 2017. Microwave and ultrasound to enhance protein extraction from peanut flour under alkaline conditions: effects in yield and functional properties of protein isolates. *Journal of Agriculture and Food Chemistry* 10(3):543-555. doi:10.1007/s11947-016-1838-3. 1-13.

In: Handbook of Chickpeas
Editors: Albert T. Lund and Noah D. Schultz
ISBN: 978-1-53616-374-2

Chapter 5

CHICKPEAS APPLICATIONS IN GLUTEN-FREE PRODUCTS

Renata Puppin Zandonadi[*] and Raquel Braz Assunção Botelho
Nutrition Department, University od Brasília, Brasília, Brazil

ABSTRACT

The gluten-free diet (GFD) is arising as a food trend due to gluten-related disorders (GRD) and also due to people who believe that they genuinely benefit from GFD. However, the GFD tend to be costly and socially restrictive since the lack of gluten in products can lead to food products without adequacy on nutritional, sensorial, and technological quality. Therefore, some studies have been attempted to develop tastier and healthier gluten-free products to help people who need to follow the GFD. In this context, chickpea - as a naturally gluten-free product, and also high in protein, iron, and fiber - has been studied to replace wheat/gluten in some types of gluten-free food to improve the quality of GFD.

Keywords: chickpea, gluten-free diet, gluten-free products, gluten-related disorders

INTRODUCTION

The demand for gluten-free products is increasing since the prevalence of gluten-related disorders is rising. In the past decades, only people who have celiac disease (CD) used to follow a restricted gluten-free diet (Sapone et al. 2012; Catassi et al. 2015;

[*] Corresponding Author's Email: renatapz@yahoo.com.br.

Zandonadi, Botelho, and Araújo 2009). The celiac disease (CD) is an autoimmune enteropathy which affects genetically susceptible individuals, in which the consumption of gluten leads to an immune reaction in the patient's intestinal mucosa with consequent inflammation and/or villous atrophy (Fasano et al. 2008). Currently, the only safe and efficient treatment for CD is the adoption of a gluten-free diet (GFD) (Sapone et al. 2012).

In the past, the CD diagnosis used to be difficult, and the demand for gluten-free products was limited to less than 1% of the global population. Nowadays, the knowledge and the diagnosis of CD are well elucidated, increasing the prevalence of diagnosed CD patients who need to follow a restrict gluten-free diet. Furthermore, other disorders related to gluten and/or wheat have arisen in the global population. The principal wheat/gluten-related disorders are the celiac disease (as we mentioned before), gluten sensitivity, and wheat allergy.

Gluten sensitivity is a reaction to gluten in which neither allergic nor autoimmune mechanisms can be identified. The diagnosis is based on exclusion criteria, since there are no markers identified until now (Sapone et al. 2012; Catassi et al. 2015). The wheat allergy consists of an immunologic reaction (mediated by IgE) to wheat proteins (Elli et al. 2015). Beyond these people who suffer from GRD, there is also a high number of consumers following a gluten-free diet without a specific diagnose believing that the gluten-restriction can be good for their health (Castillo, Theethira, and Leffler 2015; Simpson et al. 2011).

It is important to highlight that the gluten-free diet is not so easy to follow since gluten displays numerous properties in food products such as elasticity, cohesiveness, firmness, moisture, and uniformity. Therefore, gluten composes several food products, and the alternative gluten-free choices are not so feasible as gluten-containing products (Zandonadi et al. 2012). People who suffer from GRD daily face difficulties, which include high cost and poor nutrition, sensorial and technological quality of gluten-free products, unavailability of adequate food choices, and gluten contamination in gluten-free products, and social restriction (Araújo & Araújo, 2011; Araújo, Araújo, Botelho, & Zandonadi, 2010).

Therefore, some studies have attempted to develop tastier and healthier gluten-free products to help people who need to follow the GFD. In this context, chickpea (*Cicer arietinum L.*) has been studied to replace wheat/gluten in some types of gluten-free food to improve the quality of GFD.

CHICKPEA: CHARACTERIZATION

Chickpea is one of the oldest and most consumed legumes worldwide. It is the second most cultivated legume worldwide (after beans). The largest producer is India, and about 95% of the production occurs in developing countries (FAO 2017). As same as the other legumes, the chickpeas contribute to reduce food insecurity, and to improve nutrition and health since they are rich in carbohydrates (mainly starch and fiber), proteins, minerals, B complex vitamins, and antioxidants (Chillo et al. 2007; Santos et al. 2018; Gliszczyńska-Świgło, Klimczak, and Rybicka 2018).

Like other legumes, chickpea is a good source of food proteins, due to their high amounts of lysine, leucine, aspartic acid, glutamic acid and arginine, and providing a balance of essential amino acid profile when consumed with cereals and other foods rich in sulfur-containing amino acids and tryptophan (Miñarro et al. 2012). Also, their proteins present functional properties that play an essential role in gluten-free food formulation and processing. Therefore, as a gluten-free product, chickpea can be used in the GRD patients diet as a legume seed or adding it as an ingredient like chickpea flour.

CHICKPEA: APPLICATION IN GLUTEN-FREE PRODUCTS

Despite the advances in the research and development of gluten-free products, and the growth of the gluten-free market, many of the available products on the market present high prices, and lack technological, nutritional and sensory quality compared to their wheat counterparts (Santos et al. 2018; V. D. Capriles and Arêas 2016; Vanessa D. Capriles and Arêas 2014).

Several GFPs have been developed using rice flour, often combined with maize, potato, or cassava starches since they are widely available, inexpensive ingredients that are bland in flavor. However, this kind of products may lead to nutritional deficiencies in individuals who face the daily challenges from GFD, since they are not rich in fibers, minerals, vitamins, and proteins. The chickpea flour (CPF) is used as an alternative to wheat flour in gluten-free products due to its nutritional profile. CPF contains anthocyanins (1.3 mg CGE/100 g), total phenolics (45 mg GAE/100 g), and antioxidant capacity (FRAP 13 μmol GAE/100 g; ORAC 10556 μmol TE/100 g) (Rocchetti et al. 2017).

Also, CPF has the potential to be used as an ingredient in GFP due to its functional properties, such as its emulsifying and foaming characteristics, as well as its high water and oil absorption capacity (Du et al. 2014; Aguilar et al. 2015).

Products with a great amount of gluten such as bread, pasta, cakes, and cookies are a challenge to be produced with gluten-free ingredients, and, at the same time, be nutritionally balanced.

Bread

Studies of composite GFB formulations have shown promising results for people who need a gluten-free diet. A study conducted by Santos et al. (Santos et al. 2018) aimed to use chickpea in gluten-free bread (GFB) improving the nutritional and sensorial characteristics. The authors have shown that chickpea flour alone can be used for GFB production, promoting lower crumb moisture, and higher volume and crumb firmness than corn, potato or rice flour formulations. However, the interactions between chickpea and potato starch or corn starch decreased the crumb firmness and enhanced the bread volume. The most accepted GFB formulations were prepared with 75% of CPF and 25% of potato starch or corn starch. Compared to white GFB, these formulations presented almost a hundred-percent increase in protein and ash contents, and almost 150% more fiber. These blends can be used in GFB production, improving nutritional, sensory, and physical properties. The formulation used by the authors was (for the g/100 g flour): 100 g flour/starch, 100 g water, 25 g whole egg, 10.5 g whole milk powder, 6 g white cane sugar, 6 g soy oil, 2 g salt, 0.8 g dry yeast, 0.3 g xanthan gum and 0.3 g carboxymethylcellulose (CMC)(Santos et al. 2018).

Aguilar et al. (Aguilar et al. 2015) evaluated the effect of chickpea and tiger nut flours in gluten-free batters and bread to replace (partially or totally) emulsifier and/or shortening in gluten-free products. The authors compared four formulations: 100% corn starch; 7.8 g/100 g chickpea flour and corn starch; 8.6 g/100 g tiger nut flour and corn starch; 7.8 g/100 g chickpea flour 8.6 g/100 g tiger nut flour and corn starch. They also used three levels of shortening combination (5, 2.5 and 0 g/100 g) and three levels of emulsifier (2, 1 and 0 g/100 g). The results showed that CPF increased the specific volume of the bread, and decreased initial crumb firmness. The GFB prepared with CPF and tiger nut maintained the baking characteristics (crumb color and, hardness, bake loss, specific volume, crust) even when emulsifier was reduced. They had good results with the reduction of emulsifiers and shortenings, but not with the exclusion of them (Aguilar et al. 2015).

Miñarro et al. (Miñarro et al. 2012) evaluated the characteristics of GFB formulations using legume flour (chickpea, pea, carob germ) instead of soy flour. The formulations contained: corn starch + legume flour (100%); sugar (5.70%); baking powder (2.45%); shortening (4.89%); baker's yeast (4.89%); salt (2.45%); xanthan gum (1.95%); emulsifier (1.95%); water (102.77%). The chickpea bread presented the highest specific volume and the softest crumb exhibiting the best physicochemical characteristics and, in general, good acceptance as a good alternative to the soy protein. However, there was a need to use emulsifiers and shortenings with chickpea flour.

Buresová et al. (Burešová et al. 2014) studied the rheological characteristics of amaranth, chickpea, corn, millet, quinoa, and rice flour in leavened GFB. They used 300 g of flour, water according to 500 FU water absorption, 1.5% (w/w) of salt, 1.8% (w/w) of yeast, 1.86% (w/w) of saccharose and 0.005% (w/w) of ascorbic acid. The chickpea presented one of the highest values of dough extensibility (37%) compared to the wheat dough. Even it did not reach the quality of the wheat bread (100%), the bread-making quality of rice, quinoa and chickpea flours were better than amaranth, buckwheat, corn, and millet flours (Burešová et al. 2014). Also, the stickiness of chickpea GFB was comparable to wheat crumb characteristics. Comparing GFB, the highest loaf specific volumes were from chickpea, amaranth, and buckwheat gluten-free bread.

Rostamian et al. (Rostamian, Milani, and Maleki 2014) evaluated the quality of GFB prepared with chickpea flour and corn flour (together) at different proportions. They used four formulas with different concentrations of corn and chickpea flour as 20%, 40%, 60%, and 80% (A: 20% corn flour+80% chickpea flour, B: 40% corn flour+60% chickpea flour, C: 60% corn flour+40% chickpea flour, D: 80% corn flour+20% chickpea flour). Therefore they used these ingredients proportion: flour (100%); sucrose (4%); NaCl (2%); baker's yeast (3%); HPMC gum (3%); water (100%). The authors successfully produced corn and chickpea GFB. Increasing concentration of chickpea flour in formulation improved the quality parameters of gluten-free bread such as porosity, specific volume, and crust to crumb ratio. The results of texture showed that with increasing concentration of chickpea flour, there was a decreased crumb firmness. The best sample regarding quality parameters of the GFB was 80% of (w/w) chickpea flour + 20% of (w/w) corn flour + 3% (w/w) HPMC (Rostamian, Milani, and Maleki 2014).

Ouazib et al. (Ouazib et al. 2016) studied the effect of processing (germination, cooking, or toasting) of chickpea on the GF flours characteristics and their potential to obtain GFB. The authors used the GFB formulations based on % of flour basis:1.63% salt, 0.6% dry baker´s yeast, 4% sugar, 2% vegetal oil, 2% HPMC. The water amount varied depending on the water absorption of the flours. The GFB made with cooked CPF exhibited the lowest specific volume, brightest crumb, and the softest texture. The chickpea cooking decreased the content of carbohydrates, ash, and protein, although increased the protein digestibility. The authors obtained the highest specific volume of GFB with toasted chickpea flour, although crumb hardness was high. Overall, toasting and cooking led to flours that could be used for GFB with good sensory and nutritional quality(Ouazib et al. 2016). This study is the only one that used only chickpea for GFB. They compared processing types of chickpea to demonstrate the best way to produce bread.

Different from the other studies, Bird at al. (Bird et al. 2017) stated that the GF product market raised the need for alternatives to hydrocolloids as texture improving in GFB characteristics. Therefore, they studied the effect of processing on chickpea functionality with low ingredient amount, replacing hydrocolloids. The authors used these ingredients: gluten-free mix (rice flour, and corn starch, 50:50), water, canola oil, sugar, salt, xanthan gum, dry garbanzo chickpeas, and dry yeast. The authors showed that control bread was small, hard, and with low gas retention ability, depicting holes inside crumb pores. The CPF addition in low dose (2% w/w) enhanced loaf volume by 20% and reduced crumb hardness by 40%. CPF increased gas retention and increased the homogeneity of the starch-protein network.

On the other hand, the use of chickpea paste negatively affected bread quality due to loss of solubility upon cooking. Soaking and cooking water decreased significantly crumb hardness, although to a lower extent than the flour. The authors stated that chickpea products could be used as hydrocolloids substitute for GFB texture improvement, although acting with different mechanisms (Bird et al. 2017).

Based on the studies, it is possible to affirm that the use of chickpea in GFB can promote better sensory and nutritional qualities than the traditional GFB. However, the use of chickpea is recommended as an ingredient associated with other starch products or gums.

Pasta

Studies have been searching to improve the protein content of gluten-free pasta by the addition of raw materials of the vegetable source due to the importance of gluten on pasta quality. High-protein plant materials are derived mainly from legumes and pseudocereals, such as soybean, pea, amaranth, quinoa, bean, and chickpea. They can be used as isolates, flour, or in their concentrated forms (Chillo et al. 2007; Miñarro et al. 2012; Roy, Boye, and Simpson 2010).

The study from De Lima et al. (de Lima, Botelho, and Zandonadi 2019) was the first that successfully developed a GF pasta dough using only chickpea as flour source. The authors used chickpeas (65.7%) and egg whites (34.3%) in the GF pasta formulation and wheat flour (61.9%) and whole eggs (39.1%) to the control wheat pasta. The chickpea pasta presented 166.5% more dietary fiber, 53.8% more protein, and 37.3% less fat than the control. Chickpea pasta also presented good acceptance average for all attributes evaluated, suggesting that the chickpea flour can be used as a single flour to replace wheat, without using any structure agents. They also suggest that chickpea pasta can improve healthy options for individuals with gluten-related disorders.

Trevisan et al. (Trevisan, Pasini, and Simonato 2019) evaluated the *in vitro* glycemic response, and the cooking characteristics of 15 commercial GF pasta samples made using a single flour from different cereal, pseudocereal or legume flour (three from red lentil, three from green pea, two from chickpea, two from buckwheat, two from whole rice, one from white rice, two from corn). They used a durum wheat pasta sample as a control (DW). The authors showed high variation among commercial GF pasta regarding cooking properties and glycemic response. They also showed that chickpea pasta presented the lowest optimum cooking time (5 min.) compared to the other GF pasta (6 –

11 min.). The GF pasta with higher dietary fiber content, in particular, chickpea, presented the lowest glycemic index representing a healthier alternative for the GRD population than the other samples of GF pasta (Trevisan, Pasini, and Simonato 2019).

A study evaluated the quality of three gluten-free pasta (spaghetti) prepared with amaranthus flour added of chickpea and compared with traditional gluten-containing spaghetti. The authors demonstrated that the addition of chickpea presented a lower breakage susceptibility compared to the control. The cooking resistance and stickiness of chickpea spaghetti were inferior in spaghetti in base amaranthus flour compared to the wheat pasta. Regarding bulkiness, adhesiveness, and firmness, the results were statistically similar. The sensory analysis did not demonstrate differences among the spaghetti samples (Chillo et al. 2007). The produced spaghetti added of 10.7% chickpea presented high levels of protein (17.6%) and minerals (2.3%) compared with the gluten-spaghetti (12.2% and 0.81%, respectively) (Chillo et al. 2007). This study shows the importance of legumes to produce GFP to increase the amount of protein.

Flores-Silva et al. (Flores-Silva et al. 2014) developed a GF spaghetti with a low amount of glycemic carbohydrate (mixing of chickpea, unripe plantain, and maize flours). The authors used the control sample of spaghetti with 100% durum wheat semolina and eight GF samples with different percentages of chickpea (70%, 65% and 60%), unripe plantain (30%, 25%, 20% and 15%) and white maize flour (20%, 15%, 10%, 5% and 0%). They supplemented all the gluten-free formulations with 0.5% of carboxymethyl cellulose (food thickener). This value was suggested by the previous study carried out in their laboratory. Their GF spaghetti had a higher protein, fat, and ash content than the control (with semolina). The solid loss did not differ among the samples. The GF spaghetti total starch was lower than the control. The authors mentioned that lower available starch and higher resistant starch (RS) contents in the GF spaghetti could be associated with their lower glycemic index (Flores-Silva et al. 2014). They showed that it is possible to produce GF spaghetti with a reduced GI from unconventional food ingredients such as chickpea, unripe plantain, and maize flours.

Patiño-Rodríguez et al. (Patiño-Rodríguez et al. 2018) evaluated the use of unripe plantain, maize, and chickpea flour to prepare a functional gluten-free spaghetti using unripe plantain pulp and whole unripe plantain (pulp and peel). The author prepared two different samples based on the flour mixture: 25% of unripe plantain pulp flour or 25% of whole unripe plantain flour + 25% of chickpea flour + 50% of maize flour. Also, they used 1.5% (of the flours) of carboxymethylcellulose (CMC) + 32% of water. The samples presented a similar chemical composition. However, the whole flour product exhibited higher dietary fiber content. The authors did not find a difference in cooked weight or cooking loss of the spaghetti samples. The optimal cooking time was slightly higher for the pulp flour GF spaghetti. The GF spaghetti with whole flour presented higher hardness than pulp flour spaghetti. The whole flour GF spaghetti presented lower water absorption than spaghetti with pulp flour, without a difference in the diameter of the cooked spaghetti. Although the aim of the study was the use of unripe plantain in the GF spaghetti, they used chickpea as an ingredient of GF product. The authors showed that it is possible to produce a gluten-free pasta with good quality characteristics using whole unripe plantain flour associated with chickpea and maize flour (Patiño-Rodríguez et al. 2018).

Another study evaluated the use of rice and legume blend in gluten-free pasta nutritional quality (Bouasla, Ojtowicz, and Zidoune 2017). The authors used pre-cooked GF rice pasta enriched with (10 g/ 100 g, 20 g/100 g, and 30 g/100 g) legume flours (chickpea, yellow pea, and lentil). They found that the addition of legume flours decreased the GF pasta expansion ration, the hardness, cooking loss, and the lightness. Legume flours increased the yellowness, the adhesiveness, the firmness with no difference in the preparation. The samples presented acceptable scores for all sensory attributes and the overall quality. The authors mentioned that the GF pasta enriched with legumes flours (up to 30 g/100 g) presented excellent physicochemical and sensorial quality, and it can be successfully used to improve the nutritional quality of the gluten-free pasta formulation (Bouasla, Ojtowicz, and Zidoune 2017).

Camelo-Méndez et al. (Camelo-Méndez et al. 2017) studied the phenolic compounds, color and antioxidant capacity of GF pasta prepared with chickpea, unripe plantain, white maize, and blue maize flour. For all samples, the authors used 100 g of flour blended, 1.5 g of carboxymethylcellulose (CMC) and 32 mL of water. They prepared samples with six different combinations of flour based on 25%, 50% or 75% of blue maize flour or white maze flour blended with 37.5%, 25% or 12.5% of chickpea flour and the same proportion of unripe plantain flour. Despite the use of chickpea flour on the GF pasta, the authors aimed to investigate the difference of the use of white or blue maize flour. They identified 10 and 15 phenolic compounds in pasta prepared with white and blue maize flours, respectively. The different types of maize flours influenced the color and antioxidant samples characteristics. The focus of this study was not the use of chickpea flour to replace wheat. It was a base flour to the mixture with maize flours.

The studies showed that it is possible to use the chickpea as the unique GF flour or associated with other flours/starches promoting pasta with satisfactory sensory, nutritional, and technological quality.

Cake, Snacks, and Cookies

Gularte et al. (Gularte et al. 2012) aimed to evaluate the effect of different legumes (chickpea, pea, lentil, and bean) on nutritional and sensory quality, and *in vitro* protein and starch digestibility of gluten-free layer cake using the proportion of rice flour/legume flour as 50:50. They prepared GF cake samples with 350 g rice flour, 350 g legumes flour, 420 g milk, 350 g pasteurized whole eggs, 210 g sunflower oil, 630 g sucrose and

21 g baking powder. The control GF cake was prepared with 700 g rice flour. The incorporation of CPF increased the batter viscosity; however; it did not present higher specific cake volume than the control. Chickpea presented the brightest and the most yellowish crust and significantly increased the hardness and chewiness of the cakes. Also, CPF enriched cakes with protein, minerals, fat, and fiber content. Chickpea was the only legume that did not significantly affect the starch *in vitro* hydrolysis, decreasing the quickly digestible starch yielding a reduction in the glycemic index (Gularte et al. 2012). This is the only study that used chickpea to produce a cake, but it did not use chickpea exclusively. New studies should be conducted with other types of cakes, as well as the exclusive use of chickpea.

The snack foods are consumed worldwide and became a part of the eating habits of the GRD population, making them a target to improve their nutritional, sensory, and technological quality (Flores-Silva, Rodriguez-Ambriz, and Bello-Pérez 2015). Therefore, Han et al. (Han, Janz, and Gerlat 2010) aimed to develop GF legume-based cracker snacks exploiting the health-enhancing and anti-allergenic nature of legumes. The authors evaluated nine commercially available legume fractions (chickpea, yellow pea, green and red lentil, pinto, and navy bean flours and pea protein, starch, and fiber isolates) in the GF cracker formulation. They used the ingredients based on the legume GF ingredients (as 100%): water (27 - 33%); chemical leavening agent (0.6 – 0.8%); oil (20 – 27%); salt (3 - 4%); sugar (3 – 4%); hydrocolloid gums (0.6 – 0.8%). The authors showed that the nutritional and physical characteristics of these legume GF crackers were similar to the existing GF products on the market. They observed that the iron percentage of the daily value per serving in the chickpea GF crackers were three to six times higher than the existing products (Han, Janz, and Gerlat 2010).

Another study (Flores-Silva, Rodriguez-Ambriz, and Bello-Pérez 2015) aimed to develop a GF snack using chickpea, plantain, and maize blend to produce a high content of indigestible carbohydrate. The authors used three formulations of flour blend: (A) 50% unripe plantain flour, 30% chickpea flour and 20% maize flour; (B) 10% unripe plantain flour, 60% chickpea flour and 30% maize flour; and (C) 33% unripe plantain flour, 33% chickpea flour and 33% maize flour. The extrusion and deep-frying of GF flours blend produced snacks with high dietary fiber content (13.7–18.2 g/100 g) and low glycemic index (28 to 35). The GF snacks presented lower fat content (12.7 to 13.6 g/100 g) than the commercial snack. The snack with the highest proportion of unripe plantain flour showed higher slowly digestible starch (11.6 and 13.4 g/100 g) than the counterpart with the highest chickpea flour level (6 g/100 g). The overall acceptability of the GF snacks was similar to the commercial snack. The snack with the highest proportion chickpea presented the lowest total starch, slowly digested starch content, and glycemic index, but the highest resistant starch content (Flores-Silva, Rodriguez-Ambriz, and Bello-Pérez 2015).

Yamsaengsung et al. (Yamsaengsung, Berghofer, and Schoenlechner 2012) evaluated the effects of chickpea addition (0 – 100%) on the physical properties and sensory attributes of cookies based on white or whole wheat compared to buckwheat or amaranth gluten-free flour. Chickpea (0%, 20%, 40%, 60%, 80% and 100%) was added to four different types of flours: (i) white wheat flour, (ii) whole wheat flour, (iii) amaranth flour and (iv) buckwheat flour for a total of six different cookie recipes per cookie flour type. The standard cookie formulation was based on flour (150 g), shortening (vegetable fat) (62.5 g), powdered sugar (62.5 g), egg (60 g), guar gum (4.5 g), baking powder (1.5 g), salt (1.0 g) and vanillin sugar (1.0 g). The authors showed that the chickpea addition decreased the lightness in white wheat cookies while increasing it in whole wheat and amaranth cookies, and it significantly increased yellowness in all cookies. The spread factor of the gluten-free cookies was reduced by chickpea addition, and hardness was increased in white wheat and buckwheat cookies and decreased in whole wheat and amaranth cookies. The authors also showed that chickpea addition increased the acceptability of all cookies, in particular of the gluten-free cookies. In gluten-free cookies, the optimal levels of chickpea addition were 60 – 80% in amaranth and as well as buckwheat cookie formulations.

Another study proposed to develop and characterize a GF and lactose-free cookie using a mix of beet, chickpea, and quinoa flour. The cookie formulation was composed by egg (59 g); vanilla extract (15 mL); sodium bicarbonate (15 g); water (60 mL); butter (60 g); brown sugar (180 g); beet flour (60 g); chickpea flour (120 g); quinoa flour (60 g); milk-free chocolate. The GF cookie produced did not present significant difference in texture compared to their counterpart commercialized in the market. The sensory overall acceptance percentage was higher than 75%. However, the authors mentioned that the tastier's purchase intention was below 50%.

Conclusion

One of the major challenges developing GFP is the substitution of gluten and its characteristics such as elasticity, cohesiveness, firmness, moisture, and uniformity. Chickpea is a promising legume to use in GFP. Researches are increasing using this legume worldwide, and they are all recent studies. Chickpea presents high fiber and protein content that improves the quality of GFP, especially technologically and nutritionally.

For bread making, some studies demonstrated that chickpea is an excellent ingredient to use when gluten is not present. However, only one study made bread exclusively replacing wheat by chickpea in different processed forms. It was necessary to add hydrocolloids to improve texture. For gluten-free pasta, one study used 100% chickpea flour, and authors did not use any additives to improve elasticity and cohesiveness. The use of eggs helped as well as the amount of protein and fiber from chickpea. There is an interesting study that evaluated the glycemic response of GFP, and they conclude that pasta with chickpea had the best results because of its fiber content. In the literature, some researches are demonstrating more interest in evaluating the glycemic response of GFP since gluten´s substitution involves the addition of different starches which influence this response.

Fewer studies were found using chickpea in cakes or snacks. For cake development, only one study used 50% of chickpea flour, and for cookies, the best results were seen with 60 to 80% of chickpea added of amaranth or buckwheat. Authors did not test the use of 100% of chickpea flour. Other two studies evaluated chickpea for snacks, but also not exclusively. This scenario shows the promising use of chickpea to develop GFP, and more studies should be encouraged to test new formulations on a broader variety of products.

References

Aguilar, Núria., Elena, Albanell., Bego, ~ Na., Mi, ~ Narro. & Marta, Capellas. (2015). "Chickpea and Tiger Nut Flours as Alternatives to Emulsifier and Shortening in Gluten-Free Bread." *LWT - Food Science and Technology*, *62*, 225–32. https://doi.org/10.1016/ j.lwt. 2014.12.045.

Araújo, H. M. C. & Araújo, W. M. C. (2011). "Coeliac Disease. Following the Diet and Eating Habits of Participating Individuals in the Federal District, Brazil." *Appetite*, *57* (1), 105–9. https://doi.org/ 10.1016/ j.appet.2011.04.007.

Araújo, Halina Mayer Chaves., Wilma, Maria Coelho Araújo., Raquel, Braz Assunção Botelho. & Renata, Puppin Zandonadi. (2010). "Doença Celíaca, Hábitos e Práticas Alimentares e Qualidade de Vida." *Revista de Nutricao*, *23* (3), 467–74. https://doi.org/10.1590/ S1415-52732010000300014. [Celiac disease, eating habits and practices and life quality of life. Brazilian Journal of Nutrition].

Bird, Libby G., Casey, L Pilkington., Agung, Saputra. & Luca, Serventi. (2017). "Products of Chickpea Processing as Texture Improvers in Gluten-Free Bread." *Food Science and Technology International*, 690–98. https://doi.org/10.1177/1082013 217717802.

Bouasla, Abdallah., Agnieszka, W Ojtowicz. & Mohammed, Nasereddine Zidoune. (2017). "Gluten-Free Precooked Rice Pasta Enriched with Legumes Flours: Physical Properties, Texture, Sensory Attributes and Microstructure." https://doi.org/10. 1016/j.lwt.2016.10.005.

Burešová, Iva., Stanislav, Kráčmar., Petra, Dvořáková. & Tomáš, Středa. (2014). "The Relationship between Rheological Characteristics of Gluten-Free Dough and the

Quality of Biologically Leavened Bread." *Journal of Cereal Science*. https://doi.org/10.1016/j.jcs.2014.07.001.

Camelo-Méndez, Gustavo A., Pamela, C. Flores-Silva., Edith, Agama-Acevedo. & Luis, A. Bello-Pérez. (2017). "Multivariable Analysis of Gluten-Free Pasta Elaborated with Non-Conventional Flours Based on the Phenolic Profile, Antioxidant Capacity and Color." *Plant Foods for Human Nutrition*, *72* (4), 411–17. https://doi.org/10.1007/s11130-017-0639-9.

Capriles, V. D. & Arêas, J. A. G. (2016). "Approaches to Reduce the Glycemic Response of Gluten-Free Products: *In Vivo* and *in Vitro* Studies." *Food Funct.*, *7* (3), 1266–72. https://doi.org/10.1039/C5FO 01264C.

Capriles, Vanessa D. & José Alfredo, G. Arêas. (2014). "Novel Approaches in Gluten-Free Breadmaking: Interface between Food Science, Nutrition, and Health." *Comprehensive Reviews in Food Science and Food Safety*, *13* (5), 871–90. https://doi.org/10.1111/1541-4337.12091.

Castillo, N. E., Theethira, T. G. & Leffler, D. A. (2015). "The Present and the Future in the Diagnosis and Management of Celiac Disease." *Gastroenterology Report*, *3* (1), 3–11. https://doi.org/ 10.1093/ gastro/gou065.

Catassi, Carlo., Luca, Elli., Bruno, Bonaz., Gerd, Bouma., Antonio, Carroccio., Gemma, Castillejo., Christophe, Cellier., et al. (2015). "Diagnosis of Non-Celiac Gluten Sensitivity (NCGS): The Salerno Experts' Criteria." *Nutrients*, *7* (6), 4966–77. https://doi.org/10.3390/ nu7064966.

Chillo, S., Laverse, J., Falcone, P. M. & Del Nobile, M. A. (2007). "Quality of Spaghetti in Base Amaranthus Wholemeal Flour Added with Quinoa, Broad Bean and Chick Pea." https://doi.org/ 10.1016/j.jfoodeng.2007.04.022.

Du, Shuang-kui., Hongxin, Jiang., Xiuzhu, Yu. & Jay-lin, Jane. (2014). "Physicochemical and Functional Properties of Whole Legume Flour." *LWT - Food Science and Technology*, *55* (1), 308–13. https://doi.org/10.1016/J.LWT.2013.06.001.

Elli, Luca., Federica, Branchi., Carolina, Tomba., Danilo, Villalta., Lorenzo, Norsa., Francesca, Ferretti., Leda, Roncoroni. & Maria, Teresa Bardella. (2015). "Diagnosis of Gluten Related Disorders: Celiac Disease, Wheat Allergy and Non-Celiac Gluten Sensitivity." *World Journal of Gastroenterology*, *21* (23), 7110–19. https://doi.org/10.3748/wjg.v21.i23.7110.

FAO. (2017). "FAOSTAT - CHICKPEA." 2017. http://www.fao.org/faostat/en/?#data/QC/visualize.

Fasano, A., Araya, M., Bhatnagar, S., Cameron, D., Catassi, C., Dirks, M., Mearin, M. L., Ortigosa, L., Phillips, A. & FISPGHAN Celiac Disease Working Group. (2008). "Federation of International Societies of Pediatric Gastroenterology, Hepatology, and Nutrition Consensus Report on Celiac Disease." *Journal of Pediatric*

Gastroenterology and Nutrition, *47* (2), 214–19. https://doi.org/10.1097/MPG. 0b013e318 181afed.

Flores-Silva, Pamela C., Jose, De J. Berrios., James, Pan., Perla, Osorio-Díaz. & Luis, A. Bello-Pérez. (2014). "Gluten-Free Spaghetti Made with Chickpea, Unripe Plantain and Maize Flours: Functional and Chemical Properties and Starch Digestibility." *International Journal of Food Science & Technology*, *49* (9), 1985–91. https://doi.org/10.1111/ ijfs.12529.

Flores-Silva, Pamela C., Sandra, L. Rodriguez-Ambriz. & Luis, A. Bello-Pérez. (2015). "Gluten-Free Snacks Using Plantain-Chickpea and Maize Blend: Chemical Composition, Starch Digestibility, and Predicted Glycemic Index." *Journal of Food Science*, *80* (5), C961–66. https://doi.org/10.1111/1750-3841.12865.

Gliszczyńska-Świgło, Anna., Inga, Klimczak. & Iga, Rybicka. (2018). "Chemometric Analysis of Minerals in Gluten-Free Products." *Journal of the Science of Food and Agriculture*, *98* (8), 3041–48. https://doi.org/10.1002/jsfa.8803.

Gularte, Márcia Arocha., Esther, de la Hera., Manuel, Gómez. & Cristina, M. Rosell. (2012). "Effect of Different Fibers on Batter and Gluten-Free Layer Cake Properties." *LWT - Food Science and Technology*, *48* (2), 209–14. https://doi.org/10. 1016/j.lwt.2012.03.015.

Han, Jeeyup., Jennifer, A. M. Janz. & Mindy, Gerlat. (2010). "Development of Gluten-Free Cracker Snacks Using Pulse Flours and Fractions." *Food Research International*, *43*, 627–33. https://doi.org/ 10.1016/j.foodres.2009.07.015.

Lima, Bernardo Romão de., Raquel, Braz Assunção Botelho. & Renata, Puppin Zandonadi. (2019). "Gluten-Free Pasta: Replacing Wheat with Chickpea." *Journal of Culinary Science & Technology*, *17* (1), 1–8. https://doi.org/10.1080/15428052.2017. 1394952.

Miñarro, B., Albanell, E., Aguilar, N., Guamis, B. & Capellas, M. (2012). "Effect of Legume Flours on Baking Characteristics of Gluten-Free Bread." *Journal of Cereal Science*, *56* (2), 476–81. https://doi.org/ 10.1016/j.jcs.2012.04.012.

Ouazib, Meriem., Raquel, Garzon., Farid, Zaidi. & Cristina, M. Rosell. (2016). "Germinated, Toasted and Cooked Chickpea as Ingredients for Breadmaking." *Journal of Food Science and Technology*, *53* (6), 2664–72. https://doi.org/10.1007/s13197-016-2238-4.

Patiño-Rodríguez, Omar., Luis, Arturo Bello-Pérez., Pamela, Celeste Flores-Silva., Mirna, María Sánchez-Rivera. & Claudia, Andrea Romero-Bastida. (2018). "Physicochemical Properties and Metabolomic Profile of Gluten-Free Spaghetti Prepared with Unripe Plantain Flours." https://doi.org/10.1016/j.lwt.2017.12.025.

Rocchetti, Gabriele., Giulia, Chiodelli., Gianluca, Giuberti., Francesco, Masoero., Marco, Trevisan. & Luigi, Lucini. (2017). "Evaluation of Phenolic Profile and Antioxidant Capacity in Gluten-Free Flours." https://doi.org/10.1016/j.foodchem.2017.01.142.

Rostamian, Mohammad., Jafar, M. Milani. & Gisoo, Maleki. (2014). "Physical Properties of Gluten-Free Bread Made of Corn and Chickpea Flour." *International Journal of Food Engineering*, *0* (0), 467–72. https://doi.org/10.1515/ijfe-2013-0004.

Roy, F., Boye, J. I. & Simpson, B. K. (2010). "Bioactive Proteins and Peptides in Pulse Crops: Pea, Chickpea and Lentil." *Food Research International*, *43* (2), 432–42. https://doi.org/10.1016/ J.FOODRES. 2009.09.002.

Santos, Fernanda G., Camilly, Fratelli., Denise, G. Muniz. & Vanessa, D. Capriles. (2018). "Mixture Design Applied to the Development of Chickpea-Based Gluten-Free Bread with Attractive Technological, Sensory, and Nutritional Quality." *Journal of Food Science*, *83* (1), 188–97. https://doi.org/10.1111/1750-3841.14009.

Sapone, Anna., Julio, C Bai., Carolina, Ciacci., Jernej, Dolinsek., Peter, H. R Green., Marios, Hadjivassiliou., Katri, Kaukinen., et al. (2012). "Spectrum of Gluten-Related Disorders: Consensus on New Nomenclature and Classification." *BMC Medicine*, *10*, (February), 13. https://doi.org/ 10.1186/1741-7015-10-13.

Simpson, S., Lebwohl, B., Lewis, S. K., Tennyson, C. A., Sanders, D. S.& Green, P. H. (2011). "Awareness of Gluten-Related Disorders: A Survey of the General Public, Chefs and Patients." *E-SPEN, the European e-Journal of Clinical Nutrition and Metabolism*, *6* (5), e227–31. https://doi.org/10.1016/J.ECLNM.2011.08.001.

Trevisan, Sara., Gabriella, Pasini. & Barbara, Simonato. (2019). "An Overview of Expected Glycaemic Response of One Ingredient Commercial Gluten Free Pasta." https://doi.org/10.1016/j.lwt. 2019.04.013.

Yamsaengsung, Ram., Emmerich, Berghofer. & Regine, Schoenlechner. (2012). "Physical Properties and Sensory Acceptability of Cookies Made from Chickpea Addition to White Wheat or Whole Wheat Flour Compared to Gluten-Free Amaranth or Buckwheat Flour." *International Journal of Food Science & Technology*, *47* (10), 2221–27. https://doi.org/10.1111/j.1365-2621.2012.03092.x.

Zandonadi, Renata Puppin., Raquel, Braz Assunção Botelho. & Wilma, Maria Coelho Araújo. (2009). "Psyllium as a Substitute for Gluten in Bread." *Journal of the American Dietetic Association*, *109* (10), 1781–84. https://doi.org/10.1016/j.jada.2009.07.032.

Zandonadi, Renata Puppin., Raquel, Braz Assunção Botelho., Lenora, Gandolfi., Janini, Selva Ginani., Flávio, Martins Montenegro. & Riccardo, Pratesi. (2012). "Green Banana Pasta: An Alternative for Gluten-Free Diets." *Journal of the Academy of Nutrition and Dietetics*, *112* (7), 1068–72. https://doi.org/10.1016/j.jand.2012.04.002.

In: Handbook of Chickpeas
Editors: Albert T. Lund and Noah D. Schultz
ISBN: 978-1-53616-374-2

Chapter 6

CHICKPEAS APPLICATIONS IN DAIRY-FREE PRODUCTS

Luana Rincon[1], Priscila Farage[1], Renata Puppin Zandonadi[1] and Raquel Braz Assunção Botelho[1]
[1]Nutrition Department, University od Brasília, Brasília, Brazil

ABSTRACT

The adverse reactions to dairy products are prevalent worldwide in the population. Dairy products reactions include allergies and lactose intolerance, reaching almost 80% of the population. There is also the demand for dairy-free products by vegans. Therefore, the market of the dairy-free product tends to continue increasing. There are some milk substitutes based on cereals, legumes, and nuts or modified cow milk, as lactose-free dairy products.

Some of these products present limitations to replace dairy products because of their allergenic potential; some of them present high glycemic or low protein content; and others are not entirely accepted by consumers.

In this context, chickpea has been studied as a dairy-product substitute since it represents a good source of protein, iron, and fiber.

Keywords: chickpea, plant-based milk, dairy-free products, allergy to cow`s milk, lactose intolerance

INTRODUCTION

Production and commercialization of plant-based milk have grown exponentially in the last few years (Sethi et al. 2016). The demand for cow's milk substitutes has expanded mainly due to pathological reasons, such as the increase of cow's milk protein allergy (CMPA) and lactose intolerance. Moreover, changes in lifestyle and feeding habits, like the adoption of a vegetarian diet, have also contributed to the increasing the demand for such products in the market (Jeske et al. 2017).

CMPA comprises the most prevalent food allergy among children younger than three years old. Manifestations include gastrointestinal symptoms (reflux, diarrhea, constipation, among others) and extraintestinal symptoms (urticaria, respiratory manifestations, among others). Non-IgE-mediated allergy is the most common type of CMPA. The prevalence of CMPA varies between 2 - 6% in children, and 0.1 - 0.5% in adults (Makinen et al. 2016). Treatment for CMPA comprises the complete removal of cow's milk and all dairy products from the diet (Koletzko et al. 2012). As a consequence, plant-based milk is often suggested by doctors and nutritionists for the replacement of cow's milk (Singhal et al. 2017).

Lactose intolerance is another condition associated with the consumption of cow's milk. Intolerance to lactose, milk's main carbohydrate, is caused by the absence or reduction of the enzyme lactase, which is produced in the gastrointestinal system (Vanga, Raghavan 2018).

Lactose intolerance (LI) is characterized by multiple gastrointestinal symptoms, such as abdominal discomfort, nausea, abdominal pain, and flatulence, which are associated with poor digestion of lactose. Lactose poor digestion occurs when there is not enough lactase to properly hydrolyze lactose into glucose and galactose. Therefore, LI is a different condition from CMPA, though symptoms may be similar. While LI is caused by low production of lactase in the gastrointestinal tract, CMPA results from an abnormal immune response to the ingestion of cow's milk protein. The estimated prevalence of poor lactose digestion is 80% among Afro-Americans, 50% among Americans with Mexican origins and 15% among Caucasians. However, poor lactose digestion does not always manifest as LI. The real prevalence of LI is not yet known (Bailey et al. 2013), although some studies suggest prevalence of about 30% in the world population, which may vary according to the ethnicity, possibly reaching 50-80% in South Americans, Africans and Asians (Vassilopoulou, Efthymiou 2016; Misselwitz et al. 2013; Bailey et al. 2013).

CMPA and LI are the main reasons for the adoption of a dairy-free diet or low dairy-diet. Even though the use of exogenous lactase and/or consumption of lactose-free milk products is a possibility for LI patients, the complete exclusion of milk and dairy products is necessary for individuals who are allergic to cow's milk protein and individuals who wish to adopt a vegetarian diet (Bailey et al. 2013).

Plant-based milk is water-soluble extracts produced from oilseeds, cereals, pseudocereals, seeds and/or legumes. The production of these beverages depends on the raw material used and on a series of steps which may be generally described according to the following procedures: (i) the intact or grounded vegetable raw material is soaked for a few hours and then processed with water; (ii) the extract is filtered for the removal of insoluble residues. Other ingredients such as oils, flavors, sugar, and stabilizers may be added. Processes regarding stability, homogenization, and pasteurization are performed at the end, generating liquid extracts which may be colloidal suspensions or emulsions (Jeske et al. 2017; Makinen et al. 2015).

In the United States, plant-based milk produced from almond, cashew nut, coconut, hazelnut, oats, rice, and soy are the ones more commonly found in the market (Singhal et al. 2017). In general, the appearance of these beverages is similar to cow's milk, except for rice and oats plant-based milk. However, the taste of these beverages differs a lot from what is considered the usual taste of Occidental population (Makinen et al. 2016).

Nutritional properties of plant-based milk will depend on the quality and type of raw material used and also the processes to which raw materials are subjected (Makinen et al. 2016). Some of these beverages contain very low quantities of protein, calcium, and iron (Makinen et al. 2016) when compared to cow's milk (Singhal et al. 2017). Thus, low protein content along with low protein quality comprises one of the disadvantages of most plant-based milk (Jeske et al. 2017). Half of the samples analyzed by Jeske et al. (2017) contained less than 0.5% protein, and only soy extract reached similar values as to cow's milk (3.7%). In the study by Makinen et al. (2015), soy extract also displayed protein content (2.5%) comparable to cow's milk (3.32%). However, it is important to mention that about 14% of CMPA patients also present reactions to soy, which jeopardizes substitution of cow's milk by soy extract in these cases (Jeske et al. 2017).

Therefore, the market urges the development of new cow's milk substitutes with low allergenicity, good nutritional quality, and adequate protein content which are, at the same time, sensorial pleasure and accessible to consumers.

In this sense, ingredients derived from legumes become a versatile, attractive alternative for the development of new products. Legumes comprise a food group known by its high protein and fiber content and low glycemic index (Chandra-Hioe et al. 2016). Chickpea (Cicer arietinum L.) is a largely consumed legume around the world, due to its nutrients content and high acceptance (Rachwa-Rosiak et al. 2013).

Chickpea: Characterization

Chickpea is a worldwide consumed legume, which presents a high amount of protein, carbohydrates, fibers, and minerals such as zinc, calcium, and magnesium. It is cultivated in temperate and semi-arid regions, especially in Asia, Europe, Australia, and North

America. The world's larger producer is India, which provides about 66% of global production (Rachwa-Rosiak et al. 2013). Chickpea, a grain derived from the family Fabaceae with two variations (kabuli and desi), represents one of the most important legumes in the world (Rachwa-Rosiak et al. 2013; Gupta et al. 2016) and the second most consumed one, following only soy (Nascimento et al. 2016). The Desi variation is characterized by small size dark grains with a rough appearance cultivated mainly in semi-arid regions. The Kabuli variation comprises bright slightly bigger grains, and it is cultivated in temperate regions. Differences in appearance and chemical composition depending on the local of production and cultivation conditions (Rachwa-Rosiak et al. 2013). Chickpea is a versatile culinary ingredient which may be used to produce fermented products, fried and roasted food, salted or sweet. It is mainly consumed in India, Spain, and countries bathed by the Mediterranean Sea (Gupta et al. 2016). The grains may also be grounded for the production of flour, and then used for the preparation of soup, pasta, and bread (Nascimento 2016).

In regards to nutritional composition, chickpeas stand out due to its high content of good quality protein, carbohydrates, fibers, minerals (potassium, calcium, magnesium, copper, iron and zinc) and vitamins (carotenoids, thiamin and niacin) (Rachwa-Rosiak et al. 2013; Simsek et al. 2015; Malunga et al. 2014; Nascimento 2016). Moreover, results from the study by Johnson et al. (2005) point out that chickpeas display a low glycemic index (Rachwa-Rosiak et al. 2013; Johnson et al. 2005).

Thus, chickpea represents a good option for vegetarians due to its nutritional quality, and it may be consumed in its original form (whole grain) or without the bran, after heat treatment (Simsek et al. 2015).

The protein content of raw chickpea grains varies from 20.9% to 25.3%. The main proteins found in chickpea are globulins and albumins and, at a lower proportion, glutenins and prolamins. However, composition and protein amount vary according to variety and grain species, weather, and year season (Rachwa-Rosiak et al. 2013). There is no allergen officially registered in chickpea, as oppose to other legumes, such as peanut and soy (Cabanillas et al. 2017).

Isoleucine, leucine, lysine, methionine, phenylalanine, threonine, tryptophan, and valine are the most abundant amino acids in chickpea (Nascimento 2016). It has been reported that chickpea proteins display higher antioxidant capacity than white beans proteins, a feature primarily related to the amino acids profile. Legumes generally present essential amino acids like lysine but lack amino acids such as methionine and cysteine (Simsek et al. 2015). Chickpea protein quality resembles soy's (Malunga et al. 2014).

Moreover, chickpea contains polyunsaturated fatty acids, which comprises about 66% of total lipids, followed by monounsaturated fatty acids (19%), and saturated fatty acids (15%). Another compound present in chickpea is starch, a polymer with an important role regarding physical-chemical properties and digestibility of chickpea (Simsek et al. 2015).

Despite these positive nutritional features of chickpea, it is important to highlight that legumes contain antinutritional factors such as phytates and trypsin inhibitors, which affect nutrients bioavailability and also influence protein extraction. These factors may be inhibited with the application of procedures such as fermentation, germination, heat treatments, and the use of chelating agents (Jeske et al. 2017). Besides those procedures, soaking of the grains followed by water discard previous to cooking reduces the amount of anti-nutritional factors, including flatulence-causing compounds. Moreover, this technique speeds the cooking process (Nascimento 2016).

Chickpea in the Production of Plant-Based Milk

There are no studies in the literature regarding neither pure chickpea water-soluble extract nor chickpea extract associated with other compounds. Due to the protein content of this legume, the development of plant-based milk produced from chickpea becomes an important and innovative challenge for the food industry. Nutritional and sensorial evaluations of such beverage are necessary in order to provide consumers new alternatives for the substitution of cow's milk and soy extract.

Nonetheless, chickpea is darker and displays a more yellow tone when compared to cow's milk. Moreover, it has a characteristic legume flavor (Jeske et al. 2017). Thus, a combination of chickpea extract with other ingredients might be an alternative for the production of more pleasant beverages. In this sense, a coconut is an interesting option of a vegetal extract since it exhibits a similar color to cow's milk and desirable sensorial features. The combination of chickpea and coconut may be a potential strategy for sensorial improvement since coconut is generally well accepted by the population, which would generate a beverage with a pleasant taste, aroma, and color. Sensorial quality is one of the most important factors in the decision to buy a product (Jaekel et al. 2010). Therefore, it cannot be despised.

It is important to emphasize that the use of chickpea to produce vegetal milk was not previously studied. However, some studies show the application of the chickpea in the vegetarian diet as a good source of nutrients such as protein and iron replacing animal-based foods. This topic will be discussed in another chapter of this bool.

References

Bailey, R. K., Fileti, C. P., Keith, J., Tropez-Sims, S., Price, W. and Allison-Ottey, S. D. (2013). Lactose Intolerance and Health Disparities Among African Americans and

Hispanic Americans: An Updated Consensus Statement. *Journal of the National Medical Association,* 105(2), 112 - 127. doi:10.1016/s0027-9684(15)30113-9.

Cabanillas, B., Jappe, U. and Novak, N. (2017). Allergy to Peanut, Soybean, and Other Legumes: Recent Advances in Allergen Characterization, Stability to Processing and IgE Cross-Reactivity. *Molecular Nutrition and Food Research,* 62(1), 1700446.doi:10.1002/mnfr.201700446.

Chandra-Hioe, M. V., Wong, C. H. M. and Arcot, J. (2016). The Potential Use of Fermented Chickpea and Faba Bean Flour as Food Ingredients. *Plant Foods for Human Nutrition,* 71(1), 90 - 95. doi:10.1007/s11130-016-0532-y.

Gupta, R. K., Gupta, K., Sharma, A., Das, M., Ansari, I. A. and Dwivedi, P. D. (2016). Health Risks and Benefits of Chickpea (Cicer arietinum) Consumption. *Journal of Agricultural and Food Chemistry,* 65(1), 6 - 22. doi:10.1021/acs.jafc.6b02629.

Koletzko, S., Niggemann, B., Arato, A.; Dias, J. A.; Heuschkel, R.; Husby, S., Mearin, M. L., Papadopoulou, A., Ruemmele, M., Staiano, A.; Schappl, M. G., Vandenplas, Y. (2012) Diagnostic Approach and Management of Cow's-Milk Protein Allergy in Infants and Children: ESPGHAN GI Committee Practical Guidelines. *JPGN*, 55 (2), 221 - 229.

Jaekel, L. Z., Rodrigues, R. S. and Silva, A. P. (2010). Avaliação físico-química e sensorial de bebidas com diferentes proporções de extratos de soja e de arroz. *Ciên Tecnol. Aliment,* 30(2), 342 - 348.

Jeske, S., Zannini, E. and Arendt, E. K. (2017). Evaluation of Physicochemical and Glycaemic Properties of Commercial Plant-Based Milk Substitutes. *Plant Foods Hum. Nutr.*, 72(1), 26 - 33.

Makinen, O. E., Uniacke-Lowe, T., O'mahony, J. A., Arendt, E. K. (2015). Physicochemical and acid gelation properties of comercial UHT-treated plant based milk substitutes and lactose free bovine milk. *Food Chemistry*, 68, 630 - 638.

Makinen, O. E., Wanhalinna, V., Zannini, E., Arendt, E. K. (2016). Foods for Special Dietary Needs: Non-dairy Plant-based Milk Substitutes and Fermented Dairy-type Products. *Journal Critical Reviews in Food Science and Nutrition*, 56(1), 339 - 349.

Malunga, L. N., Bar-El, S. D., Zinal, E., Berkovich, Z., Abbo, S. and Reifen, R. (2014). The potential use of chickpeas in development of infant follow-on formula. *Nutrition Journal,* 13(1). doi:10.1186/1475-2891-13-8.

Misselwitz, B., Pohl, D., Frühauf, H., Fried, M., Vavricka, S. R. and Fox, M. (2013). Lactose malabsorption and intolerance: pathogenesis, diagnosis and treatment. *United European Gastroenterology Journal,* 1(3), 151 - 159. doi:10.1177/2050640613484463.

Nascimento, W. N. (2016). Hortaliças Leguminosas (1 ed). Brasília: Embrapa.

Rachwa-Rosiak, D., Nebesny, E. and Budryn, G. (2013). Chickpeas—Composition, Nutritional Value, Health Benefits, Application to Bread and Snacks: A Review.

Critical Reviews in Food Science and Nutrition, 55(8), 1137 - 1145. doi:10.1080/10408398.2012.687418.

Simsek, S., Herken, E. N. and Ovando-Martinez, M. (2015). Chemical composition, nutritional value andin vitrostarch digestibility of roasted chickpeas. *Journal of the Science of Food and Agriculture,* 96(8), 2896 - 2905. doi:10.1002/jsfa.7461.

Singhal, S., Baker, R. D. and Baker, S. S. (2017). A Comparison of the Nutritional Value of Cow's Milk and Nondairy Beverages. *J. Pediatr. Gastroenterol. Nutr.,* 64(5), 799 - 805.

Sethi, S., Tyagi, S. K. and Anurag, R. K. (2016). Plant-based milk alternatives an emerging segment of functional beverages: a review. *J. Food Sci. Technol.,* 53(9), 3408 - 3423.

Vanga, S. K. and Raghavan, V. (2018). How well do plant based alternatives fare nutritionally compared to cow's milk? *J. Food Sci. Technol.*, 55 (1), 10 - 20.

Vassilopoulou, E. and Efthymiou, D. (2016). Milk hypersensitivities: where is the grey line regarding their dietary management? *European Annals of Allergy and Clinical Immunology,* 48(5), 164 - 173.

In: Handbook of Chickpeas
Editors: Albert T. Lund and Noah D. Schultz
ISBN: 978-1-53616-374-2

Chapter 7

CHICKPEA APPLICATIONS IN A VEGETARIAN DIET

Bernardo Romão de Lima, Shila Minari Hargreaves, Renata Puppin Zandonadi and Raquel Braz Assunção Botelho
University of Brasília, Nutrition Department,
Brasília, Brazil

ABSTRACT

Vegetarianism has become an increasingly common practice around the world. A vegetarian diet excludes any foods containing meat, fish, and poultry. Since these foods are sources of vital nutrients such as protein, iron, calcium, zinc, Vitamin B12, and magnesium, adequate nutritional replacements in the diet are highlighted. Chickpeas (*Cicer arietinum*) present in their composition interesting amounts of protein, fibers, iron, zinc, and others, arousing interest by vegetarians. Therefore, the use of chickpeas has been studied as an important ingredient on a vegetarian diet to contribute to nutritional quality.

VEGETARIANISM: CONCEPT AND CONTEXT

Vegetarianism consists of a set of practices that culminate in the exclusion of meat, fish, and poultry from the dietary habit (Lynch, Johnston, & Wharton, 2018). While in vegetarian diets, consumers can maintain animal products such as dairy and eggs, plant-based and vegan diets allow only plant foods while for vegans, the restriction extends to products and brands (Chen, Chaudhary, & Mathys, 2019; Lynch et al., 2018).

The number of vegetarians varies around the world, in oriental countries, higher prevalence is found, possibly because more widespread religions condemn practices involving the sacrifice and consumption of animals or meat in general (Jathar, Inamdar-Deshmukh, Rege, & Satoskar, 1975; Shridhar et al., 2014). However, in occidental countries, the number of vegetarians tend to increase geometrically, given the advent of companies and marketing strategies involving this specific subject (Chen et al., 2019; Lynch et al., 2018).

There are several motivations for adherence to vegetarianism, such as animal liberation, environmental impact, religion, or search for healthier lifestyles (Barnard et al., 2019; Ginsberg, 2018).

As for animal liberation purposes, the suffering involved in the raising and slaughter of animals for consumption as food or even for testing of cosmetic and hygiene products serves as a primary motivation for the conversion into a vegetarian lifestyle (Le, Sabaté, Singh, & Jaceldo-Siegl, 2018; Lynch et al., 2018). The vegetarian lifestyle advocates that no animal should be harmed under any circumstance, therefore, both the sacrifice of animals for human consumption and the use of animals for testing of products is unjustified (Leitzmann, 2014).

With educational purposes, several international non-profit organizations disseminate information and orchestrate political movement regarding the theme, such as People for Ethical Treatment of Animals (PETA), Mercy for Animals (MFA) and in some instances, Greenpeace ®.

The current environmental context also serves as a motivation for vegetarians. Currently, the production and maintenance of animals in agriculture are responsible for the largest consumption of potable water and the second largest emission of polluting gases (Chen et al., 2019; Lynch et al., 2018). Also, the deforestation involved in the creation of inputs for beef cattle accounts for the majority of the destruction of native vegetation distributed throughout the world (Foley et al., 2011; Lynch et al., 2018).

In this manner, besides practices regarding better management of land use for planting and raising livestock, reducing the meat consumption and therefore the number of resources involving cattle raising might be useful to reduce the unfavorable impacts of animal-based agriculture (Foley et al., 2011; Lynch et al., 2018).

Some religions advocate the total or partial exclusion of animal-based products. In India, most of the population is composed of vegetarians, given that the common religion condemns the sacrifice of cattle for feeding, while the allowed animals present higher costs, and are inaccessible for most of the population (Jathar et al., 1975; Leitzmann, 2014). In this context, Buddhism, common in countries of Eastern origin, preaches that animals are endowed with their own "spiritual energy" and that in this sense should not be consumed (Leitzmann, 2014).

As for health benefits, vegetarian diets have proven to exert multiple desirable effects regarding longevity, sports performance, cardiac health and prevention and maintenance of diabetes (Barnard et al., 2019; Chang, Chiu, Chang, Lin, & Lin, 2019; Lynch et al., 2018; Shridhar et al., 2014). Potentially, given that plant-based foods are naturally lower in calories, saturated fats, and refined sugars, diseases regarding the unbalance of these nutrients might also be prevented (Barnard et al., 2019; Gili et al., 2019).

However, nutritional shortcomings regarding protein, calcium, zinc, folate and vitamin B12 are more prone to happen in vegetarian individuals, given that foods with either higher quantities of these nutrients or better bioavailability are usually animal-based (Lynch et al., 2018). In this manner, to achieve long-life success within a vegetarian diet, nutritional orientation is often needed. Also, to prioritize plant-based foods with better nutritional quality is necessary as well (Gili et al., 2019; Leitzmann, 2014).

Nutritional Quality of Chickpea for Vegetarians

Legumes are considered relevant sources of macro and micronutrients in vegetarian diets. They are seen as potential substitutes for animal origin food, mainly due to their high protein content (Erbersdobler, Barth, & Jahreis, 2017). Moreover, they are rich in fibers and antioxidants, and their consumption is associated with health-promoting effects. Chickpeas, which are included in this food category, are known for their high nutritional quality. Regular addition of chickpeas in vegetarian diets can contribute to a more balanced diet and bring health benefits (Bar-El Dadon, Abbo, & Reifen, 2017).

Table 1 presents the macronutrient content of chickpeas. Carbohydrates account for the majority of the grain's dry-weight basis, being most of it represented by starch. The starch digestibility is one of the critical factors that influence glycemic response to determine food intake, mainly due to the proportion of resistant starch, which represents the undigestible and fermentable starch fraction. In raw chickpeas, resistant starch accounts for 24.15% of the total starch, but cooking the grains can increase this proportion to 25.63% and up to 30.20%, depending on the cooking method. The high resistant starch proportion, along with high fiber and protein content in chickpeas, is responsible for their low estimated glycemic index, which ranges from 38.67 to 41.28 (Simsek, Herken, & Ovando-Martinez, 2016).

Moreover, resistant starch is associated with better gut health because of its prebiotic effect. The colon bacteria ferment it into short-chain fatty acids, which are known as improving the microbiome, decrease gut inflammation, and prevent large bowel cancer (Birt, Brown, & Topping, 2000). It has been shown that vegetarians have different gut bacteria profile when compared to omnivores, which is associated with their higher production of short-chain fatty acids in the gut and lower survival of putrefactive proteolytic bacteria (Zimmer et al., 2012). Including chickpeas as a source of fiber and resistant starch in vegetarian diets can, therefore, contribute to the maintenance of gut´s health.

Besides starches, chickpeas are also a source of oligosaccharides, a group of non-digestible fermentable carbohydrates. The total content ranges from 3.87 to 6.98g/100g of chickpea on dry matter, and they include raffinose, stachyose, ciceritol, and verbascose (Rachwa-Rosiak, Nebesny, & Budryn, 2015b). Because of the gut bacteria fermentation of oligosaccharides, eating chickpeas could cause abdominal discomfort. However, different cooking methods, as well as germination, can significantly reduce or eliminate the oligosaccharides content in chickpeas, improving its digestibility (Alajaji & El-Adawy, 2006; El-Adawy, 2002). Moreover, oligosaccharides intake stimulates good bacteria growth in the gut, leading to a healthier microbiome. Since they are non-digestible carbohydrates, their presence in food slows glucose release in the blood and, therefore, contributes to type II diabetes control. Other potential benefits associated with oligosaccharides intake include improving immunity and cardiovascular health and reducing constipation (Qiang, YongLie, & QianBing, 2009). Therefore, vegetarians can benefit from the oligosaccharides content in chickpeas. If there is excessive abdominal discomfort after the intake, germinating the seeds before cooking can be a good strategy to improve tolerability.

The high fiber content is another positive aspect of chickpeas since it can prevent constipation by promoting regular bowel movement. Moreover, a higher intake of dietary fiber is associated with a lower risk of developing diabetes, cardiovascular diseases, and colon-rectal cancer. The higher intake of legumes because of their high fiber content, among other beneficial components, is one of the aspects from vegetarian diets that contribute to their lower risk of non-communicable diseases (Marsh, Zeuschner, Saunders, & Zeuschner, 2012). Fiber on the gut has the potential benefit to the microbiome, which results in lower systemic inflammation levels. Another positive result obtained by high fiber consumption is the increased satiety effect correlate with less energy intake and obesity control. In fact, comparison between meals based in meat or vegetable protein (from legumes) showed that a vegetable protein based meal (with equivalent calories, fat and protein amounts) was more satiating when compared to an animal protein based meal, and this difference is attributed to its higher fiber content (Kristensen, Bendsen, Christensen, Astrup, & Raben, 2016). Therefore, the vegetarian diet seems to be a good strategy to improve overall health and to help control energy intake, contributing to obesity prevention, especially when the diet is enriched with fiber-rich foods such as chickpeas.

Table 1. Chemical composition of raw chickpeas[a]

	g/100g (dry-weight basis)
Total carbohydrates	52-71
Sugars	3-5
Starch	30-56
Amylose	23-28
Dietary fiber	6-15
Insoluble fiber	10-18
Soluble fiber	1-8
Proteins	19-27
% of total proteins	
Albumins	8-12
Globulins	53-60
Glutenin	18-24
Prolamin	3-7
Total Fats	2-7
% of total fats	
Palmitic	14
Stearic	2
Oleic	25
Linoleic	57
Linolenic	2
Others	1

[a]Reference: (Hall, Hillen, & Robinson, 2017).

The protein content of chickpeas ranges from 19 to 27% (of dry-weight matter), and its regular intake can significantly contribute to increase protein levels on a vegetarian diet. Moreover, prolamin content is very low (Table 1); it does not form gluten. Chickpea is considered gluten-free and suitable for celiac (Mlyneková, Chrenková, & Formelová, 2016). However, despite high protein concentration found in legumes, there is a concern in the vegetarian diet related to the lower biological value of vegetable proteins because of more reduced digestibility and incomplete essential amino acids profile in most of them. Modest differences can be observed in the amino acid content of chickpea after cooking. Since legumes are not consumed in the raw form, the amino acid content is displayed in Table 2, according to what is found in boiled chickpea.

In the case of chickpeas, total sulfur amino acids and valine are present in slightly lower amounts when compared to recommendations (Table 2). Legumes, in general, have a similar amino acid profile. However, it is already well established that, as long as different protein sources are consumed during the day and caloric intake is adequate, having limited amino acids in one type of food does not negatively interfere in the total amino acid pool. It is possible, then, to have chickpea as a good protein source in vegetarian diets, as long as other protein sources are consumed throughout the day (Baines, 2013a).

Table 2. Amino acid content of boiled chickpea seeds[a]

Amino acid	**g/16gN**	**FAO/WHO (1973)**
Isoleucine	4.1	4.0
Leucine	7.4	7.0
Lysine	7.5	5.5
Cystine	1.1	-
Methionine	1.4	-
Total sulfur amino acids	2.5	3.5
Tyrosine	3.6	-
Phenylalanine	5.8	-
Total aromatic amino acids	9.4	6.0
Threonine	4.3	4.0
Tryptophan	1.0	1.0
Valine	3.6	5.0
Total essential amino acids	39.8	36.0
Histidine	3.2	-
Arginine	10.1	-
Aspartic acid	11.2	-
Glutamic acid	18.5	-
Serine	4.5	-
Proline	4.3	-
Glycine	4.5	-
Alanine	3.9	-
Total non-essencial amino acids	60.2	-
Leucine/isoleucine ratio	1.8:1	1.8:1

[a]Reference: (Alajaji & El-Adawy, 2006).

On the other hand, chickpeas are rich in isoleucine, lysine, total aromatic amino acids, and tryptophan. Having good sources of lysine is important for vegetarians because cereals, which compose the basis of most diets worldwide, are lysine deficient (Marinangeli & House, 2017). Therefore, vegetarians can be at risk of lysine deficiency when their diet is mainly based in cereals, like wheat, rice, or corn. However, this issue can be solved by only including a good source of lysine, like chickpeas, on their diet. Tryptophan, also found in chickpeas in a good proportion, is used by the body to produce niacin (vitamin B3) and the neurotransmitter serotonin, responsible for regulating mood and sleep, modulate appetite and improve memory (Badawy, 2017; Berger, Gray, & Roth, 2009).

Moreover, having a balanced gut microbiome has been shown to influence the production of serotonin and other neurotransmitters, such as dopamine and GABA, and it can positively impact the brain-gut axis (Strandwitz, 2018). Besides providing tryptophan, chickpeas are also rich in fibers and resistant starch, which have a beneficial effect on the gut microbiome. These factors can contribute to positive neuronal effects, such as improving mood and reducing the risk of depression. Some studies show that vegetarians have a better mood and less anxiety and stress levels when compared to omnivores (Beezhold, Johnston, & Daigle, 2010; Beezhold, Radnitz, Rinne, & DiMatteo, 2014).

The protein digestibility of chickpeas varies depending on the cooking method, being 88.52% after boiling, 89.40% after microwaving, and 89.96% after autoclaving (pressure cooking) (Alajaji & El-Adawy, 2006). Similarly, other legumes' protein digestibility varies from 89.4 to 95.9% (Erbersdobler et al., 2017). The protein digestibility of chickpeas values is similar to the ones observed in animal proteins, such as chicken breast (86.9%), salmon fillet (87.5%) or beef loin (89.1%) (Faber et al., 2010). Therefore, regardless of the cooking method, it can be considered that chickpeas are a good protein source for vegetarians, being well digested after cooking, similarly to other vegetable or animal protein sources.

Despite the lower presence of some essential amino acids in chickpea when compared to the recommendations (FAO/WHO, 1973), it is already known that the intake of different protein sources throughout the day is sufficient to provide all the essential amino acids in adequate amounts to support vegetarian's health. Therefore, vegetarians can use chickpeas as a reliable source of protein in their diets, as long as other sources of protein, such as cereals, seeds, nuts, and vegetables, are also included (Baines, 2013a).

The lipid content of chickpeas ranges from 2 to 7% of the dry matter, much lower when compared to animal protein sources, such as beef loin (19.7%), pork loin (19.7%) and chicken breast (12.7%), and comparable to some fish, such as salmon (7.9%) and Pollock (4.8%)(Qiang et al., 2009). Chickpeas also provide less proportion of saturated fat (16% of total fat content, which represents 0.3% to 1.1% of dry matter) when compared to red meat (4.8% in pork loin; 5.8% in beef loin) and poultry (3.2% in the chicken breast) (Faber et al., 2010). In contrast, monounsaturated fat accounts for 25% of total fat content (0.5% to 1.8% of dry matter) and poly-unsaturated fat, for most of the fat content (59%), which represents 1.2% to 4.1% of dry matter. A higher proportion of unsaturated to saturated fat is associated with better health outcomes, lowering the risk for cardiovascular diseases and type 2 diabetes (Hooper, Martin, Abdelhamid, & G, 2015; Summers et al., 2002). Therefore, having chickpeas as a regular diet component can bring health benefits to those who follow a vegetarian diet.

Chickpea is also an important source of minerals, in which the concentrations vary depending on growing conditions and environment. One hundred grams of dry chickpeas provide between 82 to 272mg of calcium, which represents 8 to 27% of the daily recommended intake. Vegetarians, especially vegans, have been shown to have lower intakes of calcium (Tucker KL, 2014), being chickpea a good option to help achieve the daily recommendations. Iron, a potential limiting nutrient in vegetarian diets (Haider,

Schwingshackl, Hoffmann, & Ekmekcioglu, 2016), is also present in chickpeas in high concentrations, ranging from 4.6 to 7.5mg in 100g (equivalent to 57.5-93.7% of daily recommendations for men and 25.6 to 41.7% for women) (Hall et al., 2017). The absorption of non-heme iron is lower when compared to heme iron present in meat.

On the other hand, high consumption of heme iron is associated with higher inflammation levels and chronic diseases (Haider et al., 2016). Therefore, using vegetable sources, such as chickpeas, to achieve iron needs can be considered a healthier option when compared to animal products. An efficient way to increase non-heme iron absorption and guarantee optimal iron levels is to include vitamin C sources in the same meal, on a proportion of 20mg to vitamin C for every 3mg of iron (Teucher, Olivares, & Cori, 2004), which is common on a balanced vegetarian diet.

Zinc, together with iron, is a nutrient that requires attention in vegetarian diets. Because of the high content of phytates and polyphenols found in vegetarian diets, zinc bioavailability tends to be lower, which requires special care to achieve optimal levels in the diet (Baines, 2013b). Chickpeas provide 3.3 to 4.4mg of zinc per 100g of dry weight, and it can be considered a good source when compared animal products such as eggs (1.2mg/100g), pork fillet (2.4mg/100g) and beef fillet (7.8mg/100g) (Baines, 2013b).

Chickpea is also a good source of magnesium (147-195mg/100g), potassium (994-1264mg/100g) and phosphorus (394-452mg/100g). These minerals do not seem to be limited in vegetarian diets (Rizzo, Drph, Drph, & Fraser, 2013). It is important, however, to guarantee good sources of these nutrients when planning a vegetarian diet. The intake of chickpea can, therefore, help to achieve an optimal intake of minerals.

Cooking can reduce the mineral content in chickpeas, depending on the mineral and the cooking method. The following losses have been observed for boiling, autoclaving and microwave cooking, for each mineral, respectively: 20.8%, 10.0% and 9.6% for zinc; 11.8%, 8.0% and 5.4% for iron; and 29.5%, 25.6% and 25.6% for calcium. Therefore, microwave cooking can be considered a good choice for chickpeas to reduce cooking losses and guarantee higher minerals intake on a vegetarian diet (Alajaji & El-Adawy, 2006). In contrast, germinating chickpeas has only a slight effect on mineral losses and even an increase in the content of iron (2.5%) and zinc (14.1%). Moreover, despite small nutrient losses, cooking also helps to reduce anti-nutrients, such as phytates, which helps to increase all minerals' absorption in the gut (Alajaji & El-Adawy, 2006; El-Adawy, 2002).

Chickpeas also present proper amounts of B-complex vitamins. The folate content varies between 42 and 537μg/100g. Raw chickpeas also provide 453μg/100g of thiamine, 173μg/100g of riboflavin, and between 1602 and 3900 μg/100g of niacin (Alajaji & El-Adawy, 2006; El-Adawy, 2002; Hall et al., 2017). However, B vitamins are susceptible to heat, and cooking chickpeas can significantly reduce their content. Folate, for example, can be reduced by 47% during boiling and 38% during pressure cooking. The other vitamins can reduce by 40–96%, depending on the type and cooking method, being microwave cooking the method which results in better vitamins retention, followed by autoclaving. Boiling presents higher vitamin losses (El-Adawy, 2002; Hall et al., 2017). Germination, on the other hand, was shown to increase some vitamins content, such as riboflavin (16.1% increase) and pyridoxine (3.6% increase). Vitamin C, for example, is absent in raw chickpeas, but the germinated ones have 20mg/100g (Hall et al., 2017), which is important to the iron absorption in vegetarian diets.

Chickpeas can provide a vast range of antioxidants, which are known to be beneficial to health promotion and non-communicable diseases. Chickpeas tocopherols' contentranges from 5 to 12mg/100g (79% of which being of γ-tocopherol), and their total carotenoid content is 0.9–3.1 mg/100g, of which around 29μg are from β-carotene. Moreover, chickpeas contain a good level of phytosterols, when compared to other legumes. Phytosterols are known for their potential health benefits due to the serum cholesterol-lowering effect (Duester, 2001). One hundred grams of chickpeas provides 205mg of phytosterols, being mostly (81%) β-sitosterol (Hall et al., 2017). For example, avocados are known for their high β-sitosterol content presenting on average 76mg for every 100g (Duester, 2001). Therefore, chickpeas can be considered a reliable source of β-sitosterol, mainly because its content is not affected by cooking (Kaloustian et al., 2008).

Legumes are also related to health benefits due to their phenolic content. Soybeans, for example, are rich in isoflavones (such as daidzein and genistein), a group of phenolic compounds with antioxidant and anti-inflammatory properties, which can bring potential benefits to cardiovascular health, obesity control, diabetes, and cancer prevention (Camargo et al., 2019; Champ, 2002). However, there has been increased concern regarding soybean consumption, mainly due to the environmental impact of its large-scale production, genetic modification, and allergenicity (Bakshi, 2003; Savage, Kaeding, Matsui, & Wood, 2010). Chickpeas are rich in isoflavones as well (153 to 340 mg/100g), in similar amounts when compared to soybeans (165 to 336 mg/100g) (Camargo et al., 2019), being, therefore, a good and alternative source of isoflavones for those who need or intend to restrict soy intake. Besides isoflavones, chickpeas are also a source of other phenolic compounds, such as hydroxybenzoic acids, hydroxycinnamic acids, and catechins, which contribute to their antioxidant activity (Hall et al., 2017; Mekky et al., 2015).

Legumes are rich in anti-nutritional factors, which interfere in digestion and can reduce nutrients absorption. They include enzyme inhibitors, haemagglutinins (lectins), oxalate, saponins, phytic acid, and phenolic compounds (Champ, 2002). Most of them, however, can be deactivated during cooking. Haemagglutinin is eliminated by cooking, while trypsin inhibitor is reduced by at least 80%. Cooking also reduced tannins (48.04–50.10%), phytic acid (28.93–41.32%) and saponins (43.96–51.65%) in chickpeas (El-Adawy, 2002). Moreover, following a vegetarian diet can lead to changes in the microbiota that result in better nutrient absorption. A vegetarian diet can shape the gut microbiota by increasing the number of bacteria capable of degrading phytate (Markiewicz, Honke, Haros, Światecka, & Wróblewska, 2013). Therefore, these adaptations can reduce the potential negative effects that phytate consumption could have on a vegetarian diet.

Despite limiting nutrients absorption, anti-nutrient factors are also associated with better health outcomes, as previously discussed regarding phenolic compounds. Besides them, protease inhibitors (such as trypsin inhibitor), phytic acid, lignans, and saponins have also been shown to have anticarcinogenic properties. Blood glucose control has been associated with phytic acid, lectins, amylase inhibitor, and tannins, and blood lipid control, with saponins, phytic acid, isoflavones, and phytosterols, as well as the fiber content found in legumes (Champ, 2002).

Chickpeas present a great nutrition profile, primarily because of their protein and fiber contents. Anti-nutrients factors have been associated with properties that can improve health. Therefore, this legume can be included in diets, and more importantly, in a vegetarian diet as a preparation or as an ingredient of different recipes to vary nutrient intake.

Chickpeas as an Ingredient of Vegetarian and Vegan Products

Given its high protein and starch levels, chickpeas can be successfully implemented as a replacement for animal-based ingredients - such as dairy and eggs – it can also be used to enrich the nutritional value of different food products (Mustafa, He, Shim, & Reaney, 2018; Patil, Kote, & Mulimani, 2009; Wallace, Murray, & Zelman, 2016).

In their starch composition, chickpeas present 29.2% of amylose and 31.9% of resistant starch, large oval, and small round granules with accentuated capacity to get hydrated and swollen with also optimized hot viscoelasticity (Polesi et al., 2011). Its gelatinization peak temperature is around 64ºC, comparable to egg white's coagulation one, that is around 60ºC (Polesi et al., 2011).

Due to these rheological proprieties, chickpea starch is implemented in the market as a "vegan egg replacer", a product based of chickpea flour in combination with other stabilizing agents, such as xanthan and guar gums and isolated rice protein (Jendrikus, Francisco, Oliveira, & Francisco, 2014).

In bakery products, such as cake batters, chicken eggs are responsible for structure, tenacity and especially growth capacity (Montebello, Araújo, & Botelho, 2014), and with its absence, these characteristics are often impaired. When in contact with the liquids of the cake batter, the "vegan egg replacer" starches get hydrated and viscoelastic and, in the cooking process, chickpea's proteins get coagulated, thus forming a structure able to retain most of the gas produced, resulting in satisfactory egg-free bakery products (Jendrikus et al., 2014).

Aquafaba (produced with chickpea's waste cooking water) has shown to improve the technological aspects of egg-free bakery products as well. In the cooking process, given the high temperature and pressure needed to cook chickpea grains efficiently, the external fiber structure collapses resulting in extravasation of starch and protein fractions into the cooking water (Montebello et al., 2014; Rachwa-Rosiak, Nebesny, & Budryn, 2015a).

In a similar way to egg white's albumin, the protein fractions in *aquafaba* are capable of forming a stable structure when in physical stress - such as a food mixer rotation -, that can retain the incorporated air bubbles, resulting in a foam-like ingredient (Montebello et al., 2014). In preparations, this incorporated air in *aquafaba* contributes to texture, height, and final color, thus, being a satisfactory replacement for egg whites (Buhl, Christensen, & Hammershøj, 2019; Mustafa et al., 2018).

Mustafa *et al.* (2018) produced sponge cakes with *aquafaba* as an egg white replacer, utilizing ten different brands of canned cooked chickpeas as an *aquafaba* source. Different stabilities were found within the used brands. However, every sponge cake made with *aquafaba* presented satisfactory sensory aspects with no difference when compared to the ones with egg whites (Mustafa et al., 2018). Cakes prepared with *aquafaba* showed to be less springy and cohesive, highlighting the need for other stabilizing agents.

Also as an egg substitute, *aquafaba* can be implemented to make desserts such as custards, pudding, mousses and ice cream and serve as an emulsion agent in the making of vegan mayonnaise (Buhl et al., 2019; Mustafa et al., 2018).

As a strategy for replacing dairy milk, the water-soluble portions of different cereals and nuts are often commercialized as plant "milk" (Mäkinen, Wanhalinna, Zannini, & Arendt, 2016). However, given the predominance of carbohydrates in most of the sources, protein levels are usually prejudiced and incomparable with traditional dairy milk.

Legumes such as soy, yellow split, and pigeon peas have been used as sources for making protein-richer dairy milk alternatives, and apart from protein, satisfactory levels of iron, calcium, magnesium, and phosphorus were found (Mäkinen et al., 2016; Patil et al., 2009). Chickpea can be a source for plant "milk". However, high levels of indigestible fermentable oligosaccharides, related to abdominal distention and reduced nutrient bioavailability were found, suggesting that chickpeas cannot be used as a single source for making this product (Patil et al., 2009).

Chickpeas might also be used to improve the nutritional quality of products. With naturally fewer carbohydrates than common-use wheat flour, and higher levels of protein, fat, dietary fiber and micronutrients, when implemented in crackers, tart crusts, cookies, cakes and muffins, chickpeas were able to raise the nutritional value of these preparations, either in total or partial replacement (Alvarez, Herranz, Fuentes, Cuesta, &

Canet, 2017; Gómez, Oliete, Rosell, Pando, & Fernández, 2008; Rachwa-Rosiak et al., 2015a). In the making of wheat or gluten-free pasta, the utilization of chickpeas has raised the total protein and dietary fiber content, being an efficient strategy to improve the nutritional quality of pasta (de Lima, Botelho, & Zandonadi, 2017; Goñi & Valentín-Gamazo, 2003; Padalino, Mastromatteo, Lecce, Cozzolino, & Del Nobile, 2013).

In addition to the use of chickpeas as replacements or enrichments, culturally this legume is very present in a vegetarian/vegan lifestyle, being base of different preparations, such as "burgers" and hummus, characterizing a healthy and important food for these individuals (Wallace et al., 2016).

References

Alajaji, S. A., & El-Adawy, T. A. (2006). Nutritional composition of chickpea (Cicer arietinum L.) as affected by microwave cooking and other traditional cooking methods. *Journal of Food Composition and Analysis*, *19*(8), 806–812. https://doi.org/10.1016/j.jfca.2006.03.015.

Alvarez, M. D., Herranz, B., Fuentes, R., Cuesta, F. J., & Canet, W. (2017). Replacement of Wheat Flour by Chickpea Flour in Muffin Batter: Effect on Rheological Properties. *Journal of Food Process Engineering*, *40*(2), 1–13. https://doi.org/10.1111/jfpe.12372.

Badawy, A. A. B. (2017). Kynurenine pathway of tryptophan metabolism: Regulatory and functional aspects. *International Journal of Tryptophan Research*, *10*(1). https://doi.org/10.1177/1178646917691938.

Baines, S. K. (2013a). Protein and vegetarian diets. *The Medical Journal of Australia*, *199*(4), S7–S10. https://doi.org/10.5694/mjao11.11492.

Baines, S. K. (2013b). Zinc and vegetarian diets. *The Medical Journal of Australia*, *199*(4), S17–S21. https://doi.org/10.5694/mjao11.11493.

Bakshi, A. (2003). Potential adverse health effects of genetically modified crops. *Journal of Toxicology and Environmental Health - Part B: Critical Reviews*, *6*(3), 211–225. https://doi.org/10.1080/10937400306469.

Bar-El Dadon, S., Abbo, S., & Reifen, R. (2017). Leveraging traditional crops for better nutrition and health - The case of chickpea. *Trends in Food Science and Technology*, *64*, 39–47. https://doi.org/10.1016/j.tifs.2017.04.002.

Barnard, N. D., Goldman, D. M., Loomis, J. F., Kahleova, H., Levin, S. M., Neabore, S., & Batts, T. C. (2019). Plant-based diets for cardiovascular safety and performance in endurance sports. *Nutrients*, *11*(1), 1–10. https://doi.org/10.3390/nu11010130.

Beezhold, B. L., Johnston, C. S., & Daigle, D. R. (2010). Vegetarian diets are associated with healthy mood states: a cross-sectional study in Seventh Day Adventist adults. *Nutrition Journal*, *9*(1), 26. https://doi.org/10.1186/1475-2891-9-26.

Beezhold, B. L., Radnitz, C., Rinne, A., & DiMatteo, J. (2014). Vegans report less stress and anxiety than omnivores. *Nutritional Neuroscience*, *18*(7), 1–10. https://doi.org/10.1179/1476830514Y.0000000164.

Berger, M., Gray, J. A., & Roth, B. L. (2009). The Expanded Biology of Serotonin. *Annual Review of Medicine*, *60*(1), 355–366. https://doi.org/10.1146/annurev.med.60.042307.110802.

Birt, A. R., Brown, I. L., & Topping, D. L. (2000). Starches, Resistant Starches, the Gut Microflora and Human Health. *Current Issues in Intestinal Microbiology*, *1*(§), 25–37..

Buhl, T. F., Christensen, C. H., & Hammershøj, M. (2019). *Aquafaba as an egg white substitute in food foams and emulsions: Protein composition and functional behavior. Food Hydrocolloids*. Elsevier Ltd. https://doi.org/10.1016/j.foodhyd.2019.05.041.

Camargo, A. C. De, Favero, B. T., Morzelle, M. C., Franchin, M., Alvarez-parrilla, E., Rosa, L. A. De, … Schwember, R. (2019). *Is Chickpea a Potential Substitute for Soybean ? Phenolic Bioactives and Potential Health Benefits*..

Champ, M. M.-J. (2002). Non-nutrient bioactive substances of pulses. *British Journal of Nutrition*, *88*(S3), 307–319. https://doi.org/10.1079/bjn2002721.

Chang, C.-M., Chiu, T. H. T., Chang, C.-C., Lin, M.-N., & Lin, C.-L. (2019). Plant-Based Diet, Cholesterol, and Risk of Gallstone Disease: A Prospective Study. *Nutrients*, *11*(2), 335. https://doi.org/10.3390/nu11020335.

Chen, C., Chaudhary, A., & Mathys, A. (2019). Dietary Change Scenarios and Implications for Environmental, Nutrition, Human Health and Economic Dimensions of Food Sustainability. *Nutrients*, *11*(4), 856. https://doi.org/10.3390/nu11040856.

de Lima, B. R., Botelho, R. B. A., & Zandonadi, R. P. (2017). Gluten-Free Pasta: Replacing Wheat with Chickpea. *Journal of Culinary Science & Technology*, *00*(00), 1–8. https://doi.org/10.1080/15428052.2017.1394952.

Duester, K. C. (2001). Avocado fruit is a rich source of beta-sitosterol. *Journal of the American Dietetic Association*. https://doi.org/10.1016/S0002-8223(01)00102-X.

El-Adawy, T. A. (2002). Nutritional composition and antinutritional factors of chickpeas (cicer arietinum L.) undergoing different cooking methods and germination. *Plant Foods for Human Nutrition*, *57*(1), 83–97. https://doi.org/10.1023/A:1013189620528.

Erbersdobler, H. F., Barth, C. A., & Jahreis, G. (2017). Legumes in human nutrition: Nutrient content and protein quality of pulses Fatty acid distribution. *Science & Research | Overview 140 Ernaehrungs Umschau International*, *10*(October). https://doi.org/10.4455/eu.2017.038.

Faber, T. A., Bechtel, P. J., Hernot, D. C., Parsons, C. M., Swanson, K. S., Smiley, S., & Fahey, G. C. (2010). Protein digestibility evaluations of meat and fish substrates using laboratory, avian, and ileally cannulated dog assays. *Journal of Animal Science*, *88*(4), 1421–1432. https://doi.org/10.2527/jas.2009-2140.

FAO/WHO. (1973). *Energy and Protein Requirements*. Geneva..

Foley, J. A., Ramankutty, N., Brauman, K. A., Cassidy, E. S., Gerber, J. S., Johnston, M., Zaks, D. P. M. (2011). Solutions for a cultivated planet. https://doi.org/10.1038/nature10452.

Gili, R. V., Leeson, S., Montes-Chañi, E. M., Xutuc, D., Contreras-Guillén, I. A., Guerrero-Flores, G. N., Pacheco, S. O. S. (2019). Healthy vegan lifestyle habits among argentinian vegetarians and non-vegetarians. *Nutrients*, *11*(1), 1–18. https://doi.org/10.3390/nu11010154.

Ginsberg, C. (2018). The Market for Vegetarian Foods. *The Vegetarian Resource Guide*, 1–9..

Gómez, M., Oliete, B., Rosell, C. M., Pando, V., & Fernández, E. (2008). Studies on cake quality made of wheat-chickpea flour blends. *LWT - Food Science and Technology*, *41*(9), 1701–1709. https://doi.org/10.1016/j.lwt.2007.11.024.

Goñi, I., & Valentín-Gamazo, C. (2003). Chickpea flour ingredient slows glycemic response to pasta in healthy volunteers. *Food Chemistry*, *81*(4), 511–515. https://doi.org/10.1016/S0308-8146(02)00480-6.

Haider, L. M., Schwingshackl, L., Hoffmann, G., & Ekmekcioglu, C. (2016). The effect of vegetarian diets on iron status in adults: A systematic review and meta-analysis. *Critical Reviews in Food Science and Nutrition*. https://doi.org/10.1080/10408398.2016.1259210.

Hall, C., Hillen, C., & Robinson, J. G. (2017). Composition, nutritional value, and health benefits of pulses. *Cereal Chemistry*, *94*(1), 11–31. https://doi.org/10.1094/CCHEM-03-16-0069-FI.

Hooper, L., Martin, N., Abdelhamid, A., & G, D. S. (2015). Reduction in saturated fat intake for cardiovascular disease (Review). *Cochrane Database of Systematic Reviews*, (6). https://doi.org/10.1002/14651858.CD011737.www.cochranelibrary.com.

Jathar, V. S., Inamdar-Deshmukh, A. ., Rege, D. ., & Satoskar, R. (1975). Vitamin B12 and Vegetarianism in India. *Acta Haematologica*, *6*(2), 103..

Jendrikus, J., Francisco, S., Oliveira, M. A., & Francisco, S. (2014). Plant-based egg substitute and method of manufacture. *United States Patent Application Publication*, *1*(19).

Kaloustian, J., Alhanout, K., Amiot-Carlin, M. J., Lairon, D., Portugal, H., & Nicolay, A. (2008). Effect of water cooking on free phytosterol levels in beans and vegetables. *Food Chemistry*, *107*(4), 1379–1386. https://doi.org/10.1016/j.foodchem.2007.09.061.

Kristensen, M. D., Bendsen, N. T., Christensen, S. M., Astrup, A., & Raben, A. (2016). Meals based on vegetable protein sources (beans and peas) are more satiating than meals based on animal protein sources (veal and pork) - A randomized cross-over meal test study. *Food and Nutrition Research*, *60*, 1–9. https://doi.org/10.3402/fnr.v60.32634.

Le, L. T., Sabaté, J., Singh, P. N., & Jaceldo-Siegl, K. (2018). The design, development and evaluation of the vegetarian lifestyle index on dietary patterns among vegetarians and non-vegetarians. *Nutrients*, *10*(5), 25–27. https://doi.org/10.3390/nu10050542.

Leitzmann, C. (2014). Vegetarian nutrition: past, present, future. *The American Journal of Clinical Nutrition*, *100*(suppl_1), 496S-502S. https://doi.org/10.3945/ajcn.113.071365.

Lynch, H., Johnston, C., & Wharton, C. (2018). Plant-Based Diets: Considerations for Environmental Impact, Protein Quality, and Exercise Performance. *Nutrients*, *10*(12), 1841. https://doi.org/10.3390/nu10121841.

Mäkinen, O. E., Wanhalinna, V., Zannini, E., & Arendt, E. K. (2016). Foods for Special Dietary Needs: Non-dairy Plant-based Milk Substitutes and Fermented Dairy-type Products. *Critical Reviews in Food Science and Nutrition*, *56*(3), 339–349. https://doi.org/10.1080/10408398.2012.761950.

Marinangeli, C. P. F., & House, J. D. (2017). Potential impact of the digestible indispensable amino acid score as a measure of protein quality on dietary regulations and health. *Nutrition Reviews*, *75*(8), 658–667. https://doi.org/10.1093/nutrit/nux025.

Markiewicz, L. H., Honke, J., Haros, M., Światecka, D., & Wróblewska, B. (2013). Diet shapes the ability of human intestinal microbiota to degrade phytate - in vitro studies. *Journal of Applied Microbiology*, *115*(1), 247–259. https://doi.org/10.1111/jam.12204.

Marsh, K., Zeuschner, C., Saunders, A., & Zeuschner, C. (2012). Health Implications of a Vegetarian Diet : A Review. *American Journal of Lifestyle Medicine*, *6*(3), 250–267. https://doi.org/10.1177/1559827611425762.

Mekky, R. H., Contreras, M. D. M., El-Gindi, M. R., Abdel-Monem, A. R., Abdel-Sattar, E., & Segura-Carretero, A. (2015). Profiling of phenolic and other compounds from Egyptian cultivars of chickpea (Cicer arietinum L.) and antioxidant activity: A comparative study. *RSC Advances*, *5*(23), 17751–17767. https://doi.org/10.1039/c4ra13155j.

Mlyneková, Z., Chrenková, M., & Formelová, Z. (2016). Cereals and Legumes in Nutrition of People with Celiac Disease. *International Journal of Celiac Disease*, *2*(3), 105–109. https://doi.org/10.12691/ijcd-2-3-3.

Montebello, N. de P., Araújo, W. M. C., & Botelho, R. B. A. (2014). *Alquimia Dos Alimentos - Série Alimentos e Bebidas - 3ª Ed. 2014*. (Senac, Ed.). Senac. Retrieved from https://www.saraiva.com.br/alquimia-dos-alimentos-serie-alimentos-e-bebidas-3-ed-2014-7827117.html.

Mustafa, R., He, Y., Shim, Y. Y., & Reaney, M. J. T. (2018). Aquafaba, wastewater from chickpea canning, functions as an egg replacer in sponge cake. *International Journal of Food Science and Technology*, *53*(10), 2247–2255. https://doi.org/10.1111/ijfs.13813.

Padalino, L., Mastromatteo, M., Lecce, L., Cozzolino, F., & Del Nobile, M. A. (2013). Manufacture and characterization of gluten-free spaghetti enriched with vegetable flour. *Journal of Cereal Science*, *57*(3), 333–342. https://doi.org/10.1016/j.jcs.2012.12.010.

Patil, A. G. G., Kote, N. V., & Mulimani, V. (2009). Enzymatic removal of flatulence-inducing sugars in chickpea milk using free and polyvinyl alcohol immobilized α-galactosidase from Aspergillus oryzae. *Journal of Industrial Microbiology and Biotechnology*, *36*(1), 29–33. https://doi.org/10.1007/s10295-008-0467-x.

Polesi, L. F., Bruder, S., Sarmento, S., Prudente, C. B., Anjos, D., Polesi, L. F., … Boralli Prudente dos ANJOS, C. (2011). Composition and characterization of pea and chickpea starches. *Braz. J. Food Technol. Campinas*, *14*(1), 74–81. https://doi.org/10.4260/BJFT2011140100010.

Qiang, X., YongLie, C., & QianBing, W. (2009). Health benefit application of functional oligosaccharides. *Carbohydrate Polymers*, *77*(3), 435–441. https://doi.org/10.1016/j.carbpol.2009.03.016.

Rachwa-Rosiak, D., Nebesny, E., & Budryn, G. (2015a). Chickpeas—Composition, Nutritional Value, Health Benefits, Application to Bread and Snacks: A Review. *Critical Reviews in Food Science and Nutrition*, *55*(8), 1135–1143. https://doi.org/10.1080/10408398.2012.687418.

Rachwa-Rosiak, D., Nebesny, E., & Budryn, G. (2015b). Chickpeas — Composition, Nutritional Value, Health Benefits, Application to Bread and Snacks: A Review. *Critical Reviews in Food Science and Nutrition*, *55*(8), 1135–1143. https://doi.org/10.1080/10408398.2012.687418.

Rizzo, N. S., Drph, K. J., Drph, J. S., & Fraser, G. E. (2013). Nutrient Profiles of Vegetarian and Nonvegetarian Dietary Patterns. *Journal of the Academy of Nutrition and Dietetics*, *113*(12), 1610–1619. https://doi.org/10.1016/j.jand.2013.06.349.

Savage, J. H., Kaeding, A. J., Matsui, E. C., & Wood, R. A. (2010). The natural history of soy allergy. *Journal of Allergy and Clinical Immunology*, *125*(3), 683–686. https://doi.org/10.1016/j.jaci.2009.12.994.

Shridhar, K., Dhillon, P. K., Bowen, L., Kinra, S., Bharathi, A. V., Prabhakaran, D., … Ebrahim, S. (2014). The association between a vegetarian diet and cardiovascular disease (cvd) risk factors in india: The indian migration study. *PLoS ONE*, *9*(10), 1–8. https://doi.org/10.1371/journal.pone.0110586.

Simsek, S., Herken, E. N., & Ovando-Martinez, M. (2016). Chemical composition, nutritional value and in vitro starch digestibility of roasted chickpeas. *Journal of the Science of Food and Agriculture*, *96*(8), 2896–2905. https://doi.org/10.1002/jsfa.7461.

Strandwitz, P. (2018). Neurotransmitter modulation by the gut microbiota. *Brain Research*, *1693*, 128–133. https://doi.org/10.1016/j.brainres.2018.03.015.

Summers, L. K. M., Fielding, B. A., Bradshaw, H. A., Ilic, V., Beysen, C., Clark, M. L., … Frayn, K. N. (2002). Substituting dietary saturated fat with polyunsaturated fat changes abdominal fat distribution and improves insulin sensitivity. *Diabetologia*, *45*(3), 369–377. https://doi.org/10.1007/s00125-001-0768-3.

Teucher, B., Olivares, M., & Cori, H. (2004). Enhancers of iron absorption: Ascorbic acid and other organic acids. *International Journal for Vitamin and Nutrition Research*, *74*(6), 403–419. https://doi.org/10.1024/0300-9831.74.6.403.

Tucker KL. (2014). Vegetarian Diets and Bone Status. *The American Journal of Clinical Nutrition*, *100*(suppl), 329–335. https://doi.org/10.3945/ajcn.113.071621.1.

Wallace, T. C., Murray, R., & Zelman, K. M. (2016). The nutritional value and health benefits of chickpeas and hummus. *Nutrients*, *8*(12), 1–10. https://doi.org/10.3390/nu8120766.

Zimmer, J., Lange, B., Frick, J.-S., Sauer, H., Zimmermann, K., Schwiertz, A., … Enck, P. (2012). A vegan or vegetarian diet substantially alters the human colonic faecal microbiota. *European Journal of Clinical Nutrition*, *66*(1), 53–60. https://doi.org/10.1038/ejcn.2011.141.

In: Handbook of Chickpeas
Editors: Albert T. Lund and Noah D. Schultz
ISBN: 978-1-53616-374-2

Chapter 8

CHICKPEA FLOUR TO IMPROVE THE NUTRITIONAL VALUE OF BAKERY PRODUCTS

Abdelrahman R. Ahmed[1,2], Haiam O. Elkatry[1,2] and Idriss Mohammed[3]

[1]Food and Nutrition Science Department, Faculty of Agricultural and Food Science, King Faisal University, Kingdom of Saudi Arabia
[2]Food and Nutrition Science, Home Economics Department, Faculty of Specific Education, Ain Shams University, Cairo, Egypt
[3]Institute for Food Technology and Food Chemistry, Department of Food Rheology, Technical University of Berlin, Germany

ABSTRACT

Legumes are very important economic crops and have a high protein content up to 40% in some species. Legumes proteins can be used to fortify some foods to increase its nutritional value and balance amino acid content. The most important of these foods are bakery products.

The main ingredient of bakery products is cereal flour, especially wheat, which contains functional proteins that contribute to obtaining a good final product through the formation of the gluten network. Nutritionally, wheat protein is rich in sulfur amino acids, such as cysteine and methionine, which contribute significantly to formation of sulfur bonds in the gluten network, but on the other hand, poor in lysine (one of essential amino acids), which caused that cereal proteins not completed nutritionally. So many researchers tend to fortify cereal proteins with other sources proteins rich in these amino acids. Soy protein, chickpeas, peas and sweet or bitter lupine were used to balance the amino acids in the final product.

Chickpea (*Cicer arietinum L.*) is one of the most important legumes after beans and peas in many countries of the world such as India, America, Australia, Asia, Europe, East

Africa and Middle East. Syria and Turkey are among the most important producing countries. Chickpea is a crop cultivated from ancient times and in different agricultural lands and climatic areas, which gives it high agricultural productivity. There are two main types of chickpeas, Desi seeds which are small size, brown, green or black color and Kabuli seeds are large size, light colors and have low fiber content. Chickpea protein is rich in lysine and poor in methionine content, making it an ideal model for the fortification of cereal proteins for balance of essential amino acids.

The rheological properties of the dough play a very important role in predicting the characteristics of the final product. The gluten network is the main component of the wheat flour dough, which is formed primarily from the gliadin and glutenin, which give the dough its ideal structure. Therefore, when wheat flour is fortified with any other component, an interfere of this component with the gluten network occurs, which affects its rheological properties. This can be measured by instruments of rheological properties, as well as studying the properties of resulting bread to determine the optimum addition which improve the nutritional value of wheat bread dough with optimum technological characteristics.

INTRODUCTION

The nutritional situation in most developing countries is still critical despite many efforts. Protein-calorie malnutrition (PCM) is a major nutritional syndrome affecting more than 170 million preschool children and nursing mothers in developing Afro-Asian countries. The present trend in population growth indicates that the Protein gap may continue to increase in the future unless well-plane measures are taken to tackle the situation. Provision of adequate proteins of animal origin is difficult and expensive. An alternative for improving nutritional status of the people is to supplement the diet with plant proteins. Attention, therefore, must be directed to the nutritional evaluation of proteins from plant species. Legumes (poor man 's meat) play an important role in human nutrition since they are rich sources of protein, calories, certain minerals and vitamins (Deshpande, 1992) In Afro- Asian diets, legumes are the major contributors of protein and calories for economic and cultural reasons. Food legumes are crops of the family Leguminosae also called Fabaceae. They are mainly grown for their edible seeds, and thus are also named grain legumes. They occupy large cropped areas worldwide. Grain legumes are used as pulse (dhal) with cereals, grown in both tropical and temperate regions of the globe. They enhance the protein content of cereal-based diets and may improve the nutritional status of the cereal-based diets. Cereal proteins are deficient in certain essential amino acids, particularly lysine (Iqbal et al., 2003). On the other hand, legumes have been reported to contain adequate amounts of lysine but are deficient in S-containing amino acids (methionine, cystine and cysteine) (Farzana and Khalil, 1999). Since the chemical composition of crops varies with crop cultivars, soil and climatic conditions of the area, it is imperative to study the chemical composition of some important food legumes (Iqbal et al., 2006). Pulse crops (peas, lentils, chickpeas and beans) are among the most important sources of protein, starch and dietary fiber in food

(Perez-Hidalgo et al., 1997). These crops contain 18.5–30% protein, 35–52% starch and 14.6–26.3% dietary fiber on a dry weight basis (Dalgetty and Baik, 2003; de Almeida Costa et al., 2006; Perez-Hidalgo et al., 1997; Wang et al., 2009).

Legume proteins can be successfully used in baked products, to obtain a protein-enriched product with improved amino acid balance. The potential use of legumes as protein-enriching agents of baked products, mainly in the form of protein flours, has been reported by several authors. Among the legume protein products tested were various soybean protein preparations (Ribotta et al., 2005), chickpea flour (Güémes-Vera et al., 2008), germinated chickpea flour (Fernandez and Berry, 1989), germinated pea flour (Sadowska et al., 2003) and lupin flour (Dervas et al., 1999; Doxastakis et al., 2002; Pollard et al., 2002).

Chickpea (*Cicer arietinum L.*) is a valuable ancient leguminous plant which grows well in different soils and climates. It is the third most important cool-season food legume after common bean (*Phaseolus vulgaris L.*) and pea (*Pisum sativum L.*), based on world production estimates (FAO, 2001). It is cultivated in the Indian subcontinent, North Africa, the Middle East, Southern Europe, Asia, the Americas and Australia. The half of the world production of chickpea is exported from Syria and Turkey (Sayer et al., 2001). The high lysine content makes chickpea an excellent enhancer of protein quality when combined with cereal grain proteins, which are low in lysine but rich in sulphur amino acids (Iqbal et al., 2006). Chickpea cultivars are broadly divided into two groups, desi and Kabuli Figure 1. Kabuli seeds are large and light-colored beans, and are characterized by their larger size, ram-head shape and low fiber content (Singh et al., 1991). The seeds of desi cultivars are small, wrinkled at beak, with brown, black or green color.

Figure 1. Kabuli vs desi chickpeas.

Information on the rheological properties of dough will be useful for predicting the potential application of the wheat flour and the quality of the end product. Since protein has a major role in the quality of bread, supplementation of wheat dough with chickpea or other grain legume flours and proteins certainly affects rheological properties of the fortified wheat flour dough and its subsequent finished products (Eliasson, 1990; Sathe et al., 1981; Singh and Ram, 1990). Knorr and Betschart (1978) suggested that the weakening effect of foreign proteins on wheat flour doughs was the result of a dilution of the gluten structure by the added protein. These effects can be measured by using physical dough testing devices to evaluate the bread making potential and performance characteristics of the fortified flour.

1. The Chickpea Market Worldwide

Legumes are multipurpose crops and are consumed either directly as food or in various processed forms or as feed in many farming systems (Kumara and Deb, 2014). The legume crops are often grown as rotation crops with cereals because of their role in nitrogen fixation. However, over the past few decades, the yields and production of legume crops have been stagnant in the developing countries. Agricultural research and development efforts in many of these countries have concentrated on increasing cereal yields and production and lowering crop losses in order to achieve food security. Due to the diverse roles played by grain legume crops in farming systems and nutritional security, the research on legume crops will have significant impacts on nutritional security and soil fertility. Since 1961, the global production of the entire grain legume crops, namely chickpea, pigeon pea, cowpea, dry bean, faba bean, lentil, has increased at the rate of more than 1% per annum. Globally, chickpea has yield levels of about 850 kg/ha. The crop yields in the developing regions are very low. However, in the case of chickpea, some developing regions have exceeded the developed countries in terms of yield levels. The yield level in South and South-East Asia has increased by 13% from 717 kg/ha in 1994–1996 to 812 kg/ha in 2008–2010, growing at an annual rate of 0.8% (Nedumaran et al., 2015). Mean annual production share of chickpeas by region from 2008 to 2017 revealed that Asia shares 83% globally (Figure 2).

Chickpea is grown all over the world in about 57 countries under varied environmental conditions. South and South-East Asia dominates in chickpea production with 80% of regional contribution. Although developed countries do not contribute much toward chickpea production, the yield is particularly high in some Eastern European countries. China also showed a high yield level at 4,537 kg/ha in 2013–2017. The International Crops Research Institute for the Semi-Arid Tropics (ICRISAT) has released high-yielding, short-duration chickpea varieties that are resistant to Fusarium wilt in Southern India. Andhra Pradesh has the highest chickpea yields averaging 1.4 metric tons per ha. Almost 80% of the chickpea area in Myanmar during 2013–2017 was covered by ICRISAT-bred chickpea varieties.

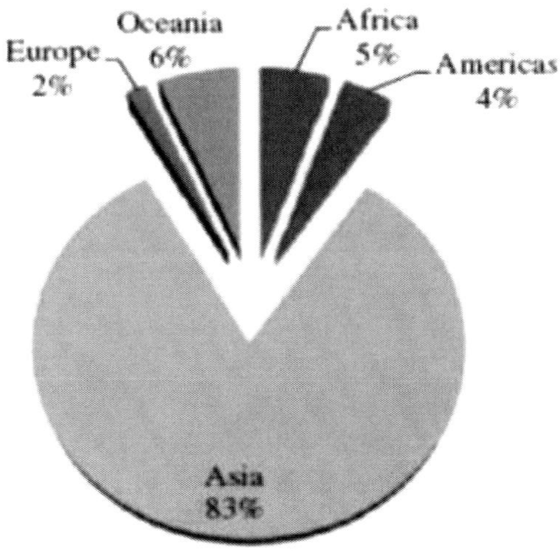

Source: Food and Agriculture Organization (FAO 2019).

Figure 2. Mean annual production share of Chickpeas by region from 2008 to 2017.

Worldwide, chickpea ranks third among the pulse crops and accounting for 11.67 million tons annually. This ranking place chickpea behind beans (25.66 million tons) and peas (11.69 million tons) with mean annual production of 11.67 million tons from 2013 to 2017 (Table 1). Taken together, annual combined production of peas and chickpea is nearly equal to that of beans, an indication of their overall importance.

Table 1. Mean annual global production of pulse crops 2008-2017

Pulse	Production (tons)
Beans	25,657,833
Peas	11,691,517.3
Chickpeas	11,672,579
Cowpeas	6,498,236.8
Faba beans (Broad beans)	4,468,240.1
Lentils	4,990,522.6
Pigeon peas	4,449,435.9
Other pulses	6,254,656.9
Total pulses	75,683,021.6

Data for other pulses were obtained by difference between total pulse production as reported by Food and Agriculture Organization (FAO., 2019) and the total of the seven pulse crops listed in this table.
Source: (FAO 2019).

These three pulses (beans, peas, and chickpeas) account for about 64.77% of global pulse production with chickpea accounting for approximately 15.42% of the total annually. Production of chickpea in terms of harvested area from 1961 to 2013 ranged from a low of 11.84 million hectares in 1961 to a high of 14.56 million hectares in 2017 (Figure 3).

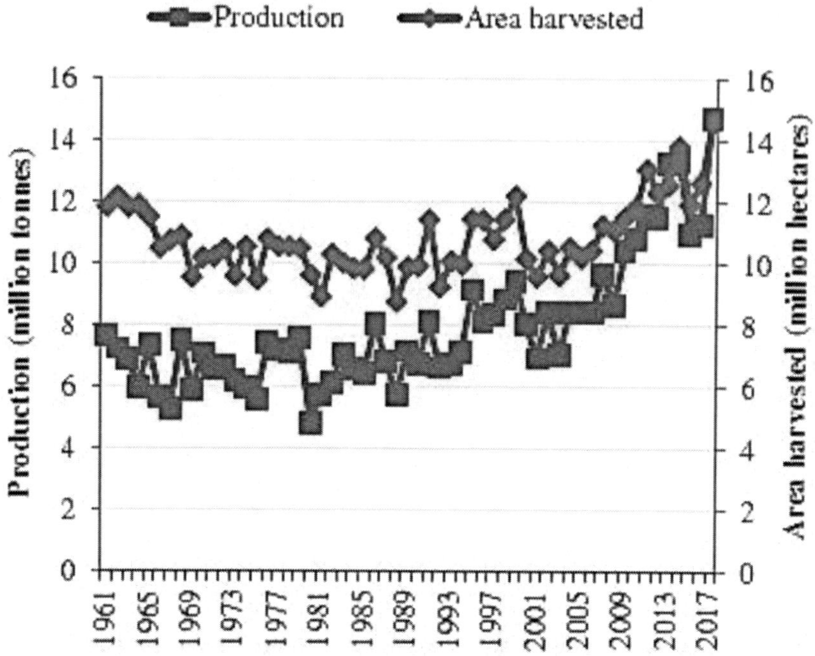

Source: Food and Agriculture Organization (FAO 2019).

Figure 3. Chickpea worldwide, area harvested (million hectares; filled diamond), and production (million tons; filled square) from 1961 to 2017.

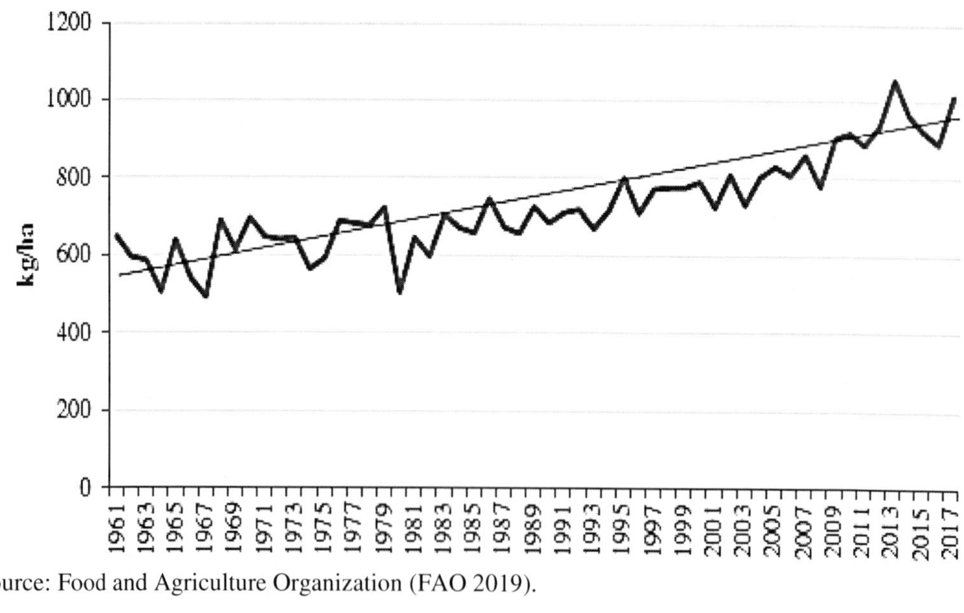

Source: Food and Agriculture Organization (FAO 2019).

Figure 4. Annual worldwide chickpea yields (kg/ha) and trend line from 1961 to 2017.

Earlier production trends from 1961 to 2013, in terms of area harvested, was somewhat static or slightly declining; however, yield increases began to have an impact on total production starting in the late 1900s. Steady increases in production appear in the early 2000s and continue to the present, and especially since 2013. Chickpea yields have been steadily increasing globally since 1961 (Figure 4) and trending to increases of over 6 kg/ha per annum.

This positive development is likely the result and benefits of sustained research programs toward the development improved germplasm and higher yielding varieties characterized by improved disease resistance and adaptation to environment. Also important are factors such as improved seed sources, seed supplies, and overall management practices. Expanded production in more productive environments also may account for these consistent yield increases. The expanded production in developed countries such as Australia, Canada, and the USA appears to have had a positive influence on mean yields worldwide. However, with most of the production centered in South Asia and India in particular, production increases in South Asia have clearly had an impact globally. Production increases have been pronounced in the most recent 10-year period (Figure 5).

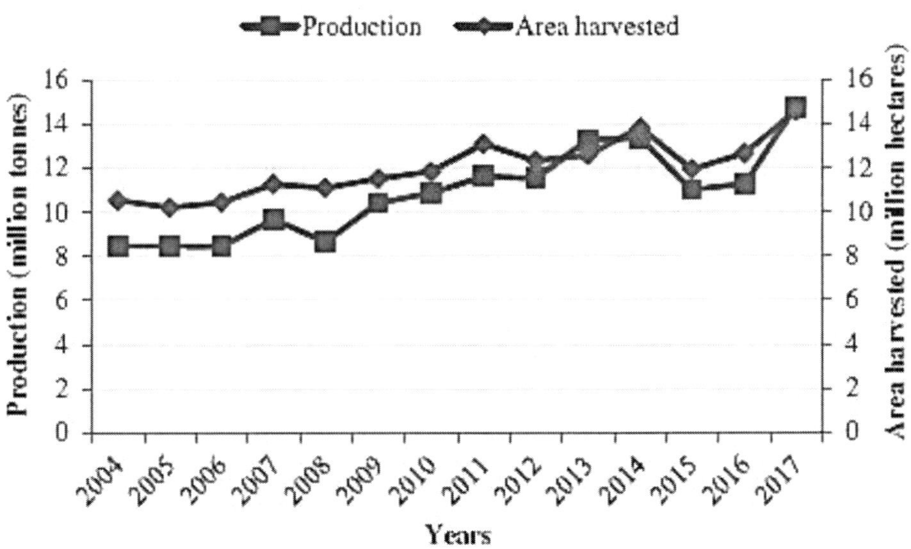

Source: Food and Agriculture Organization (FAO 2019).

Figure 5. Chickpea production worldwide, area harvested (million hectares; filled diamond), and tons (million t; filled square) from 2004 to 2017.

Area harvested and tonnage produced have expanded starting in 2004 leading up to 2017 where global production reached an all-time high of 14.78 million tons from nearly 14.56 million hectares. Gains in yield per hectare and area harvested are positively impacting overall production. Chickpea is produced in over 50 countries with India having the largest production and accounting for over 65% of total world production.

Figure 6 illustrates the dominance of India in chickpea production and the relative importance of the next most important producers. Australia, Myanmar, and Ethiopia, the next most important producers, account for 14%, 4%, and 4% of world production, respectively.

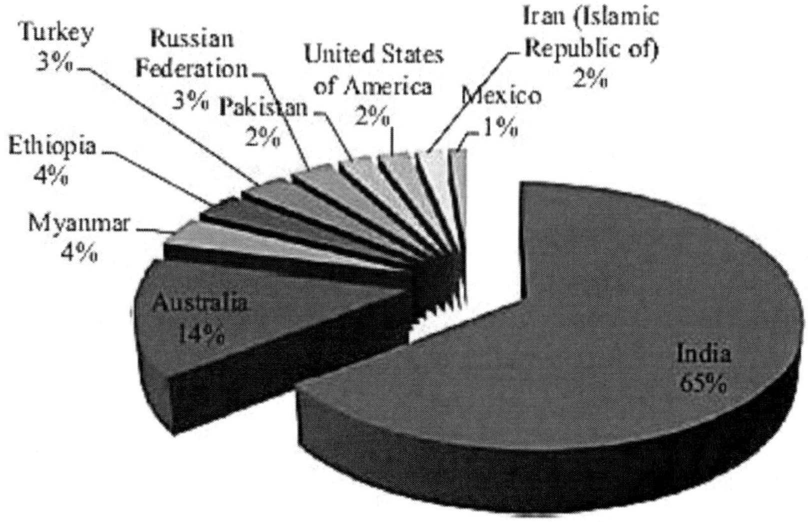

Source: Food and Agriculture Organization (FAO 2019).

Figure 6. Relative importance of the 10 leading chickpea producing countries.

Other major producing countries such as Turkey and Russia account for equal 3%; while Iran, USA, and Pakistan having an equal share of 2% of world production. Other important producing countries include Mexico, Malawi, Morocco, and Syria. Mean yields of chickpea have varied widely among producing countries and range from relatively low yields averaging 400–700 kg/ha in Iran, Malawi, Morocco, Pakistan, and Syria to relatively high yields in Ethiopia and Mexico (Figure 6). India, the largest producer has had stable mean yields of about 935 kg/ha. The high yields in Mexico are largely because most of the crop is irrigated and grown over the cool winter season.

India is the single largest producer of chickpea in the world, accounting for 65% (9.075 million tons) of the total production under chickpea (Figure 6). Australia is the second leading country over the world in 14% share. The average global chickpea yield is about 1.8 ton/ha, but the average yield in west and south Asia is only 1.46 tons/ha (Table 2). Though the production is low in other developing countries, such as Turkey, Myanmar, Ethiopia, and Mexico, the yield levels exceed 1.8 ton/ha (Table 2). The yield rates are above 2 tons/ha in Yemen and Russia. Similar or higher yield levels prevail in most developed countries (Table 2). The highest yield increase was observed in sub-Saharan Africa at about 80% between 2013 and 2017 (Table 2). West and South Asia showed a strong upward trend in the area under production and, therefore, production also follows suit from a low base. Area decline is observed in South America, and Middle-East and North Africa.

Table 2. Area harvested (hectares), mean yields (kg/ha), and total production (tons) from chickpea producing countries in important regions from 2013 to 2017

Region	Country	Area (ha)	Productivity (kg/ ha)	Production (tons)
Africa	Algeria	27,925.5	1,038.28	145,905
	Egypt	916.2	2,156.38	27,925.5
	Eritrea	8,596.6	438.62	18,855
	Ethiopia	237,182	1,943.82	2,307,096
	Malawi	62,922.4	675.98	196,405
	Morocco	62,235.2	729.34	222,488
	Sudan	7,036.2	1,755.1	61,780
	Tunisia	6,204.6	830.02	26,079
West Asia	Iran	505,829.4	472.36	1,199,901
	Iraq	374	1,687.12	2,517
	Israel	4,075.2	4,770.82	95,665
	Jordan	725.8	2,472.3	7,987
	Lebanon	3,116.2	979.68	15,275
	Kazakhstan	12,277.6	611.08	41,298
	Syria	70,233.4	654.98	231,608
	Turkey	381,728	1,229.14	2341,000
	Yemen	19,877.4	2,643.48	2,62,483
South Asia	India	8,927,600	935.34	41,827,500
	Bangladesh	6,646.8	976.0	32,291
	Myanmar	*374,765*	*1,489.7*	2,790,562
	Pakistan	959,027.4	*444.8*	2,145,445
	Nepal	9,642.6	1,043.7	50,367
Asia	China	2.763	4,534.82	62,733
	Uzbekistan	2,636.6	2,267.66	29,957
Europe	Greece	6,093.6	1,373.02	42,077
	Italy	12,577.4	1,525.44	97,779
	Portugal	1,246.8	675.76	4,411
	Russia	212,555.2	987.56	998,293
	Spain	39,346.6	930.68	182,766
North America	Canada	64,991.2	1,690.98	546,700
	Mexico	93,437.6	1,779.08	829,921
	USA	125,703.6	1,580.1	963,523
South America	Argentina	58,891.2	1,062.8	312,953
	Chile	780.6	749.54	3,144
Australia		650,528.8	1,426.08	4,876,693
World (total)		13,111,816.6	969.5	63,669,396

Source: Food and Agriculture Organization (FAO., 2019).

2. Exports and from Which Countries

Considerable quantities of chickpea (estimated at 2.4 million tons) have entered world trade as consuming countries have been unable to meet demand through domestic production. Australia nearly meets half of world export demand and provides an over 926,802 tons to the market annually in the most recent period where data are available (FAO, 2019). Exports had a value of $905.7 million (Australian) in 2016 (Table 3). India, while being the largest producer and importer of chickpea is also a major exporter ranking second only to Australia. Mexico, with a large production of high quality and large seeded Kabuli types, is the fifth most important exporter with the commodity being exported to over 50 countries worldwide with Algeria, Turkey, and Spain their most important customers. Turkey, also a major exporter with over 21,615.2 tons annually, with most of the tonnage moving to neighboring countries of Iraq, Jordan, and Saudi Arabia (Table 3).

Table 3. Annual exports from major chickpea exporting countries from 2012 to 2016

Country	Exports (tons)
Argentina	77,579.8
Australia	926,802
Canada	81,869.6
India	212,467.8
Mexico	149,691.4
Russian Federation	243,778.2
Turkey	21,615.2
USA	70,059
Ethiopia	59,768.4

Source: Food and Agriculture Organization (FAO., 2019).

3. Demand and Consumption

Chickpea is divided into two distinct classifications. The most prominent type is referred to as "Desi" and is characterized by relatively small seeds that range from light tan to black and with many variations including various markings of anthocyanin pigmentation. The relatively small seeds have rather thick seed coats and yellow cotyledons. The seeds are often decorticated (seed coat removed) and cotyledons split to form a product referred to as "dal." Dal, which can be made from most pulse crops, is used in soup making and is popular in South Asia. The less prominent type referred to as "Kabuli" is characterized by relatively large seeds that can range up to 22 mm in diameter and larger. Seed coats of Kabuli types lack pigment and are light tan and quite thin. This type is preferred in most markets outside of South Asia most likely because it is easier to produce and less expensive. Overall, world production of chickpea is predominated by the Desi type that accounts for 80% of production with the remaining 20% devoted to Kabuli types (Table 4).

Table 4. Major chickpea importing countries (mean imports from 2012 to 2016)

Country	Imports (tons)	Country	Imports (tons)
Algeria	71,675	Bangladesh	208,308.4
Sri Lanka	24,869.6	Egypt	26,834.2
France	5,856.6	Greece	3,909.4
India	590,656.8	Iran	18,131.8
Iraq	8,485.8	Jordan	33,471.8
Lebanon	13,214.6	Libya	5,472.4
Pakistan	104,035.8	Portugal	13,641.4
Saudi Arabia	43,705.8	Spain	54,618.2
Syria	21,98.6	Oman	3,689.4
Turkey	40,146	United Arab Emirates	103,909
USA	36,734.8		

Source: Food and Agriculture Organization (FAO., 2019).

4. Nutritional Content of Chickpea

Chickpea (Cicer arietinum L.) is a main pulse crop and is known for its nutritional quality, particularly in the Afro-Asian countries. It is a good source of carbohydrates containing monosaccharides (fructose, ribose, galactose and glucose), disaccharides (maltose and sucrose) and oligosaccharides (raffinose, ciceritol and verbascose stachyose). The two important galactosides of chickpea stachyose and ciceritol constitute 25% and 36–43% of total sugars respectively in chickpea seeds (Aguilera et al., 2009; Sánchez-Mata et al., 1998). Chickpea contains of all the essential amino acids except sulphur-containing amino acids. Starch is the main storage carbohydrate along with dietary fiber, oligosaccharides and simple sugars such as sucrose and glucose. Although the lipids are found in small quantity, they are high in unsaturated fatty acids such as oleic and linoleic acids. The chickpea seeds also contain P, Mg and Ca, especially K. Nutritionally important vitamins such as niacin, riboflavin, folate, thiamin and the vitamin A precursor β-carotene are also present in chickpea. It is cholesterol-free and an important source of both soluble and insoluble fiber. Chickpea seeds have a number of phenolic compounds (Wood and Grusak, 2007). The isoflavones biochanin and formononetin are two significant phenolic compounds of chickpea (Wood and Grusak, 2007). Matairesinol, genistein, diadzein and secoisolariciresinol are other phenolic compounds found in chickpea oil. The protein content of seeds of eight annual wild species of the genus Cicer ranged from 168 g/kg in Cicer cuneatum to 268 g/kg in Cicer pinnatifidum, with an average of 207 g/kg over the eight wild species (Ocampo et al., 1998). The protein quality of chickpea is better than some pulse crops such as green gram (Vigna radiata L.), black gram (Vigna mungo L.), and red gram (Cajanus cajan L.) (Kaur and Singh, 2005). In chickpea, the total dietary fiber content (DFC) is 18–22 g/100 g of raw chickpea seed (Aguilera et al., 2009), while pulse contains higher amount of DF. Soluble DFC is about 4–8 g/100 g and insoluble DFC of raw chickpea seed is 10–18 g/100 g (Dalgetty and Baik, 2003). The total DFC and insoluble DFC of desi types are higher compared with the Kabuli types, due to thicker hulls and seed coat in the Desi types (11 5% of total seed weight) compared with the Kabuli types (only 4.3–4.4% of total seed weight) (Rincón et al., 1998).

The nutritional quality of seeds can vary depending on the environmental, climate, soil nutrition, soil nutrition, soil biology, agronomic practices and stress factors (biotic and abiotic). The literature covering the proximate and chemical composition of desi and kabuli seeds is summarized in Tables 5-8. In general, the cotyledon and embryo make up most of the nutritionally beneficial part of the seed whilst the seed coat contains many of the antinutritional factors (ANFs). Desi types have a thicker seed coat than the kabuli types, reflected most obviously in the greater fiber content of desi seeds (Knights and Mailer, 1989).

4.1. Energy

Energy is often expressed as gross energy (MJ/kg) or as a caloric value (Kcal/ 100 g) and refers to the amount of energy contained in a food. Energy values for chickpea have been reported at 14–18 MJ/kg (334-437 Kcal/100 g) for desi types and 15-19 MJ/kg (357-446 Kcal/100 g) for kabuli types Table 5. The kabuli types generally have slightly higher energy values than desi types grown under identical conditions due to a smaller seed coat component. The WHO recommends a high consumption of energy-dilute foods rich in non-starch polysaccharides (NSPs) such as chickpea, other vegetables and fruits (WHO, 2003).

4.2. Protein and Amino Acid

Most legumes have high nitrogen contents, due to their ability to fix atmospheric nitrogen through a symbiotic association with soil microbes. The protein concentration of chickpea seed ranges from 16.7% to 30.6% and 12.6% to 29.0% for desi and kabuli types, respectively, and is commonly 2-3 times higher than cereal grains. Chickpea has been specifically used to treat protein malnutrition and kwashiorkor in children (Krishna-

Murti, 1975). Protein in the diet is essential in providing the body with amino acids to build new proteins for tissue repair and replacement, and to synthesize enzymes, antibodies and hormones. The amino acid composition of chickpea is well balanced, apart from the limited sulphur amino acids (methionine and cysteine) and is high in lysine. Hence, chickpea is an ideal companion to cereals, which are known to be higher in sulphur amino acids but limited in lysine Table 6.

4.3. Lipid and Fatty Acid

Chickpea exhibits higher lipid content than other pulses, with wide genotypic variation. The total lipid concentration of desi and kabuli types ranges from 2.9% to 7.4% and 3.4% to 8.8%, respectively (Table 7), which although high for pulses is low for some grain legumes (e.g., soybean and groundnut). The total lipid content of chickpea mainly comprises polyunsaturated (62-67%), mono-unsaturated (19-26%) and saturated (12-14%) fatty acids (Table 7). Hence, the small amount of lipid in chickpea is mostly of the beneficial kind (mono-unsaturated and polyunsaturated) rather than saturated fats that have been linked to heart and circulatory diseases. Essential fatty acids are those that cannot be synthesized by the human body and must be supplied through the diet. The two most important essential fatty acids are the omega-6 (linoleic) and omega-3 (linolenic) fatty acids. They are necessary for normal growth, physiological functions and cell maintenance. The major fatty acid in chickpea is linoleic acid, with desi oil containing 46-62% of the acid and kabuli oil containing 16-56%. Oleic acid, a monounsaturated fatty acid, is the next most common type with 18-23% found in desi oil and 19-32% found in kabuli oil. Linoleic acid has been shown to be hypocholesterolemic and can reduce the likelihood of atherosclerosis and coronary heart disease. The high linoleic acid levels in chickpea can explain lowered serum cholesterol levels in chickpea feeding trials (Jaya et al., 1979; Mathur et al., 1968). Conversely, palmitic acid, a saturated fatty acid, that is hypercholesterolemic and adverse to health, is found in comparatively small amounts in chickpea. The lipid content of foods is often responsible for its flavor, which in the case of chickpea may contribute to its 'nutty' taste. On the other hand, deterioration of food quality and formation of objectionable flavors can sometimes occur during storage and processing due to enzymatic oxidation of unsaturated fatty acids. This can lead to rejection of poorly stored legumes for human consumption, although the problem is more common in legumes with higher lipid concentrations such as soybean and groundnut.

Table 5. General composition and carbohydrates composition of desi and kabuli chickpeas seeds

Parameter	Unit	Desi whole seed				Kabuli whole seed			
		Minimum	Maximum	Number of cultivars	References	Minimum	Maximum	Number of cultivars	References
General									
Seed coat	%	10.1	21.69	111	1-9	4.5	9.5	31	[1, 3, 6-10]
Gross energy	MJ/kg	16.2	18.3	6	11-12	14.958	18.7	3	[11-12]
Calorific value	Cal/100 g	334	387	15	2, 11, 13	357.5	391	26	[11, 15,]
Protein	%	16.7	30.57	1898	1-6, 8, 11-13, 17	12.6	29	1749	[1, 3, 6, 8, 10-12, 14, 16, 17-18, 20, 21, 34]
Ash	%	2.04	4.2	180	1, 2, 5, 11-13, 17, 21, 22	2	3.9	156	[1, 10-12, 14, 16, 17, 21, 20, 21, 26]
Fat	%	2.9	7.42	1954	1, 2, 5, 11-13, 17, 21, 19	3.4	8.88	167	[1, 11-12, 14, 17, 21, 20, 21, 26]
Crude fibre	%	3.7	13	107	1, 2, 5, 7, 12-13, 17, 19-21	1.17	4.95	92	[1, 7, 13, 14, 17, 20, 23-26]
CHÓ, total	%	50.64	64.9	48	2, 5, 13, 18, 19, 20,	54.27	70.9	38	[11, 15, 20, 23]
Carbohydrates									
NSP, total	%	-	12.2	2	27	-	8.78	1	[36]
Total soluble sugars	%	5.33	11.8	17	3, 6, 21	6.65	7.5	2	[23]
Total reducing sugars	%	2.61	4.77	10	16	2.25	2.42	2	[23]
Total non-reducing sugars	%	1.12	1.89	10	15	4.4	5.08	2	[23]
Sugars, total	%	2.2	10.7	26	1, 24	5.5	10.85	21	[1, 11]
Starch	%	32	56.3	44	1, 3, 6, 15, 17	30	57.2	18	[1, 3, 17, 27, 28]
Amylose	% of starch	20	42	7	3, 6, 25	20.7	46.5	6	[3]
Resistant starch	%	-	3.39	1	19	0.31	16.43	2	[28]
ADF	%	9.4	16.7	37	3, 6, 7, 12, 21,	3.24	12.2	33	[3, 6, 7, 12, 21]
NDF	%	15.6	30.2	20	7, 12, 21	5.16	16	21	[7, 12, 21]
Soluble dietary fiber	%	18.4	22.7	10	7, 11	10.6	16.63	11	[7, 11]
Insoluble dietary fiber	%	-	0.0	1	19	-	-	-	-

Parameter	Unit	Desi whole seed				Kabuli whole seed			
		Minimum	Maximum	Number of cultivars	References	Minimum	Maximum	Number of cultivars	References
Dietary fiber	%	-	13.9	1	19	-	-	-	-
Hemicellulose	%	3.5	8.8	8	7,21	4	7.3	8	[7, 21]
Pectic substances	%	1.5	3.8	7	7	2.4	4.1	8	[7, 21]
Lignin	%	1.3	17.0	14	7, 12, 21	0.01	2.1	19	[7, 12, 20, 21]

1. Jambunathan and Singh (1980); 2. Agrawal and Singh (2003); 3. Saini and Knights (1984); 4. Awasthi et al. (1991); 5. Suryawanshi et al. (1998); 6. Saini et al., (1989); 7. Singh, (1984); 8. Singh and Jambunathan (1981); 9. Knights and Mailer (1989); 10. Bellido and Fuentes (1990); 11. Nutrient Data Laboratory (USDA Agricultural Research Service) (2009); 12. Petterson et al. (1997); 13. Agrawal and Bhattacharya (1980); 14. El Hardallou et al. (1980); 15. Mittal et al. (1999); 16. Vijayalakshmi et al. (2001); 17. Viveros et al. (2001): 18. Amirshahi and Tavakoll (1970): 19. de Almeida Costa et al. (2006); 20. Rossi et al. (1984); 21. Salgado et al. (2001): 22. Iqbal et al. (2006); 23. Zaki et al. (1996) ; 24. Sotelo et al. (1987); 25. Madhusudhan and Tharanathan (1995); 26. Dodok et al. (1993); 27. Perez-Maldonado et al. (1997); 28. Bravo (1999).

Table 6. Amino acids and minerals composition of desi and kabuli chickpeas seeds

Parameter	Unit	Desi whole seed				Kabuli whole seed			
		Minimum	Maximum	Number of cultivars	References	Minimum	Maximum	Number of cultivars	References
Amino acids									
Alanine	g/16g N	2.64	6.06	85	1-6	2.81	5.66	44	[1-4, 6-10]
Arginine	g/16g N	5.1	12.79	117	1-6, 11, 12	5.14	13.2	44	[1,2, 4, 6-10, 12]
Aspartic acid	g/16g N	8.36	14.19	83	1-6	8.49	14.26	44	[1-4, 6-10]
Cystine	g/16g N	0.5	1.93	121	1-5, 11, 13	0.53	1.77	39	[1,2 , 4, 8]
Glutamic acid	g/16g N	12.94	21.09	84	1-6	13.14	18.9	44	[1-4, 6-10]
Glycine	g/16g N	2.44	5.1	85	1-6	2.49	4.43	42	[1, 2, 4, 6, 7, 9, 10]
Histadine	g/16g N	1.48	5.71	115	1-6, 11-13	1.4	4.2	519	[1, 2,4, 6-10, 12, 13]
Isoleucine	g/16g N	2.15	5.82	114	1-6, 11,13	3.2	6.03	47	[1 2, 4, 6-10, 13]
Leucine	g/16g N	3.12	9.95	114	1-6, 11,13	5.73	9.09	51	[1, 2, 4, 10, 13]
Lysine	g/16g N	4.86	9.62	120	1-6, 11-13	4.85	8.71	53	[1, 2, 4, 6-10, 12, 13]
Methionine	g/16g N	0.64	2.29	175	1-6, 11-14	0.61	3.63	54	[1, 2, 4, 6-10, 12-14]
Phenylalanine	g/16g N	2.76	6.83	114	1-6, 11, 13	4,06	8.41	51	[1, 2, 4, 6-10, 13]

Table 6. (Continued)

Parameter	Unit	Desi whole seed				Kabuli whole seed			
		Minimum	Maximum	Number of cultivars	References	Minimum	Maximum	Number of cultivars	References
Proline	g/16g N	2.82	4.98	83	1-3, 5,6	2.88	5.28	41	[1, 2, 6, 7,9, 10]
Serine	g/16g N	3.26	6.08	86	1-6	3.13	5.89	43	[1, 2, 4, 6, 7, 9,10]
Threonine	g/16g N	1.18	5.58	119	1-6, 11, 13, 14	2.59	4.93	48	[1, 2, 4, 6-10, 13, 14]
Tryptophan	g/16g N	0.39	1.30	80	1, 4-6, 11, 13, 14	0.31	2.36	15	[1,6, 8-10, 13, 14]
Tyrosine	g/16g N	0.71	3.81	102	1-6,11,13	1.1	4.21	46	[1, 2, 4, 6-10, 13]
Valine	g/16g N	2.03	5.55	114	1-6,11,13	3.11	6.22	47	[1, 2, 4, 6-10, 13]
Minerals									
Calcium	mg/100g	68	269	185	1,3, 5, 6, 11, 14-19	40	267	73	[1, 3, 6-10, 12, 14, 16, 17, 20]
Magnesium	mg/100g	107.4	168	95	1,3,5,6,16,17,19	10	239	29	[1, 3, 6, 9, 10, 16, 17]
Phosphorus	mg/100g	169	860	180	1,3,5,6,11,12, 14-17, 19	159	930	78	[1, 3, 6-10, 12, 14, 16, 17, 20]
Potassium	mg/100g	230	1272	142	1,3, 5,6, 16	220	1333	43	[1, 3, 6, 8-10,16]
Sodium	mg/100 g	1.0	101	74	1,3, 5, 6, 18	2.06	64	28	[1, 3, 6, 8-10, 18]
Sulphur	mg/100 g	160	220	84	1	160	200	23	[1]
Copper	mg/100 g	0.31	11.6	114	1,3, 5, 6, 18	0.27	1.42	27	[1, 3, 6, 9, 10,16, 17]
Iron	mg/100 g	3	12	113	1,3, 5,6, 18, 19	3.2	14.3	62	[1, 3, 6-10, 12, 14, 16, 17, 20]
Manganese	mg/100 g	1.78	5.16	80	1,3,5,6	0.09	9.4	31	[1, 3, 6, 9, 10, 16, 17]
Molybdenum	mg/100 g	0.03	0.13	20	1	0.07	0.13	9	[1]
Zinc	mg/100 g	2.2	20	101	1,3, 5, 6, 18	2	5.4	31	[1, 3, 6, 9, 10, 16, 17]
Cobalt	mg/100 g	11.0	35.0	8	1	6.0	41.0	5	[1]
Selenium	mg/100 g	5.0	100.0	15	1.3	0.5	10.0	8	[1, 3]

Petterson ef al. (1997); 2. Rossi et al. (1984); 3. Nutrient Data Laboratory (USDA Agricultural Research Service) (2009); 4. Salgado et al. (2001): 5. Iqbal et al. (2006); 6. Avancinl et al. (1993); 7. El-Hardallou et al. (1980); 8. FAO (2001) ; 9. Madhusudhan and Tharanathan (1995); 10. Dodok et al. (1993); 11. Singh et al., (1995); 12. Sharma et al., (1999); 13. Batterham (1989); 14. Siddhuraju et al. (1996); 15. Agrawal and Bhattacharya (1980); 16. Jambunathan and Singh (1981); 17. Viveros et al. (2001); 18. Awasthi et al. (1991); 19. Kaur and Singh, (2005); 20. Zaki et al. (1996).

Table 7. Fatty acids composition and vitamins of desi and kabuli chickpeas seeds

Parameter	Unit	Desi whole seed				Kabuli whole seed			
		Minimum	Maximum	Number of cultivars	References	Minimum	Maximum	Number of cultivars	References
Fatty acids									
Myristic (14:0)	% in oil (of total FAs)	0.2	0.2	2	1,2	0.01	0.2	11	[1-5]
Palmitic (16:0)	% in oil	9.1	11.5	6	1,2	3.3	10.8	12	[1-5]
Palmitoleic (16:1)	% in oil	-	0.3	1	2	0.002	0.2	4	[1,2,5]
Stearic (18:0)	% in oil	0.3	1.8	6	1,2	0.05	5.3	12	[1-5]
Vaccenic (18:1, 7)	% in oil	1.2	1.4	4	1	0.7	1.3	9	[1]
Elaidic (18:1, 9t}	% in oil	1	1	1	1	0.7	1.0	2	[6]
Oleic (18:1, 9c}	% in oil	17.6	23.3	5	1	17.37	32	11	[5,6]
Linoleic (18:2)	% in oil	45.95	61.5	13	1,2	16.4	70.4	14	[1-5]
Linolanic (18:3)	% in oil	2.16	4.8	6	1,2	0.3	3.3	12	[1-5]
Total 18:1		0.6	0.8	5	1	0.7	0.9	6	[6]
Arachidie (20:0)	% in oil	0.5	0.6	5	1	0.5	0.7	4	[6]
Eicosenoic (20:1)	% in oil	0.4	0.8	5	1	0.5	0.6	3	[6]
Behenic (22:0)	% in oil	0.1	0.1	1	1	0.3	0.5	4	[6]
Lignoceric (24:0)	% in oil	-	13.4	1	2	12	14	2	[1-3]
Total saturated	% in oil	-	28.8	1	2	5	29	2	[1-3]
Tatal manounsaturated	% in oil	-	29.0	1	2	19	29.0	2	[1-3]
Total polyunsaturated	% in oil	-	57.6	1	2	62	67	2	[1-3]
Vitamins									
Vit A (β-carotene equivalent)	µg/100g	-	40	1	2	9.6	49	9	[2, 7, 8]
Vit C (ascorbic acid) -	Mg/100g	-	4.0	3	2	-	-	-	
Vit E (tocopherals and tocotrienols)	Mg/100g	-	0.82	1	2	0.82	13.7	8	[2, 8]
Thiamine (B1)	Mg/100g	-	0.48	42	2	0.41	0.58	11	[2, 3, 5]
Riboflavin (B2)	Mg/100g	-	0.21	48	2	0.106	1.58	7	[2, 3, 5]
Niacin (B3)	Mg/100g	-	1.54	43	2	1.5	1.76	5	[2, 3]
Pantothenic acid (B5)	Mg/100g	-	1.59	20	2	-	0.83	1	[2]

Table 7. (Continued)

Parameter	Unit	Desi whole seed				Kabuli whole seed			
		Minimum	Maximum	Number of cultivars	References	Minimum	Maximum	Number of cultivars	References
Vit B6	Mg/100g	-	0.54	26	2	-	0.49	1	[2]
Folate (B9)	Mg/100g	150	557	16	2,9, 10	347	437	1	[2, 11]
Vit K	Mg/100g	-	9.0	1	2	-	9.1	1	[2]

Petterson et al. (1997); 2. Nutriant Data Laboratory (USDA Agricultural Research Service) (2009) 3. FAO (2001), 4. Madhusudhan and Tharanathan (1995); 5. Dodok et al. (1993), 6. Biliaderis et al. (1980); 7. Abbo et al. (2005); 8. Atienza et al. (1998); 9. Dang et al. (2000); 10, Babu and Srikantia (1976); 11. Hefni, and Witthöft (2014).

Table 8. Antinutritional factors, phytosterol and phytoestrogens content and vitamins of desi and kabuli chickpeas seeds

Parameter	Unit	Desi whole seed				Kabuli whole seed			
		Minimum	Maximum	Number of cultivars	References	Minimum	Maximum	Number of cultivars	References
Antinutritional factors									
Oligosaccharides	%	1.2	3.9	82	1,2	0.4	2.8	35	[2, 3]
Phytate	%	0.37	1.13	63	2	0.25	0.99	14	[2, 3]
Tannins (total)	%	0.36	0.72	53	2,4	0.12	0.51	39	[2, 3, 5]
Tannins (condensed)	%	0.01	0.09	47	2,4	0	0.04	37	[2, 4]
Trypsin inhibitor (TIA)	mg/g	1.16	15.7	129	1,2,4,6,7	1.39	12,1	62	[2, 4-8]
Chymotrypsin inhibitor (CTIA)	mg/g	2.40	13.19	77	1, 2,7	3	10.74	24	[2, 5-7]
Polyphenols	mg/g	0.84	6.00	19	4,7,9,10	0.02	2.2	10	[4, 7, 10]
Saponins	mg/g	17.7	56.0	3	4,11, 12	0.7	21.9	1	[4, 13]
Phytoestrogens (isoflavones)									
Formononetin	μg/100g	-	215.0	1	14	94.3	126.0	2	[14]
Biochanin A	μg/100g	-	838.0	1	14	1420	3080	2	[14]
Daidzein	μg/100g	-	11.4	1	14	34.2	192.0	2	[14]
Geristein	μg/100g	-	76.3	1	14	69.3	214.0	2	[14]
Coumestrol	μg/100g	-	5.0	1	14	0.00	0.0	2	[14]
Phytoestrogens (lignans)									

Parameter	Unit	Desi whole seed				Kabuli whole seed			
		Minimum	Maximum	Number of cultivars	References	Minimum	Maximum	Number of cultivars	References
Secoisolariciresinol	µg/100g	-	8.4	1	14	6.7	8.1	2	[14]
Matairesinol	µg/100g	-	0.0	1	14	0.0	0.0	2	[14]
Phytesterols	µg/100g	-	39	1	15	35	40	2	[15, 16]

1, Saini (1989); 2. Petterson et al. (1997); 3. Viveros et al. (2001); 4. Salgado et al. (2001); 5. Zaki et af, (1996); 6. Batterham (1989); 7. Singh and Jambunathan (1981), 8. Sotelo et al. (1987); 9. Agrawal and Bhattacharya (1980); 10. Avancini et al. (1992); 11. Fenwick and Oakentull (1983); 12. Kerem et al. (2005); 13. Adebowale et al. (2005); 14. Mazur (1998); 15. Nutrient Data Laboratory (USDA Agricultural Research Service) (2009); 16. Sanchez-Vioque et al. (1998).

4.4. Carbohydrates

Carbohydrates are the major nutritional component in chickpea, with 51-65% in desi type and 54-71% in kabuli type. The major classes of carbohydrates are monosaccharides, disaccharides, oligosaccharides and polysaccharides.

4.5. Monosaccharides and Disaccharides

Monosaccharides are single sugar units, whereas disaccharides consist of 2 monosaccharides joined by a glycosidic bond. Free monosaccharides and disaccharides are the most readily available sources of energy; however, only small amounts are present in dried, mature pulse seeds. Most sugars are found as components of glycosides, oligosaccharides and polysaccharides. The most common monosaccharides in chickpea seeds are glucose (0.7%), fructose (0.25%), ribose 0.1%) and galactose (0.05%) (Sanchez Mata et al., 1998). The most abundant freely occurring disaccharides in chickpea are sucrose (1-2%) and maltose (0.6%).

4.6. Oligosaccharides

Oligosaccharides are generally defined as polymeric sugars made up of 2-4 monosaccharides. Legume seeds are rich sources of the raffinose family of oligosaccharides (RFO) in which galactose is present in a-D-1-6-linkage. They are often considered ANFs because they are neither hydrolysed nor absorbed by the human digestive system, passing into the large intestine where they undergo fermentation by colonic bacteria with the release of gases causing flatulence. Oligosaccharides, therefore, may be classified as part of dietary fiber. Chickpea contains 0.4-2.8% and 1.2-3.9% oligosaccharides in desi and kabuli seeds, respectively (Table 8). Similar amounts are also found in beans (*Phaseolus vulgaris*), which are less than the content in peas (Pisum sativus), but more than that in faba beans (*Vicia faba*) and lentils (*Lens culinaris*) (Hedley, 2000). The most important oligosaccharides in chickpea are raffinose, stachyose, ciceritol and verbascose, reported in amounts of up to 2.2%, 6.5%, 3.1% and 0.4%, respectively (Table 8). Interestingly, ciceritol, which was discovered only in 1983, does not belong to the raffinose family (Quemener and Brillouet, 1983). It is most abundant in chickpea - hence the name - and is present in only a few other pulses such as lentil, lupin and kidney bean. Quemener and Brillouet (1983) also showed that, unlike the RFO, ciceritol does not contribute significantly to flatulence. However, the ingestion of large quantities of chickpea has been known to cause flatulence.

4.7. Polysaccharides

Polysaccharides are high-molecular-weight polymers of monosaccharides. They are generally classified by their function in plants, either as an energy reserve or by providing structural support, although there is some overlap. The two main storage polysaccharides in legumes are galactomannans and starch, of which chickpea contains only starch.

4.8. Starch

Starch is the principal carbohydrate constituent in chickpea seeds (30-57%) and is the main dietary energy source derived from chickpea. A high proportion of the human requirement for energy is met by starch from seeds and tubers. Starch is composed of two types of glucose polymer (amylose and amylopectin), with differing properties, occurring together in a starch 'granule'. Amylose is an essentially linear molecule and comprises 20-41% and 23-47% in desi and kabuli types, respectively, depending on the analytical method used (Table 5). The remaining 59–80% and 53-77%, respectively, is amylopectin, a highly branched molecule. Comparatively, cereal starches contain less amylose (20– 25%) and more amylopectin (75–80%) than pulses. Legume starches also differ from cereal and tuber starches in terms of starch granule structure and crystallinity, referred to as C-type starch. The type of starch (A, B or C) and the proportion of amylose to amylopectin and their respective characteristics are responsible for many of the differences in seed, flour and dough behavior during processing. On ingestion, starch is cleaved by pancreatic a-amylase in the duodenum and is hydrolysed by enzymes in the small intestine to produce glucose, an energy source (Hedley, 2000). Legume starch has been reported to be less digestible than cereal starch, possibly due to less amylopectin, higher levels of ANFs, starch granule structure differences and/or cell wall components blocking enzyme access to the granules (Carré et al., 1998).

Starch can also be divided according to digestibility as soluble, insoluble or resistant. Digestibility varies according to plant species and genotype, chemical and physical granule structure and method of food preparation. Soluble and insoluble starches are completely digested and absorbed, differing only in the speed at which these processes occur. Starch was thought to be fully digested until Englyst et al. (1992) recognized a portion of undigested starch (usually from wholegrain or processed foods) that passed into the large intestine. It was named resistant starch and defined as 'the sum of starch and products of starch degradation not absorbed in the small intestine of healthy humans' (Asp and Björck, 1992). As such, resistant starch is now classified as a component of dietary fiber. Kabuli seeds have been reported to contain 16.43% resistant starch and desi seeds 3.4–10.3% (Table 5).

4.9. Non-Starch Polysaccharides

NSPs are polysaccharides that are not part of starch and can be classified as soluble or insoluble (Choct, 1997). Soluble NSP includes hemicellulose and pectic substances. Hemicelluloses and pectic substances are usually present in amorphous matrixes around cellulose or can act as binding agents between cells in the middle lamella. Hemicellulose contents range from 3.5% to 9% and pectic substances from 1.5% to 4% in chickpea (Table 5). Soluble NSP is digested slowly due to its hygroscopic and gummy nature. Insoluble NSP includes cellulose, some insoluble hemicellulose and other polysaccharides, and is indigestible. Cellulose is present in the cell walls of plants and contributes to most of the rigidity and strength of plant structures (often referred to as crude fibre). Desi types contain 4–13% cellulose whilst kabuli types have less (1-5%), primarily due to a thinner seed coat.

Bravo (1999) measured 3.41% soluble NSP (mainly uronic acid and arabinose), 5.37% insoluble NSP (mainly arabinose and glucose) and 8.78% total NSP in kabuli seeds. In chickpea the total NSP fraction is dominated by the monosaccharide residues of glucose and arabinose followed by xylose and mannose, mainly in the form of insoluble NSP from the seed coat (Farrell et al., 1999).

4.10. Dietary Fiber

Monosaccharides, disaccharides and most of the starch are generally classified as available carbohydrates. These are enzymically digested into smaller sugar units within the small intestine and are a rapidly available energy source. Resistant starch and NSP (including oligosaccharides) are classified as unavailable carbohydrates and constitute dietary fiber. Kamath and Belavady (1980) reported chickpea as having more non-available carbohydrates (sum of non-cellulosic polysaccharides, cellulose and lignin) compared to pigeon pea (*Cajanus cajan*), black gram (*P. mungo*) and mung bean (*P. aureus*).

Dietary fiber includes unavailable carbohydrates and lignin (not a carbohydrate, but a major constituent of plant cell walls). It is often difficult to compare dietary fibre contents as numerous methods of calculation are reported in the literature. Legume seeds are generally characterized by their relatively high content of dietary fiber compared to other grains. Much of this fraction comes from the seed coat; hence kabuli types have lower dietary fibre content (11-16%) than desi types (19-23%). A high content of dietary fiber in diets has been shown to have a negative effect on nutrient digestibility in animals, especially monogastrics (Choct et al., 1999). NSP is generally the largest fraction of dietary fiber in grain legume seeds, the rest comprising resistant starch and lignin. NSP and resistant starch are fermented to varying degrees in the colon. Lignin is not a carbohydrate but is a polyphenolic constituent of plant cell walls. Lignin's role in plant cells is to cement and anchor NSP and strengthen cell walls to prevent biochemical degradation and physical damage. Lignin is classified as a constituent of dietary fibre as it is indigestible and not fermented in the colon. Dietary fiber, as a proportion of total carbohydrate, constitutes -25% in peas and 27% in faba beans (Gdala, 1998). In chickpea, kabuli types have 20-21% and desi types 35-37% (Table 5). This large difference between desi and kabuli types is mainly due to differing seed coat contents, with desi types having seed coat with 2-3 times higher levels of lignin than kabuli types.

Fiber can also be calculated as acid detergent fiber (ADF) and neutral detergent fiber (NDF). This terminology is usually seen in reference to stock-feeds, more so than for human nutrition. The ADF consists of cellulose, lignin, bound protein and acid-insoluble ash portions, which are all indigestible. The NDF consists of the ADF and hemicellulose. The NDF is used to predict dry matter intake whilst the ADF is inversely correlated with digestible energy (Table 5).

Fiber was traditionally considered an antinutritive factor that diluted the energy content and decreased the nutrient availability of food. This is still considered a problem for intensive monogastric livestock production where rapid growth is preferred, but is also a problem in malnourished populations where a high-fiber diet can inhibit the absorption of other nutrients leading to deficiency-related diseases. On the other hand, a high-fiber diet has many health benefits in affluent societies, especially where obesity, heart disease and diabetes are a growing problem. Dietary fiber is fermented slowly by microflora in the large intestine to provide energy, producing beneficial short-chain fatty acids and gaseous by-products (Southgate, 1992).

4.11. Minerals

Chickpea plants obtain minerals from their soil environment and partition these to their seeds. Roots utilize specific and/or selective transport proteins to obtain all the minerals (usually as ions) that are essential for plant growth and development (i.e., Ca, Mg, K, P, S, CI, B, Fe, Mn, Zn, Cu, Ni and Mo) (Grusak and DellaPenna, 1999). Together these minerals constitute what is referred to as the ash fraction of the seed. For the plant, these minerals are important for many metabolic activities including photosynthesis, respiration, chlorophyll synthesis, cell division and various responses to biotic and abiotic stress. Although most of the essential plant minerals are also essential for humans, there are certain minerals required by humans (e.g., Na, I, Se and Cr) that plants do not require. Fortunately, these additional minerals can be absorbed by plants and accumulated in seeds and other edible tissues. Chickpea seeds can thus supply several minerals essential for humans. Although the concentrations of any given mineral vary depending on genotype and environmental constraints (Table 5), chickpea contributes significantly to several minerals in the human diet.

4.12. Vitamins

Vitamins are essential for normal bodily growth and many metabolic processes. They can be obtained from food and are consumed as the vitamin themselves or as precursors that require conversion into vitamin-active compounds in the body. Chickpea contains several water-soluble vitamins such as the B-complex vitamins (including folate) and vitamin C, as well as several lipid-soluble vitamins such as vitamin A (found as provitamin A carotenoids), vitamin E (tocopherols and tocotrienols) and vitamin K.

4.13. Water-Soluble Vitamins

The B-complex vitamins include vitamin B, (thiamine), vitamin B2 (riboflavin), vitamin B3 (niacin), vitamin B6 and pantothenic acid (Table 7). Thiamine (vitamin B1) is present in small amounts in a wide variety of plant and animal foods; however, the process of refining cereal and pulse products removes much of the thiamine content. Riboflavin (vitamin B2) is also present in small amounts in chickpea but is not active until conversion after absorption in the small intestine. Niacin (vitamin B3) is associated with protein; hence, high protein foods such as chickpea are rich sources of niacin. Vitamin B6 exists in 3 major chemical forms (pyridoxine, pyridoxal and pyridoxamine). Grain legumes, including chickpea, are among the richest sources of pyridoxine. Pantothenic acid (vitamin B5) is widely distributed in food and is important in the metabolism and breakdown of carbohydrates, proteins and fats in the body. Vitamin B6 performs a wide variety of functions in the body. It is essential for synthesizing enzymes and neurotransmitters required for protein metabolism and normal nerve communication, respectively. Vitamin B6 is also important for immune system function, blood glucose control, red blood cell metabolism and the conversion of tryptophan to niacin. Folic acid (or folate - the anion form) is also known as vitamin Bg. Folacin is a provitamin that is converted to the active form (folic acid) for use by the body. Folic acid is required for RNA synthesis and DNA replication and is very important for new cell growth and maintenance. Chickpea contains 150–557 ug/g folate (Table 7). There is evidence that folic acid may help prevent cancers induced by DNA damage (Coppen and Bolander-Gouaille, 2005).

Ascorbic acid (vitamin C) is readily available from fruits and vegetables. Chickpea contains 4mg/100 g vitamin C, although it is easily destroyed by processes such as cooking and prolonged storage. Vitamin C is a strong antioxidant and is required to produce collagen in connective tissues (including skin, mucous membranes, teeth and bones), synthesis of neurotransmitters and the synthesis of carnitine (important for fatty acid metabolism).

4.14. Lipid-Soluble Vitamins

Vitamin A (retinol) helps maintain healthy skin and mucous membranes, immune system regulation, cell division and differentiation, bone growth, reproduction and vision. Whilst chickpea does not contain vitamin A per se (only found in animal foods), they are rich sources of carotenoids. Carotenoids are primarily responsible for the yellow color of chickpea cotyledons and some carotenoids can be readily converted to vitamin A after consumption. Abbo et al. (2005) found that chickpea seeds contain a higher concentration of carotenoids than the engineered carotene-boosted, proof-of-concept 'golden rice'. There are more than 600 carotenoids identified from plants, of which only 10% are precursors for vitamin A. Of these, ß-carotene has been reported to be the most efficient precursor of vitamin A (Paiva and Russell, 1999). Chickpea contains up to 49 g/100g ß-carotene (Table 7) existing in both the cotyledon and seed coat (Atienza et al., 1998). In addition to serving as sources of vitamin A, some carotenoids have been shown to act as antioxidants in laboratory studies. Antioxidants help eliminate free radicals in the body, thereby protecting cells from potential damage. Free radicals are by-products of oxygen metabolism in the body that have been suggested as contributors to the development of many chronic diseases (Paiva and Russell, 1999). B-carotene has been shown to be protective against coronary heart disease (Gey et al., 1993). Chickpea contains other carotenoids that do not have vitamin A activity, such as lutein and zeaxanthin, but which are believed to have other health-promoting properties such as the prevention of atherosclerosis and age-related macular degeneration (Trumbo et al., 2001).

Vitamin E is fat-soluble and exists in at least eight different forms of either tocopherol or tocotrienol. Of these eight, only a-, B- and y-tocopherols and a- and B-tocotrienols are widespread in nature. The most important sources of vitamin E in the diet are the plant seed oils. Chickpea is a reasonable source as the seed contains ~3-9% lipids (Table 5) and up to 13.7 mg/100 g vitamin E (Table 6). The most important characteristic of vitamin E is as an antioxidant, more powerful than vitamins A or C. It has been reported to aid in the prevention of atherosclerosis and cardiovascular disease (by limiting oxidation of low-density lipoprotein (LDL) cholesterol and preventing formation of blood clots), cancer (by protecting cell membranes from free radical damage and blocking the formation of carcinogens in the stomach) and cataracts. Vitamin E also plays a role in immune function, DNA repair and other metabolic processes. Much of the literature has focused on a-tocopherol, the most active form and a powerful antioxidant in humans. y-tocopherol has also been shown to be potentially protective against chronic conditions such as inflammation. In addition, the natural admixture of tocopherol forms in natural vegetable oils has an additive and synergistic relationship that provides a higher antioxidant effect than any single form on its own and a broader range of health benefits (Saldeen and Saldeen, 2005).

Vitamin K (phylloquinone) is required for the modification of certain proteins required for blood coagulation, bone metabolism and vascular health. Chickpea contains a low concentration of vitamin K compared to leafy vegetables, but higher than fruits and most animal products (USDA, 2009).

5. Uses for Chickpea

Chickpea is a highly nutritious and an inexpensive source of protein that is estimated at 24% and ranges from 15% to 30% (Hulse, 1994) depending on variety and environmental conditions. Chickpea also has estimated 60–65% carbohydrates, 6% fat and is a good source of minerals and essential B vitamins. There are numerous uses for Desi and Kabuli types. They can be boiled, eaten raw as a fresh vegetable, roasted, dehulled to make dal or processed into flour that can be added to bread. Dal made mostly from Desi types, used in making a soup that is served with rice in most areas of South Asia providing a nutritious combination of a cereal grain and a pulse crop. It is well known that the amino acids of pulses, particularly those containing sulfur amino acids compensate for those that are limiting in the cereals.

Chickpeas (*Cicer arietinum L.*), commonly known as garbanzo beans, are an old-world pulse (i.e., edible seeds) in the legume family (DeSalvo et al., 2016) and have traditionally been incorporated into many culinary creations because of their nut-like flavor and versatile sensory applications in food (Deosthale, 1982). The origin of the chickpeas is thought to have been Levant and ancient Egypt (Davidson, 2014), which is logical since the plant prefers temperate and semiarid regions. India is the world's leading producer of chickpeas. Worldwide over 14.2 metric tons of chickpeas were harvested in 2014 according to the Food and Agriculture Organization (FAO) of the United Nations (FAO., 2019). Pulses are unique in comparison to other plant foods in that they contain higher proportions of protein (17%–30% by dry weight). The main proteins found in chickpeas, like other legumes, are albumins and globulins. Smaller amounts of glutelins and prolamines are also present (Saharan and Khetarpaul, 1994).

In Western culture, chickpea consumption is somewhat driven through the intake of hummus. Traditional hummus is a dip or spread made from cooked, mashed chickpeas, blended with tahini, olive oil, lemon juice, and spices. A variety of other forms of hummus—or bean-based dips labeled as hummus that do not follow the traditional hummus recipe—exist on the market, each containing unique ingredients which may or may not contribute to nutrient intakes and/or have benefits beyond basic nutrition. For instance, commercial processing of hummus may alter the profile and bioavailability of some nutrients. Over the past decade, the hummus market has grown to over $530M in U.S. sales alone as of 2013, increasing by over 25% since 2010. This could be in part due to increased consumer recognition of pulses (e.g., chickpeas) and their derived products or, in scientific terms, their high nutritional quality.

6. Improvement of Nutritional Properties of Bakery Products

Wheat flour is used as the major ingredient in most of bakery products and has been implicated in various health problems in small segments of the population. Highly refined wheat flour is usually used for bread production. Although, wheat is naturally a good source of proteins (8-12%), vitamins such as Vitamin E, minerals such as Iron, Zinc, and dietary fibers, substantial proportions of these nutrients are lost during milling and refining of the wheat grains for flour production. Excessive processing and milling lead to significant loss of fibers due to removal of the outer layer of the wheat (Muhammad Anjum et al., 2006). Also, wheat, like many cereals, lacks essential amino acids such as Lysine (Khetarpaul and Goyal, 2009). To overcome this problem, one solution is to fortify flour with substances that can supplement the essential nutrients, compensate for the lost nutrients during processing and milling, and reduce the risk of serious nutrition-related diseases such as cancer, diabetes, and cardiovascular diseases. Therefore, to provide nutritious food to consumers, one solution is to use less refined wheat flour and to fortify bread with substances that can compensate for the nutrient loss of wheat flour during processing.

Bread is consumed in all parts of the world and is one of the oldest foods known throughout history. Wheat is a major ingredient in bread. To meet the requirements of modern lifestyle, bread is often fortified with various substances to improve its nutrient content and taste. The United States of America is one of the leading countries in the fortification of food. For example, Folic acid has been added to flour, and niacin has been added to bread in the United States since 1938. Other countries have also done many recent studies on fortification. Folic acid is used for fortification of bread in Australia to prevent folic acid deficiency in the population, especially young girls and women. Similarly, flax seeds are added to bread formulation to increase its dietary fiber content and for supplementing it with omega-3 fatty acids. Furthermore, fruit bits and milk solids are added to bread to improve taste. Adding nutritional value to bread is one of the ways to provide healthy food to consumers. Nutrients are also added to get desired texture, physical and chemical properties, and to increase the shelf-life of the bread. Since, bread is consumed on a regular basis throughout the world, enrichment of bread with fiber and protein can potentially benefit people of all age groups in having healthy diet and in overcoming various nutritional problems. Moreover, the American Dietetic Association (2013) has reported lower than recommended intake of dietary fiber among US children and adults and has expounded the beneficial role of fibers in controlling diabetes. These facts again support the fortification of bread with fibers and other nutrients so as to ensure daily provision of healthy food to children and diabetics (Lafrance et al., 1998). Cereals being a relatively cheap source of protein and fibers, are an economical choice for fortification of bread. This makes them affordable and nutritious food for use with low income families and food relief programs.

Consumption of low-glycemic index (GI) foods, have been shown to improve glucose tolerance in human subjects. The estimated cost of diabetes in the US is $245 Billion, and it is expected to rise by 53% to more than $622 billion dollars between the years 2015-2030 (Rowley et al., 2017). While the consumption of low glycemic response foods (LGR) has increased in recent years (Riccardi et al., 2008), there is a need for a more diverse range of such foods in the market that are also affordable. A potential solution to manage diabetes cost is to consume foods that have a low GI. Low GI diets are more expensive than the higher GI equivalents, which affect the consumer buying behavior and food choice (Cleary et al., 2012). Malnutrition is a complex problem and has become a major problem in different countries around the world. Different programs and potential solutions have been suggested as answers to the problem of malnutrition, but most of these are long-term solutions. One of the solutions is the development of cereal and cereal-based products for food supplementations. These products are used throughout the world as inexpensive energy and protein sources.

Chickpea flour is excellent source for fortification of wheat flour in order to enhance its nutritional content. Chickpea is known to be of excellent nutritional quality. It is rich in carbohydrates, proteins, vitamins and minerals. It is low in fat and sodium content, and as such, it is beneficial for diabetics and hypertensive individuals. It is cholesterol-free and a significant source of both soluble and insoluble fiber. In the scientific literature, it has been reported that chickpea confers various health benefits such as lowering of glycemic index (GI) of diabetic patients, increase in satiety, cancer prevention, and protection against cardiovascular disorders due to its high dietary fiber content (Jukanti et al., 2012). Chickpea seeds are eaten as fresh, boiled, canned, roasted, or fried products. It is ground into powder and used for making various fried snacks. In the Middle Eastern and Mediterranean countries, it is also used as an ingredient in bread making.

7. Chickpea Fortifications

Adding nutritional value to bread is one of the ways to provide healthy food to consumers. Nutrients are also added to get desired texture, physical and chemical properties, and to increase shelf-life of the bread. Since, bread is consumed on a regular basis throughout the world. Fortification of bread with fiber and protein can potentially be beneficial for people of all age groups in having healthy diet and in overcoming various nutritional problems. Moreover, the American Dietetic Association (2013) has reported lower than recommended intake of dietary fiber among US children and adults and have advocated the beneficial role of fibers in controlling diabetes. These facts again support the fortification of bread with fibers to ensure providing daily healthy food to children and diabetics (Lafrance et al., 1998). Chickpeas being a relatively cheap source of protein and fibers, are an economical choice for fortification of bread. This makes them affordable for use in feeding low income families and for use in food relief. Table 9, presents some studies that have been summarized which used chickpeas for the fortification purpose.

Table 9. Chickpeas fortification studies

Title of the study	Author/year	Brief description of study
Lactobacillus plantarum ITM21B fermentation product and chickpea flour enhance the nutritional profile of salt reduced bakery products	Di Biase et al., (2019)	The use of chickpea flour was validated at industrial pilot plant producing salt reduced sliced bread and puccia type bread. The resulting products showed improved nutritional profile and a sensory quality comparable to the company's products containing salt
Effect of Processed Chickpea Flour Incorporation on Sensory Properties of Mankoushe Zaatar	Dandachy et al., (2019)	Hence, conventionally processed chickpea flour can be used as a fortifier to improve the nutritional quality of bakery products without negatively affecting their sensory properties
Effects of fermented-chickpea liquor (chickpea yeast) on whole-grain wheat flour bread properties	Sayaslan & Şahin, (2018)	Fermented-chickpea liquor incorporation into whole-grain wheat flour bread formulations at 30 and 45% levels, respectively, increased loaf volume about 10 and 25%, decreased crumb hardness approximately 10 and 15%, and enhanced antioxidant activity (about 10%) of whole-grain wheat flour breads.
Optimization of the Level of Chickpea Sourdough and Baking Powder in Cake Formulation by Response Surface Methodology: Effects on Physicochemical, Sensory and Antioxidant Properties	Ertop and Hayta, (2018)	The aim of this study was to optimize the amounts of chickpea sourdough (A) and baking powder (B) in cake formulation. The cake also acquired slightly sour, sweetish and soft sensory profiles. DPPH radical scavenging activity and physicochemical properties were also improved by the use of the chickpea sourdough.
Shelf-life, physicochemical, and nutritional properties of wheat bread with optimized amount of dried chickpea sourdough and yeast by response surface methodology	Hendek Ertop and Coşkun, (2018)	The aim of this study was to optimize of the instant active dry yeast (A) and the dried chickpea sourdough (B) amount in bread formulation. The bread texture; antioxidant activity, mineral content, microbial shelf life were improved, and the volume and crust thickness of the bread increased by the usage of chickpea sourdough.

8. Chickpeas and Health

Pulses have been used for their nutritional qualities for thousands of years (Jukanti et al., 2012). The interest in pulses as food and their potential impact on human health have been revived, during the past two to three decades. It is also reported that many pulses overcome the risk of chronic diseases and optimize health. Therefore, chickpea is considered as a 'functional food' along with its role in providing protein and fiber. Chickpea contains different vitamins, minerals and several bioactive constituents (phenolic, phytates, enzyme inhibitors, and oligosaccharides, etc.) that could help to reduce the risk of chronic diseases.

8.1. Diabetes and Blood Pressure

Chickpea have a higher quantity of resistant starch and amylose. The starch of chickpea is more resistant to digestion in the small intestine, which lowers availability of glucose (Pittaway et al., 2008). There are several other studies which relate to the use of Chickpea in treatment of diabetes and blood pressure (summarized in the Table 10).

Table 10. Use of Chick pea diet in diabetes and blood pressure

Title of the study	Author/ year	Brief description of study & results
Pasta added with chickpea flour: chemical composition, in vitro starch digestibility and predicted glycemic index.	Osorio-Diaz P et al., (2008)	Protein, ash, lipid, and fiber content increased while total starch decreased with the chickpea flour level in the composite pasta. The starch hydrolysis index (HI) decreased as chickpea flour in the pasta increased, reflecting the slow and low digestion of the starch in chickpea. Predicted glycemic index was lower in spaghetti added with chickpea flour than in durum wheat control pasta.
Chickpeas may influence fatty acid and fiber intake in an ad libitum diet, leading to small improvements in serum lipid profile and glycemic control.	Pittaway et al., (2008)	Incorporating chickpeas in the habitual ad libitum intake of 45 healthy participants for 12 weeks resulted in reduced serum total cholesterol, fasting insulin concentration. This may benefit in a more hypercholesterolemic and hyperglycemic population.
The effect of yellow pea protein and fiber on short-term food intake, subjective appetite and glycemic response in healthyyoung men	Smith et al., (2012)	Yellow pea consumption suppressed mean pre-meal BG compared to control. In conclusion, protein is the component responsible for the short-term effects of yellow peas in the regulation of glycaemia.
Chickpea flour ingredient slows glycemic response to pasta in healthy volunteers	Goni and Valentin-Gamazo (2003)	Incorporation of 25% of chickpea flour into wheat pasta significantly lowered starch hydrolysis than in white bread. Chickpea flour, evidently provide a food with a low glycemic response and could help in achieving a wider range of low-GI foods.

8.2. Weight Loss/Obesity

Dietary fiber may influence body-weight regulation by physiologic mechanisms involving intrinsic, hormonal, and colonic effects. Ultimately, these mechanisms act to decrease food intake by promoting satiation (lower meal energy content) or satiety (longer duration between meals) or by influencing metabolic fuel partitioning (increased fat oxidation and decreased fat storage). Therefore, it is concluded that fiber-rich diets, contain non-starch fruits, vegetables, whole grains, nuts and legumes, and may be effective in the prevention and treatment of obesity in children (Pereira and Ludwig, 2001). The use of low-GI foods resulted in an increase in cholecystokinin (hunger suppressant) and increased satiety (Swinburn et al., 2004). Chickpea is a low-GI food, and therefore may helpful in the reduction of obesity as well as in weight-loss. Presented Table 11, summarizes the reports related to chickpea diet and weight loss/obesity.

Table 11. Use of Chick pea diet in Obesity/weight loss

Title of the study	Author/ year	Brief description of study & results
Dietary chickpea reverses visceral adiposity, dyslipidaemia and insulin resistance in rats induced by a chronic high-fat diet.	Yang, et al., (2007)	Chickpea supplementation in the diet prevented increased body weight and weight of epididymal adipose tissues. Chickpea is reported to decrease fat accumulation in obese subjects. This aids in improving fat metabolism and could be helpful in correcting obesity-related disorders
Chickpea supplementation in an Australian diet affects food choice, satiety and bowel function	Murty et al., (2010)	Chickpea supplementation in the diet resulted in increased satiation and fullness.

8.3. CVD, CHD and Cholesterol Control

Table 12. Use of Chick pea diet in CVD, CHD and cholesterol control

Title of the study	Author/ year	Brief description of study & results
A hypocaloric diet enriched in legumes specifically mitigates lipid peroxidation in obese subjects	Crujeiras et al. (2007)	Fiber-rich chickpea-based pulse diet has been shown to reduce the total plasma cholesterol levels in obese subjects.
A pulse-based diet is effective for reducing total and LDL cholesterol in older adults	Abeysekara et al. (2012)	Pulse-based diet is effective for reducing LDL-C and total cholesterol in older adults and that's why reduces the risk of CVD.
Non-soy legume consumption lowers cholesterol levels: a meta-analysis of randomized controlled trials	Bazzano et al. (2011)	Pulse-rich diet decreases total and LDL cholesterol.
Dietary Supplementation with Chickpeas for at Least 5 Weeks Results in Small but Significant Reductions in Serum Total and Low- Density Lipoprotein Cholesterols in Adult Women and Men	Pittaway et al. (2006)	Inclusion of chickpeas in an intervention diet results in lower serum total and low-density lipoprotein cholesterol levels as compared with a wheat supplemented diet.

Foods that contain high amounts of soluble fiber reduced the total cholesterol of serum and LDL-cholesterol (LDL-C) and have an inverse correlation with CHD mortality (Kushi et al., 1999). Chickpea seeds are a comparatively high source of DF and bioactive compounds (e.g., saponins, phytosterols, and oligosaccharides) as well as low glycemic index (GI), therefore chickpea reduced the risk of CVD (Duranti, 2006). Table 12 presents some research that supports the health benefits of chickpea and heart disease.

8.4. Other Health Benefits

Chickpea seeds contain sterols, tocopherols and tocotrienols exhibit anti-ulcerative, anti-bacterial, anti-fungal, antitumour and anti-inflammatory properties (Murty, et al., 2010). Carotenoids such as lutein and zeaxanthin, present in chickpea seeds, play a role in senile or age-related macular degeneration (Mozaffarieh et al., 2003). Vitamin A is significant in numerous developmental processes in humans such as cell division, bone growth and most importantly, vision. In traditional medicine, the chickpea seeds have been used as tonics, stimulants, and aphrodisiacs. They are also used as appetizers, for thirst quenching and reducing burning sensation in the stomach (Aguilera, et al. 2009). In addition to these applications, chickpea seeds are also used for treating skin ailments, ear infections, blood enrichment, and liver and spleen disorders. For over 2500 years, the Uygur people of China have used chickpea in herbal medicine for treating diabetes and hypertension (Zhang et al., 2007).

9. Antioxidant in Chickpea

The consumption of legumes has been associated with decreasing incidence of diseases, a feature that relates to their high content of antioxidant phenolics, low lipid content, and low glycemic index. Legumes such as chickpeas is seen as staple food and it's nutritious and improve health, known as the meat of the poor people because of its high protein and fiber content. Chickpea is now presented as a staple food for vegetarians and for people affected by nutrition related health problems, such as diabetes, obesity, and over-weight. Such a trend is caused by a general feature of pulses, namely, their appreciable content in slowly digestible carbohydrates (Silva-Cristobal et al., 2010). Phenolic compounds are abundant in legumes and their flours. Phenolics from legume flours are potentially safer, and therefore may be preferred alternatives for inhibition of carbohydrate breakdown and control of glycemic index of food products. Therefore, utilization of legume flours in the development of functional foods with increased therapeutic value would be a significant step toward disease prevention and management through diet. Chickpea is an excellent source of protein, dietary fiber, a variety of micronutrients and phytochemicals with potential health benefits. Chickpea with lower a-amylase and higher a-glucosidase inhibitory activities could be used as food ingredients and in composite flours for the delay absorption of dietary carbohydrates in the meal, leading to suppression of an increase in postprandial blood glucose level without adverse effects. Due to favorable flour functionality and phytochemical associated health benefits, chickpea offers enormous potential for the production of legume composite flours (Sreerama et al., 2012). Legumes contain other bioactive compounds beside phenolics such as vitamins and carotenoids that might also behave as antioxidant. Carotenoids are fat-soluble pigments (Jayawardena et al., 2015). Epidemiological studies have shown a positive correlation between ingestion of vegetables and fruits containing carotenoids and prevention of several chronic diseases. The health-promoting properties of carotenoids are due to their free radical scavenging activity through the stabilization of single oxygen by its conjugate double bounds (Quesada et al., 2011).

10. Effect of Chickpea Supplementation on the Rheological Properties of Bakery Products

The rheological properties of the dough play a very important role in predicting the characteristics of the final product. The gluten network is the main component of the wheat flour dough, which is formed primarily from the gliadin and glutenin, which give the dough its ideal structure. Therefore, when wheat flour is fortified with any other component, an interfere of this component with the gluten network occurs, which affects its rheological properties. This can be measured by instruments of rheological properties, as well as studying the properties of resulting bread to determine the optimum addition which improve the nutritional value of wheat bread dough with optimum technological characteristics.

The farinograph was invented in the 1930s—the first of the physical dough testing instruments (Janssen et al., 1996) and it is the most popular of the empirical rheological instruments. The farinograph is one of the most widely used recording dough mixers. The two Z-shaped blades of the farinograph mixer rotate at constant speeds and subject the dough to mixing at constant temperature. The farinograms generated from testing a flour are analyzed to obtain quantitative information on arrival time (time required for the top of the curve to cross the 500 Brabender Unit (BU) line), peak time (time required for dough to reach maximum consistency without breakdown), departure time or time to breakdown (time lapsed until the top of the curve permanently drops below the 500 BU line), stability (time difference between arrival time and departure time), mixing tolerance index (MTI, the difference in BU's between the peak time and peak time plus 5min), and water absorption. Farinographs are commonly used for determining the water absorption of flour, especially in industrial settings (Mondal and Datta, 2008). From the 1st step in the breadmaking process (blending of flour and water with other ingredients) to the final step (baking), the ingredients used undergo several physical and chemical changes (Faridi and Faubion, 1990) such as evaporation of water, formation of porous structure, volume expansion, protein denaturation, starch gelatinization, and crust formation take place during bread baking. Crumb structure of cereal products like bread is a very important factor determining the sensorial quality as may be quantified for example as texture or crispness as well as storage and staling properties (Regier et al., 2007).

The supplementation of wheat flour with chickpea flour would greatly improve the protein nutritional quality of bread. This high protein content in the chickpea supplemented breads would be of nutritional importance in most developing countries, such as Africa and Asia, where many people can hardly afford high portentous foods because of their expensive costs. Chickpea cultivars are broadly divided into two groups, desi and Kabuli. Kabuli seeds are large and light-colored beans, and are characterized by their larger size, ram-head shape and low fiber content (Singh et al., 1991). The seeds of desi cultivars are small, wrinkled at beak, with brown, black or green color. The influence of additives of non-bakery crops on the pasting behavior is related mainly to their different (in comparison to wheat) starch and protein complex. The addition of legumes to wheat flour delays the pasting temperature and deteriorates the baking properties of bread. The unique bread making properties of wheat flour can be attributed mainly to the ability of its gluten proteins to form a viscoelastic network when mixed with water. The reduction of viscoelastic properties of wheat flour dough, after substitution by legume flour, reduces bread making potential. Knorr and Betschart (1978) suggested that the weakening effect of foreign proteins on wheat flour doughs was the result of a dilution of the gluten structure by the added protein. The farinograph is one of the most widely used recording dough mixers. The two Z-shaped blades of the farinograph mixer rotate at constant speeds and subject the dough to mixing at constant temperature. The farinogram generated from testing a flour are analyzed to obtain quantitative information on arrival time (time required for the top of the curve to cross the 500 Brabender Unit (BU line), peak time (time required for dough to reach maximum consistency without breakdown), departure time or time to breakdown (time lapsed until the top of the curve permanently drops below the 500 BU line), stability (time difference between arrival time and departure time), mixing tolerance index (MTI, the difference in BU's between the peak time and peak time plus 5 min), and water absorption. Farinograph is commonly used for determining the water absorption of flour, especially in industrial settings (Mondal and Datta, 2008).

Full understanding of the rheological behavior of flour dough is of great importance from the practical point of view. Dough rheology directly affects the baking performance of flours, and rheological analyses have been made in order to optimize dough formulation. Although dough rheology has long been investigated, there remains a significant lack of understanding. This lacks progress is due to the complexity of this biological system (Masi et al., 2001).

REFERENCES

Abbo, S., Molina, C., Jungmann, R., Grusak, M., Berkovitch, Z., Reifen, R., . . . Reifen, R. (2005). Quantitative trait loci governing carotenoid concentration and weight in seeds of chickpea (Cicer arietinum L.). *Theoretical and Applied Genetics, 111*. 195-185,(2).

Abeysekara, S., Chilibeck, P. D., Vatanparast, H., & Zello, G. A. (2012). A pulse-based diet is effective for reducing total and LDL-cholesterol in older adults. *British Journal of Nutrition, 108*(S1), S103-S110.

Adebowale, Y., Adeyemi, A., & Oshodi, A. (2005). Variability in the physicochemical, nutritional and antinutritional attributes of six Mucuna species. *Food chemistry, 89*(1), 37-48.

Agrawal, K., & Singh, G. (2003). Physico-chemical and milling quality of some improved varieties of chickpea (Cicer arietinum). *Journal of food science and technology, 40*(4), 439-442.

Agrawal, P., & Bhattacharya, L. (1980). Proximate composition of seeds of improved varieties of Bengal gram (Cicer arietinum Linn.). *Seed Research, 8*(1), 5-14.

Aguilera, Y., Martín-Cabrejas, M. A., Benítez, V., Mollá, E., López-Andréu, F. J., & Esteban, R. M. (2009). Changes in carbohydrate fraction during dehydration process of common legumes. *Journal of Food Composition and Analysis, 22*(7-8), 678-683.

Amirshahi, M., & Tavakoli, M. (1970). *Protein content of different varieties of five species of pulse crops.* Paper presented at the Improving plant protein by nuclear techniques. Proceedings of a symposium, Vienna, 8-12 June 1970. Jointly organized by the IAEA and FAO.

Asp, N. G., & Björck, I. (1992). Resistant starch. *Trends in Food Science & Technology, 3*, 111-114.

Association, A. D. (2013). Economic costs of diabetes in the US in 2012. *Diabetes care, 36*(4), 1033-1046.

Atienza, J., Sanz, M., Herguedas, A., Alejos, J. & Jimenez, J. (1998). Note. β—Carotene, α—tocopherol and γ—tocopherol contents in dry legumes. Influence of cooking Nota. Contenido de β—caroteno, α—tocopherol y γ—tocopherol en legumbres secas. Influencia de la cocción. *Food science and technology international, 4*(6), 437-441.

Avancini, S., Sales, A., De Aguirre, J., & Mantovani, D. (1993). Chemical composition and nutritional value of chickpea cultivars grown in the state of Sao Paulo. *COLETANEA-INSTITUTO DE TECNOLOGIA DE ALIMENTOS, 22*, 145-145.

Awasthi, C., Abidi, A., & Chowdhury, A. (1991). Studies on the nutritional quality of different varieties of chickpea. *Indian Journal of Agricultural Research, 25*(1), 21.

Babu, S., & Srikantia, S. (1976). Availability of folates from some foods. *The American journal of clinical nutrition, 29*(4), 376-379.

Batterham, E. S. (1989). Use of chickpeas (Cicer arietinum L.) in animal breeding. *Nutrition Reviews, 56*, 317–333.

Bazzano, L. A., Thompson, A. M., Tees, M. T., Nguyen, C. H., & Winham, D. M. (2011). Non-soy legume consumption lowers cholesterol levels: a meta-analysis of randomized controlled trials. *Nutrition, metabolism and cardiovascular diseases, 21*(2), 94-103.

Bellido, L., & Fuentes, M. (1990). Cooking quality of chickpea. *Options Mediterraneennes-Serie Seminire, 9*, 113-125.

Biliaderis, C., Maurice, T., & Vose, J. (1980). Starch gelatinization phenomena studied by differential scanning calorimetry. *Journal of food science, 45*(6), 1669-1674.

Bravo, L. (1999). Effect of processing on the non-starch polysaccharides and in vitro starch digestibility of legumes/Efecto del procesado en el contenido de polisacáridos no amiláceos y la digestibilidad in vitro del almidón de legumbres. *Food science and technology international, 5*(5), 415-423.

Carré, B., Melcion, J., Widiez, J., & Biot, P. (1998). Effects of various processes of fractionation, grinding and storage of peas on the digestibility of pea starch in chickens. *Animal Feed Science and Technology, 71*(1-2), 19-33.

Choct, M. (1997). Feed non-starch polysaccharides: chemical structures and nutritional significance. *Feed milling international, 191*(June issue), 13-26.

Cleary, J., Casey, S., Hofsteede, C., Moses, R., Milosavljevic, M., & Brand-Miller, J. (2012). Does a low glycaemic index (GI) diet cost more during pregnancy? *Nutrients, 4*(11), 1759-1766.

Coppen, A., & Bolander-Gouaille, C. (2005). Treatment of depression: time to consider folic acid and vitamin B12. *Journal of Psychopharmacology, 19*(1), 59-65.

Crujeiras, A. B., Parra, D., Abete, I., & Martínez, J. A. (2007). A hypocaloric diet enriched in legumes specifically mitigates lipid peroxidation in obese subjects. *Free radical research, 41*(4), 498-506.

Dalgetty, D. D., & Baik, B. K. (2003). Isolation and characterization of cotyledon fibers from peas, lentils, and chickpeas. *Cereal chemistry, 80*(3), 310-315.

Dandachy, S., Mawlawi, H., & Obeid, O. (2019). Effect of Processed Chickpea Flour Incorporation on Sensory Properties of Mankoushe Zaatar. *Foods, 8*(5), 151.

Dang, J., Arcot, J., & Shrestha ,A. (2000). Folate retention in selected processed legumes. *Food chemistry, 68*(3), 295-298.

Davidson, A. (2014). *The Oxford companion to food*: OUP Oxford.

de Almeida Costa, G. E., da Silva Queiroz-Monici, K., Machado Reis, S. M. P., & Costa de Oliveira ,A. (2006). Chemical composition, dietary fibre and resistant starch contents of raw and cooked pea, common bean, chickpea and lentil legumes. *Food chemistry, 94*(3), 327-330.

Deosthale, Y. (1982). *Food processing and nutritive value of legumes.* Paper presented at the Pulse production, constraints and opportunities: proceedings of Symposium on Increasing Pulse Production in India--Constraints and Opportunities, October 1982, New Delhi, India/edited by HC Srivastava...[et al.].

Dervas, G., Doxastakis, G ,.Hadjisavva-Zinoviadi, S., & Triantafillakos, N. (1999). Lupin flour addition to wheat flour doughs and effect on rheological properties. *Food chemistry, 66*(1), 67-73.

DeSalvo, K. B., Olson, R., & Casavale, K. O. (2016). Dietary guidelines for Americans . *Jama, 315*(5), 457-458.

Deshpande, S. (1992). Food legumes in human nutrition: a personal perspective. *Critical Reviews in Food Science & Nutrition, 32*(4), 333-363.

Di Biase, M., Bavaro, A. R., Lonigro, S. L., Pontonio, E., Conte, A., Padalino, L . . ,. .Valerio, F. (2019). Lactobacillus plantarum ITM21B fermentation product and chickpea flour enhance the nutritional profile of salt reduced bakery products. *International journal of food sciences and nutrition*, 1-13.

Dodok, L., Modhir, A. A., Hozova, B ,.Halasova, G., & Polacek, I. (1993). Importance and utilization of chickpea in cereal technology. *Acta alimentaria, 22*(2), 119-129.

Doxastakis, G., Zafiriadis, I., Irakli, M., & Tananaki, C. (2002). Lupin, soya and triticale addition to wheat flour doughs and their effect on rheological properties. *Food chemistry, 77*(2), 219-227.

Duranti, M. (2006). Grain legume proteins and nutraceutical properties. *Fitoterapia, 77*(2), 67-82. doi:10.1016/j.fitote.2005.11.008.

El Hardollou, S. B., El Tinay, A. H., & Nour ,A. A. M. (1980). Chemical characteristics of some legumes grown in the Sudan. *Sudan Journal of Food Science and Technology, 12,* , 35–42.

Eliasson, A. C. (1990). Rheological properties of cereal proteins in Dough rheology and baked product texture. In H .Faridi & J. Faubion (Eds.), (pp. 67-110). Reinhold: Van Nostrand.

Englyst, H. N., Kingman, S. M., & Cummings, J. H. (1992). *Eur. J. Clin. Nutr., 46*(Suppl. 2), S33.

Ertop, M. H., & Hayta, M. (2018). Optimization of the Level of Chickpea Sourdough and Baking Powder in Cake Formulation by Response Surface Methodology: Effects on Physicochemical, Sensory and Antioxidant Properties. *Food Science and Technology Research, 24*(4), 697-706.

FAO. (2001). *Bulletin of Statistics*. Retrieved from

FAO. (2019). FAOSTAT Statistical Database of the United Nation Food and Agriculture Organization (FAO) statistical division. Rome.

Faridi, H., & Faubion, J. M. (1990). *Dough rheology and baked product texture*. New York: Van Norstrand Reinhold.

Farrell, D., Perez-Maldonado ,R., & Mannion, P. (1999). Optimum inclusion of field peas, faba beans, chick peas and sweet lupins in poultry diets. II. Broiler experiments. *British Poultry Science, 40*(5), 674-680.

Farzana, W., & Khalil, I. A. (1999). Protein quality of tropical food legumes. *Journal of Science and Technology, 23*, 13-19.

Fenwick, D. E., & Oakenfull, D. (1983). Saponin content of food plants and some prepared foods. *Journal of the Science of Food and Agriculture, 34*(2), 186-191.

Fernandez, M. L., & Berry, J. W. (198 .(9Rheological properties of flour and sensory characteristics of bread made from germinated chickpea. *International Journal of Food Science & Technology, 24*(1), 103-110.

Gdala, J. (1998). Composition, properties, and nutritive value of dietary fibre of legume seeds. A review. *Journal of Animal and Feed Sciences, 7*(2), 131-149.

Gey, K. F., Moser, U., Jordan, P., Stähelin, H., Eichholzer, M., & Lüdin, E. (1993). Increased risk of cardiovascular disease at suboptimal plasma concentrations of essential antioxidants: an epidemiological update with special attention to carotene and vitamin C. *The American journal of clinical nutrition, 57*(5), 787S-797S.

Goni, I., & Valentin-Gamazo, C. (2003). Chickpea flour ingredient slows glycemic response to pasta in healthy volunteers. *Food chemistry, 81*(4), 511-515.

Grusak, M. A., & DellaPenna, D. (1999). Improving the nutrient composition of plants to enhance human nutrition and health. *Annual review of plant biology, 50*(1), 133-161.

Güémes-Vera, N., Peña-Bautista ,R. J., Jiménez-Martínez, C., Dávila-Ortiz, G., & Calderón-Domínguez, G. (2008). Effective detoxification and decoloration of Lupinus mutabilis seed derivatives, and effect of these derivatives on bread quality and acceptance. *Journal of the Science of Food and Agriculture, 88*(7), 1135-1143. doi:10.1002/jsfa.3152.

Hedley, C. L. (2000). *Carbohydrates in grain legume seeds: improving nutritional quality and agronomic characteristics*: CABI.

Hefni, M., & Witthöft, C. M. (2014). Folate content in processed legume foods commonly consumed in Egypt. *LWT-Food Science and Technology, 57*(1), 337-343.

Hendek Ertop, M., & Coşkun, Y. (2018). Shelf-life, physicochemical, and nutritional properties of wheat bread with optimized amount of dried chickpea sourdough and yeast by response surface methodology. *Journal of Food Processing and Preservation, 42*(7), e13650.

Hulse, J. (1994). Nature, composition, and utilization of food legumes. In *Expanding the production and use of cool season food legumes* (pp. 77-97): Springer.

Iqbal, A., Khalil, I. A., Ateeq, N., & Khan, M. S. (2006). Nutritional quality of important food legumes. *Food chemistry, 97*(2), 331-335.

Iqbal, A., Khalil, I. A., & Shah, H. (2003). Nutritional yield and amino acid profile of rice protein as influenced by nitrogen fertilizer. *Sarhad Journal of Agriculture (Pakistan).*

Jambunathan, R., & Singh, U. (1980). *Studies on Desi and Kabuli chickpea (Cicer arietinum) cultivars. 1. Chemical composition.* Paper presented at the Proceedings of the international workshop on chickpea improvement.

Jambunathan, R., & Singh, U. (1981). Studies on desi and kabuli chickpea (Cicer arietinum L.) cultivars. 3. Mineral and trace elements composition. *Journal of Agricultural and Food Chemistry, 29*(5), 1091-1093.

Janssen, A. M ,.van Vliet, T., & Vereijken, J. M. (1996). Fundamental and Empirical Rheological Behaviour of Wheat Flour Doughs and Comparison with Bread Making Performance. *Journal of Cereal Science, 23*(1), 43-54.

Jaya, T., Venkataraman, L., & Krishnamurthy, K .(1979) .Influence of germinated legumes on the levels of tissue cholesterol and liver enzymes of hypercholesterolemic rats: Influence of ungerminated and germinated chickpea and greengram (whole). *Nutrition Reports International (USA).*

Jayawardena, N., Watawana, M. I., Jayathilaka, R. T., & Waisundara, V. Y. (2015). The antioxidant and starch hydrolase inhibitory activity of ten spices in an in vitro model of digestion: bioaccessibility of anthocyanins and carotenoids. *Evidence-Based Complementary and Alternative Medicine, 2015.*

Jukanti, A. K., Gaur, P. M., Gowda, C., & Chibbar, R. N. (2012). Nutritional quality and health benefits of chickpea (Cicer arietinum L.): a review. *British Journal of Nutrition, 108*(S1), S11-S26.

Kamath, M. V., & Belavady, B .(1980) .Unavailable carbohydrates of commonly consumed Indian foods. *Journal of the Science of Food and Agriculture, 31*(2), 194-202.

Kaur, M., & Singh, N. (2005). Studies on functional, thermal and pasting properties of flours from different chickpea (Cicer arietinum L.) cultivars. *Food chemistry, 91*(3), 403-411.

Kerem, Z., German-Shashoua, H., & Yarden, O. (2005). Microwave-assisted extraction of bioactive saponins from chickpea (Cicer arietinum L). *Journal of the Science of Food and Agriculture, 85*(3), 406-412.

Khetarpaul, N., & Goyal, R. (2009). Effect of composite flour fortification to wheat flour on the quality characteristics of unleavened bread. *British Food Journal, 111*(6), 554-564.

Knights, E., & Mailer, R. (1989). Association of seed type and colour with establishment, yield and seed quality in chickpea (Cicer arietinum). *The Journal of Agricultural Science, 113*(3), 325-330.

Knorr, O., & Betschart, A. A. (1978). The relative effect of an inert substance and protein concentrates upon loaf volume breads. *Lebensm. Wiss. Technol., 11*, 198-207.

Krishna-Murti, C. (1975). Biochemical studies on Bengal gram. *Biochem Rev Bangalore.*

Kumara, C. D., & Deb, U. (2014). *Proceedings of the "8th International Conference viability of small farmers in Asia.* Paper presented at the International Conference on Targeting of Grain Legumes for Income and Nutritional Security in South Asia, Savar, Bangladesh.

Kushi, L. H., Meyer, K. A., & Jacobs Jr, D. R. (1999). Cereals, legumes, and chronic disease risk reduction: evidence from epidemiologic studies. *The American journal of clinical nutrition, 70*(3), 451s-458s.

Lafrance, L., Rabasa-Lhoret, R., Poisson, D., Ducros, F., & Chiasson, J. L. (1998). Effects of different glycaemic index foods and dietary fibre intake on glycaemic control in type 1 diabetic patients on intensive insulin therapy. *Diabetic Medicine, 15*(11), 972-978.

Madhusudhan, B., & Tharanathan, R. N. (1995). Legume and Cereal Starches — Why Differrences in Digestibility? Part 1: Isolation and Composition of Legume (Greengram and Bengalgram) Starches. *Starch - Stärke, 47*(5), 165-171. doi:10.1002/star.19950470502.

Masi, P., Cavella, S., & Piazza, L. (2001). An interpretation of the rheological behavior of wheat flour dough based on fundamental tests. In P. Chinachoti & Y. Vodovotz (Eds.), *Bread staling*. New York: CRC press Boca Raton Boston. (Reprinted from: NOT IN FILE).

Mathur, K., Khan, M., & Sharma, R. (1968). Hypocholesterolaemic effect of Bengal gram: a long-term study in man. *British Medical Journal,* 1. 30, (5583).

Mazur, W. (1998). 11 Phytoestrogen content in foods. *Baillière's Clinical Endocrinology and Metabolism, 12*(4), 729-742.

Mittal, M., Nigam, P. K., Rao, S. K., & Rao, S. (1999). Effect of environment on seed quality composition in chickpea *Legume Research, 22* 211–221.

Mondal, A., & Datta, A. (2008). Bread baking _ A review. *Journal of Food Engineering, 86*(4), 465-474.

Mozaffarieh, M., Sacu, S., & Wedrich, A. (2003). The role of the carotenoids, lutein and zeaxanthin, in protecting against age-related macular degeneration: a review based on controversial evidence. *Nutrition journal, 2*(1), 20.

Muhammad Anjum, F., Rauf, M., Issa Khan, M., & Hussain, S. (2006). Preparation of low calorific fiber rich cakes by wheat bran supplementation. *Nutrition & Food Science, 36*(6), 438-444.

Murty, C. M., Pittaway, J. K., & Ball, M. J. (2010). Chickpea supplementation in an Australian diet affects food choice, satiety and bowel health. *Appetite, 54*(2), 282-288.

Nedumaran, S., Abinaya, P., Jyosthnaa, P., Shraavya, B., Rao, P., & Bantilan, C. (2015). Grain Legumes Production, Consumption and Trade Trends in Developing Countries; Working Paper Series No. 60.

Ocampo, B., Robertson, L. D., & Singh, K. B. (1998). Variation in seed protein content in the annual wild Cicer species. *Journal of the Science of Food and Agriculture, 78*(2), 220-224.

Osorio-Diaz, P., Agama-Acevedo, E., Mendoza-Vinalay, M., Tovar, J., & Bello-Perez, L. A. (2008). Pasta added with chickpea flour: Chemical composition, in vitro starch digestibility and predicted glycemic index pasta adicionada con harina de garbanzo: Composición química, digestibilidad in vitro del almidóny predicción del índice glucémico. *CYTA-Journal of Food, 6*(1), 6-12.

Paiva, S. A., & Russell, R. M. (1999). β-carotene and other carotenoids as antioxidants. *Journal of the American college of nutrition, 18*(5), 426-433.

Pereira, M. A., & Ludwig, D. S. (2001). Dietary fiber and body-weight regulation: observations and mechanisms. *Pediatric Clinics of North America, 48 (4) 969-980.*

Perez-Hidalgo, M. A., Guerra-Hernandez, E., & Garcia-Villanova, B. (1997). Dietary Fiber in Three Raw Legumes and Processing Effect on Chick Peas by an Enzymatic-Gravimetric Method. *Journal of Food Composition and Analysis, 10*(1), 66-72.

Petterson, D., Sipsas, S., & Mackintosh, J. B. (1997). *The chemical composition and nutritive value of Australian pulses*: Grains Research and Development Corporation.

Pittaway, J., Ahuja, K., Cehun, M., Chronopoulos, A., Robertson, I., Nestel, P., & Ball, M .(2006) .Dietary supplementation with chickpeas for at least 5 weeks results in small but significant reductions in serum total and low-density lipoprotein cholesterols in adult women and men. *Annals of Nutrition and Metabolism, 50*(6), 512-518.

Pittaway ,J. K., Robertson, I. K., & Ball, M. J. (2008). Chickpeas may influence fatty acid and fiber intake in an ad libitum diet, leading to small improvements in serum lipid profile and glycemic control. *Journal of the American Dietetic Association, 108*(6), 100. 1013-9.

Pollard, N. J., Stoddard, F. L., Popineau, Y., Wrigley, C. W., & MacRitchie, F. (2002). Lupin flours as additives: Dough mixing, breadmaking, emulsifying, and foaming. *Cereal chemistry, 79*(5), 662-669.

Quemener, B., & Brillouet, J.-M. (1983). Ciceritol, a pinitol digalactoside form seeds of chickpea, lentil and white lupin. *Phytochemistry, 22*(8), 1745-1751.

Quesada, S., Azofeifa, G., Jatunov, S., & Jiménez, G. (2011). Carotenoids composition, antioxidant activity and glycemic indexof two varieties of bactris gasipaes. *Emirates Journal of Food and Agriculture*, 482-489.

Regier, M., Hardy, E. H., Knoerzer, K., Leeb, C. V., & Schuchmann, H. P. (2007). Determination of structural and transport properties of cereal products by optical scanning, magnetic resonance imaging and Monte Carlo simulations. *Journal of Food Engineering, 81*, 485-491.

Ribotta, P. D., Arnulphi, S. A., Leôn, A. E., & Anôn, M. C. (2005). Effect of soybean addition on the rheological properties and breadmaking quality of wheat flour. *Journal of the Science of Food and Agriculture, 85*(11), 1889-1896.

Riccardi, G., Rivellese, A. A., & Giacco, R. (2008). Role of glycemic index and glycemic load in the healthy state, in prediabetes, and in diabetes. *The American journal of clinical nutrition, 87*(1), 269S-274S.

Rincón, F., Martínez, B., & Ibáñez, M. V. (1998). Proximate composition and antinutritive substances in chickpea (Cicer arietinum L) as affected by the biotype factor. *Journal of the Science of Food and Agriculture, 78*(3), 382-388.

Rossi, M., Germondari, I., & Casini, P. (1984). Comparison of chickpea cultivars: chemical composition, nutritional evaluation, and oligosaccharide content. *Journal of Agricultural and Food Chemistry, 32*(4), 811-814.

Rowley, W. R., Bezold, C., Arikan, Y., Byrne, E., & Krohe, S. (2017). Diabetes 2030: insights from yesterday, today, and future trends. *Population health management, 20*(1), 6-12.

Sadowska, J., Blaszczak, W., Fornal, J., Vidal-Valverde, C., & Frias, J. (2003). Changes of wheat dough and bread quality and structure as a result of germinated pea flour addition. *Eur Food Res Technol, 216*(46), 50.

Saharan, K., & Khetarpaul, N. (1994). Protein quality traits of vegetable and field peas: Varietal differences. *Plant Foods for Human Nutrition, 45*(1), 11-22.

Saini, H. S. (1989). Chickpea quality. In: Brinsmead, R.B. and Knights, E.J. (eds) Australian Chickpea Workshop. Australian Institute of Agricultural Science, Brisbane.

Saini, H. S., Huisman, J., Van der Poel, T. F. B., & Liener, I. E .(1989) .*Legume seed oligosaccharides.*

Saini, H. S., & Knights, E. J. (1984). Chemical constitution of starch and oligosaccharide components of'desi'and'kabuli'chickpea (Cicer arietinum) seed types. *Journal of Agricultural and Food Chemistry, 32*(4), 940-944.

Saldeen, K., & Saldeen, T. (2005). Importance of tocopherols beyond α-tocopherol: evidence from animal and human studies. *Nutrition Research, 25*(10), 877-889.

Salgado, P., Lallès, J.-P., Toullec, R., Mourato, M., Cabral, F., & Freire, J. P .(2001) . Nutrient digestibility of chickpea (Cicer arietinum L.) seeds and effects on the small intestine of weaned piglets. *Animal Feed Science and Technology, 91*(3-4), 197-212.

Sánchez-Mata, M. C., Peñuela-Teruel, M. J., Cámara-Hurtado, M., Díez-Marqués, C & ,. Torija-Isasa, M. E. (1998). Determination of mono-, di-, and oligosaccharides in legumes by high-performance liquid chromatography using an amino-bonded silica column. *Journal of Agricultural and Food Chemistry, 46*(9), 3648-3652.

Sathe, S. K., Ponte, J .G., Jr., Rangnekar, P. D., & Salunkhe, D. K. (1981). Effects of addition of Great Northern bean flour and protein concentrates on rheological properties of dough and baking quality of bread. *Cereal Chem, 58*(2), 97-100.

Sayaslan, A., & Şahin, N. (2018). Effects of fermented-chickpea liquor (chickpea yeast) on whole-grain wheat flour bread properties. *Quality Assurance and Safety of Crops & Foods, 10*(2), 183-192.

Sayer, S., Turhan, M., & Sundaram, G. (2001). Analysis of chickpea soaking by simultaneous water transfer and water_starch reaction. *Journal of Food Engineering, 50*(2), 91-98.

Sharma, S., Bajwa, U. H., & Nagi, H. P. S. (1999). Rheological and baking properties of cowpea and wheat flour blends. *Journal of the Science of Food and Agriculture, 79(5) 657-662.*

Siddhuraju, P., Vijayakumari, K., & Janardhanan, K. (1996). Chemical composition and protein quality of the little-known legume, velvet bean (Mucuna pruriens (L.) DC.). *Journal of Agricultural and Food Chemistry, 44*(9), 2636-2641.

Silva-Cristobal, L., Osorio-Díaz, P., Tovar, J., & Bello-Pérez, L. (2010). Chemical composition, carbohydrate digestibility, and antioxidant capacity of cooked black bean, chickpea, and lentil Mexican varieties Composición química, digestibilidad de carbohidratos ,y capacidad antioxidante de variedades mexicanas cocidas de frijol negro, garbanzo, y lenteja. *Cyta–Journal of Food, 8*(1), 7-14.

Singh, B., Sekhon, K. S., & Singh, N. (1995). Suitability of full fat and defatted rice brain obtained from Indian rice for use in food products. *Plant Foods Hum Nutr, 47*(3), 191-200.

Singh, K. B., Subrahmanyam, N., & Kumar, J. (1991). Cooking quality and nutritional attributes of some nely developed cultivars of chickpea (Cicer arietinum) L. *Journal of the Science of Food and Agriculture, 55*(1), 37-46.

Singh, P. N., & Ram, H. (1990). Effect of phosphorous and sulphur application on protein and amino acid contents in chickpea. *Indian Journal of Pulses Research, 3*(1), 36-39.

Singh, U. (1984). Dietary fiber and its constituents in desi and kabuli chickpea (Cicer arietinum L.) cultivars. *Nutrition reports international, 29*(2), 419-426.

Singh, U., & Jambunathan, R. (1981). Studies on desi and kabull chickpea (Cicer arietinum L.) cultivars: levels of protease inhibitors, levels of polyphenolic compounds and in vitro protein digestibility. *Journal of food science, 46*(5), 1364-1367.

Smith, C. E., Mollard, R. C., Luhovyy, B. L., & Anderson, G. H. (2012). The effect of yellow pea protein and fibre on short-term food intake, subjective appetite and glycaemic response in healthy young men. *British Journal of Nutrition, 108*(S1), S74-S80.

Sotelo, A., Flores, F., & Hernández, M. (1987). Chemical composition and nutritional value of Mexican varieties of chickpea (Cicer arietinum L.). *Plant Foods for Human Nutrition, 37*(4), 299-306.

Southgate, D. A. T. (1992). Starch and Dietary Fibre. Health Education Authority, London.

Sreerama, Y. N., Sashikala, V. B., & Pratape, V. M. (2012). Phenolic compounds in cowpea and horse gram flours in comparison to chickpea flour: Evaluation of their antioxidant and enzyme inhibitory properties associated with hyperglycemia and hypertension. *Food chemistry, 133*(1), 156-162.

Suryawanshi, R., Snehalatha Reddy, N., & Sawate, A. (1998). Physico-chemical characteristics and acceptability of snacks of different varieties of chickpea (Cicer arietinum). *Journal of food science and technology, 35*(2), 179-182.

Swinburn, B. A., Caterson, I., Seidell, J. C., & James, W. (2004). Diet, nutrition and the prevention of excess weight gain and obesity. *Public health nutrition, 7*(1a), 123-146.

Trumbo, P., Yates, A. A., Schlicker, S., & Poos, M. (2001). Dietary reference intakes: vitamin A, vitamin K, arsenic, boron, chromium, copper, iodine, iron, manganese, molybdenum ,nickel, silicon, vanadium, and zinc. *Journal of the Academy of Nutrition and Dietetics, 101*(3), 294.

USDA. (2009). National Nutrient Database for Standard Reference. In *Nutrient values and weights are for edible portion*: Release 22.

Vijayalakshmi, V., Kumar, J., & Rao, T. N. (2001). Inheritance of protein content in chickpea. *Legume Research-An International Journal, 24*(1), 28-31.

Viveros, A., Brenes, A., Elices, R., Arija, I., & Canales, R. (2001). Nutritional value of raw and autoclaved kabuli and desi chickpeas (Cicer arietinum L.) for growing chickens. *British Poultry Science, 42*(2), 242-251.

Wang, N., Hatcher, D. W., Toews, R., & Gawalko, E. J. (2009). Influence of cooking and dehulling on nutritional composition of several varieties of lentils) Lens culinaris). *LWT - Food Science and Technology, 42*(4), 842-848.

WHO. (2003). *Report of a WHO/FAO expert consultation 2003. Diet Nutrition and Prevention of Chronic Diseases;* . Retrieved from Geneva.

Wood, J. A., & Grusak, M. A. (2007). Nutritional Value of Chickpea. In S. S. Yadav, B. Redden, W. Chen, & B. Sharma (Eds.), *Chickpea Breeding and Management* (pp. 101-142). Wallingford: CAB International. (Reprinted from: NOT IN FILE).

Yang, Y., Zhou, L., Gu, Y., Zhang, Y., Tang, J., Li, F., . . . Chen ,M. (2007). Dietary chickpeas reverse visceral adiposity, dyslipidaemia and insulin resistance in rats induced by a chronic high-fat diet. *British Journal of Nutrition, 98*(4), 720-726.

Zaki, A. M., Fetouh, S. A. A., Mohany, H., & E.M.rsi, E. A. (1996). Biochemical changes of some chemical constituents in chickpea seeds during germination and processing. *Annals of Agricultural Science, Moshtohor, 34*(3), 995-1022.

Zhang, T., Jiang, B., & Wang, Z. (2007). Nutrition and application of chickpea (in Chinese) .*Cereals Oils, 7*, 10-20.

In: Handbook of Chickpeas
Editors: Albert T. Lund and Noah D. Schultz ISBN: 978-1-53616-374-2

Chapter 9

Health Effects of Starch and Undigested Carbohydrates Associated to Chickpea-Based Foods

Ada K. Milán-Noris[1], PhD, Alvaro Montoya-Rodríguez[1], PhD, Julian De la Rosa-Millan[2], PhD, and Sergio O. Serna-Saldivar[2,*], PhD

[1]Laboratorio de Nutracéuticos (18), Facultad de Ciencias Químico Biológicas, Universidad Autónoma de Sinaloa, Culiacán Sinaloa, México
[2]Tecnológico de Monterrey, Escuela de Ingeniería y Ciencias, Centro de Biotecnología-FEMSA, Monterrey, México

Abstract

Pulses are good sources of protein, slow digesting starch, insoluble and soluble dietary fibers, oligosaccharides and relevant antioxidants mainly of phenolic nature. In addition, they commonly contain less than 10% lipids rich in polyunsaturated fatty acids. These nutrients and phytochemicals have been related to relevant health benefits especially in terms of prevention of chronic diseases as diabetes, cardiovascular and cancers. Chickpea is the third most produced pulse worldwide, after common beans and peas. This crop is classified as Desi or Kabuli. The first yield seeds with thick and pigmented seed coats and the second thin and cream-white seed coat pigmentation. Carbohydrates in chickpea constitute more than 60% of the total seed weight and are divided into digestible and non-digestible. The amounts of carbohydrate fractions vary according to the class (Desi or Kabuli genotypes). Starch is the main carbohydrate and its

* Corresponding Author's Email: sserna@tec.mx.

content varies from 41 to 50%. Unlike cereal grains, the chickpea starch is rich in slowly digestible (SDS) and resistant (RS) fractions. In chickpea seeds, RS is 20-30 % of total starch and the rest is comprised of SDS and rapidly digestible starch (RDS). As a result, the i*n vitro* starch digestibility and glycemic index in chickpea foods is significantly lower compared to cereal starches favoring diabetic patients because they release glucose slowly into the blood stream and leftover starch readily fermented in the hind gut. Besides RS, the non-digestible carbohydrates are oligosaccharides, insoluble dietary fiber and soluble dietary fiber rich in β-glucans. The oligosaccharides reported in chickpea are raffinose, stachyose, verbascose and ciceritol. Chickpea non-digestible carbohydrates play an important role in cholesterolemia, glucose homeostasis, cancer prevention and the modulation of microbiota composition. Hence, this chapter aims to review and analyze current information related to: 1) starch digestion and glycemic index of raw/processed chickpea flours and starch isolates, 2) undigested carbohydrates content in raw/processed chickpea and 3) health effects of chickpea digestible and undigestible carbohydrates.

Keywords: chickpea, starch, starch digestion, fiber, glycemic index, oligosaccharides, prebiotics

INTRODUCTION

In recent years, the use of pulses as food ingredients has increased mainly due to their high availability with good nutritional and phytochemical profiles. According to the FAO [1] these crops that were planted on more than 59 million Ha and produced more than 62 million tons in 2017, have good agronomic flexibility because they thrive under different environmental conditions, poor soils and low water availability [2]. Pulses have become the staple crops of people inhabiting in drought regions of the globe, mostly in the middle east and India [3]. Among pulses, chickpeas (*Cicer arietinum* L) are the third most planted pulse after common beans and peas. The world annual production in 2017 was estimated in 14.77 million tons. Of this output, India, Australia and Nyanmar are the top producers with 9, 2 and 0.52 million tons, respectively [1]. The chickpea seeds are classified as either Kabuli, meaning "from Kabul" which possesses a thin and cream-white seed coat pigmentation or Desi (from country or native) types which have thick and pigmented seed coats [2, 4]. This trend responds to an increasing demand of chickpea in diverse industries, being the food sector their main market [5]. Recent studies have shown several benefits derived from its consumption, due to this, historically, chickpea have been used as ingredient for preparation of puree (humus), pasta, soups and specialty breads [6]. The uses of this seed have evolved to the production of whole flours because they are highly convenient and have comparatively longer shelf life than other seeds. Besides flours have higher bioavailability of their bioactive and antioxidant compounds which are known to benefit human health [7, 8]. In this aspect, when chickpeas are processed to flours it is important to keep in mind that even when the seed structure was disrupted, these maintain most of their native cotyledon arrangement, and that these are complex systems in which proteins, starch, fibers and other seed components coexist in different proportions. Additionally, their arrangement and molecular conformations can differ widely due to their agronomic characteristics and their post-harvest handling [9, 10]. Chickpeas also contain relative high amounts of good quality protein (18–22%) compared with other plant sources. The protein and starch contents in chickpea flour have an important influence in the physicochemical, pasting and thermal properties that could be determinant during product development of functional foods [11, 12].

Nowadays, pulse protein isolates have gained great attention due to its protein quality, bioavailability and its relative ease of isolation, which may allow to manufacture a variety of novel food ingredients [13]. Most of the recent literature in chickpea-related research focuses in the production or in taking advantage of protein rich ingredients. Consequently, current advances have shown that the application of pulse proteins in food formulations greatly depends on their behavior within the food system that is influenced by both extrinsic and intrinsic factors. The first one is related with their isolation conditions whereas the second to their molecular characteristics [14, 15].

The raise in the protein content in chickpea enriched products may promote or enhance molecular and structural interactions among seed components. For instance, there is a strong interaction among cotyledon components like starch granules, soluble and insoluble dietary fibers, along with oligosaccharides and phenolics [10, 14]. Therefore, the functionality of chickpea ingredients may differ due to the processing conditions and native molecular characteristics of these fractions. Therefore, it is important to understand the subtle differences in chickpeas and their potential effects in food systems.

STARCH DIGESTION IN CHICKPEA AND PROCESSING EFFECTS

Starch is the main source of carbon in pulses and is stored inside highly organized granules comprised of amylose and amylopectin chains. The first type of starch contains glucose residues linked by α (1-4) bonds whereas the second also contains branches linked by α (1-6) glycosidic bonds. The starch is usually categorized as type A in cereals, type B in tubers and type C in pulses. These polymorphs differ in the form that side amylopectin chains are packed in the lamella of starch granules. Type C is an intermediate of A and B in structure and packing density [2, 16]. From the structural point of view, chickpea starches often present type C polymorphism that is a shared feature in most pulse seeds. The special internal crystalline arrangement of the starch granules differentiate from other plant sources [17, 18]. At molecular level, there are studies that have reported the presence of small and disperse branches in amylose molecules as well as a high ratio of long to short chains in the amylopectin moieties [18-20]. These features influence pasting viscosity, thermal stability, molecular interactions and digestion properties of chickpea starches [20, 21]. Regarding viscosity, the starch outstands from other pulse starches because it reaches higher peak and final viscosities, and the resulting gels tend to present less syneriesis during refrigeration as well as greater stability during freeze-thaw cycles [20, 22]. In the same sense, due to their particular starch granule architecture, its gelatinization temperatures and enthalpies are notably higher compaed to other pulses, ranging form Tp = 60 to ≈70°C and ΔH = 9 to ≈12. In terms of the important phenomena of retrogradation, these starches appear to reorganize at greater extent than other sources [21, 23]. These characteristics make them very useful to develop new food ingredients with diverse functionalities [24, 25]. The average content of starch in chickpea seeds varies from 41 to 50% of the total weight [2] and its digestibly is mainly affected by the amount of resistant starch (RS) in samples. The prebiotic RS content is influenced by several factors such as starch granules structure and cooking and cooling conditions that enhance starch retrogradation [3].

Starch digestion can be divided as rapidly digestive (RDS), slowly digestible (SDS) and resistant starch (RS) [26]. RDS is the amorphous and dispersed starch and its digestion can lead to an elevation of blood glucose [17]. SDS is completed digested in the small intestine but is comparatively released more slowly compared to RDS. This fraction is desirable for the management of diabetes and is known to cause more satiety. RS is not digestible but can act as prebiotic in the hindgut or colon. RS is readily fermented by the microbiota present in the hind gut [26].

The starch digestion rates have been widely quantified using the Englyst method. This assay determines the RDS, SDS, and RS starch fractions and is considered as the most reproducible and repeatable. The test consists in the simulation of physiological conditions of starch digestion in the intestine using a mixture of pancreatin, amyloglucosidase and invertase at 37°C [18, 26]. RDS fraction is determined as the glucose released within the first 20 min of reaction whereas SDS from 20 to 120 min and the RS is the indegested starch after 120 min of reaction [26]. Nevertheless, some authors question the accuracy of the *in vitro* test because there are not good correlation with *in vivo* starch digestion assays [18].

Table 1 depicts values of *in vitro* starch digestibility. It is important to mention that not all the investigations evaluated the Englyst starch fractions and therefore the information related to RS and other starch fractions such as SDS and RDS are difficult to combine. However, the information can show the tendencies of *in vitro* starch digestion as affected by processing. Importantly, the RS fraction is known to account for 35% of total starch (TS) in chickpeas, although some authors reported RS fractions higher than 60% of the TS. The RS content in raw chickpeas varied from 1.9 to 39%. The high variation is credited to the genetical background and the digestion assay [19, 27-34].

Although, there is not a clear relationship between chickpea processing and digestion rates, cooking usually leads to higher RDS values due to the increase of available starch molecules after starch gelatinization.

Table 1. *In vitro* starch digestibility of chickpea foods

Process	Chickpea Type	RS (%)	SDS (%)	RDS (%)	SD (%)	Ref
Raw	N.S.	1.9				[27]
Raw	N.S.	18.1			33.7	[34]
Raw	Kabuli	19.8				[28]
Raw	N.S.	4.1			27.7	[29]
Raw	Kabuli (2 cv)	3.1 to 4.7	30.3 to 30.7	11.1 to 12.4		[19]
Raw	Desi	6.4	27.1	9.4		[19]
Raw	N.S.	39.9	5.4	2.2		[30]
Raw	N.S.	31.6	10.2	4.9		[31]
Raw	N.S.				9.2	[32]
Raw	Kabuli (3 cv)	3.8 to 7.2				[33]
Dehulled	N.S.				24.4	[32]
Soaked	N.S.	12.1			36.5	[34]
Soaked	Kabuli	15.1				[28]
Soaked	N.S.				40.6	[105]
Soaked	N.S.				7.4 to 13.3	[39]
Cooked	N.S.	0.6	14.7	35.2		[31]
Cooked	N.S.	10.6	14.6	2.6		[30]
Cooked	N.S.	4.4			40.6	[106]
Cooked	N.S.	2.2				[107]
Cooked	Kabuli	8.7	3.8	25.7		[24]
Cooked	Desi (9 cv)	11.6 to 14.1	4.2 to 9.9	18.6 to 26.9		[24]
Cooked	Desi	3.9				[35]
Cooked	Kabuli	4.2				[35]
Cooked	Kabuli (3 cv)	5.7 to 7.4				[33]
Cooked	N.S.	25.6 to 28.2	11.8	2.8 to 3.4		[36]
Cooked (Pr)	N.S.				29.0	[38]
Cooked (Pr)	N.S.				26.6	[39]
Cooked (Pr)	N.S.	5.5 to 7.3				[33]
Cooked (MW)	N.S.	6.6 to 7.6				[33]
Cooked (MW)	N.S.				30.2	[38]
Cooked (R)	N.S.	17 to 30	11.8 to 12.2	2.9 to 14.4		[36]
Cooked (T)	N.S.	3.6				[37]
S/Cooked	N.S.	9.7			39.1	[34]
S/Cooked	Kabuli	9.1				[28]
S/Cooked	N.S.				33.0	[40]
S/Cooked (Pr)	N.S.				13 to 21.9	[39]
S/Cooked/D	N.S.	9.8			39.4	[34]
S/Cooked/D	Kabuli	9.2				[28]
S/Cooked (A)	N.S.				38.6 to 38.8	[40]
Germinated	N.S.				42.7 to 42.9	[105]
Germinated	N.S.				10.40	[32]
G/Cooked (Pr)	N.S.				25.6	[38]
G/Cooked (MW)	N.S.				25.3	[38]
G/Cooked (Pr)	N.S.				38.5 to 39.4	[105]
G/Cooked (MW)	N.S.				46.8 to 47.8	[105]
Fermented	Kabuli	7.6				[108]

N.S. not specified. RS. Resistant starch. RDS. Rapidly starch digestibility SDS. Slowly digested starch SD. Available digested starch. S. Soaked Pr. Pressure. MW. Microwave. G. Germinated D. Dehydrated. A. Autoclaved T. Toasted. R. Roasted.

In cooked chickpea flours of nine Desi cultivars, RDS, SDS and RS ranged from 18.6 to 26.96%, 4.2 to 9.9% and 11.6 to 14.1%, respectively [24]. According to the source and thermal processing, the RS fraction widely varied from 0.6 to 30% (Table 1) [28, 31, 33-40]. The reduction in starch digestion in some cereals and pulses has been associated with the content of phenolics which are known to inhibit enzymes including amylases; therefore, diminishing starch digestion rates [41]. This relationship was observed in cooked chickpea flours, in which the RS fraction correlated positively with phenolic and β-glucans contents [24].

Starch Digestibility of Chickpea Products

It is recognized that pulses are good sources of starch, dietary fiber and protein. Although starch is the main component, its isolation is difficult due to the co-sedimentation of protein and fiber with the starch granules that form a brown deposit [25, 42]. The principal methods to isolate or obtain starch are: dry milling and wet milling. Undoubtedly, wet milling is preferred because the process yields starches with the highest purity [42]. The starches associated to pulses are usually digested to lower extent compared to counterparts isolated from cereals or tubers. Therefore, the lower digestibilities of their starches lower the glycemic index [43]. These features are attributed to the starch granular morphology, the absence of pores on the surface, degree of crystallinity, type of crystalline polymorphs, molecular composition, amylose content and structure, and the amylopectin conformation [42, 44-46]. Most authors coincide that isolated starches from pulses have high SDS and RS and low RDS contents that lower starch digestibility and the release of glucose into the blood stream. Moreover, pulse starches relatively rich in RS can be supplemented as prebiotics [47, 48]. Table 2 summarizes starch digestibility values of chickpea starches. The values of RS, SDS and RDS in native chickpea starches varied from 15 to 73%, 14 to 60 % and 10.95 to 51.3%, correspondingly [20-23, 25, 49]. These variations could be attributed to the chickpea type and genotype. Nevertheless, as most starchy foods, the application and consumption of chickpea starches is after thermal processing that leds to the gelatinization of starch granules. Typically, gelatinized starch is more prone to amylolytic hydrolysis increasing both RDS and digestibility [26, 43]. Milan Moris et al. [37] assayed digested starch fractions of nine cooked Desi chickpeas and concluded that the RDS and SDS contents of cooked pastes increased from 56 to 59.2% and from 33.2 to 35.4%, respectively and reduced RS contents from 9.2 to 6.4%. The RS of the cooked paste is still considered good especially when compared to cereal starches [25]. Likewise, Ma et al. (2017) evaluated the digestion fractions in raw and cooked pulses starches and concluded that the RDS fraction increased to 88.3% in the cooked chickpea starch whereas the RS fraction decreased to 9.8%. The RS of the cooked chickpea was comparatively higher than lentils and beans starches [50]. Besides, in the evaluation of cooked Kabuli chickpea starches, the RS and SDS diminished after thermal processing whereas RS and SDS values ranged from 25 to 36% and 4 to 26% in cooked samples [22]. The relative high content of RS in cooked chickpea starches can act as potent prebiotic and modulator of the hind gut microbiota and as ingredients to low the glycemic index especially in type I and II diabetics [47, 48].

Table 2. Starch digestion rates of chickpea starches

Type		Processing		Ref
		Raw	Cooked	
Kabuli	RDS (%)	17.60	56.8	[25]
	SDS (%)	31.1	34.7	
	RS (%)	51.2	8.4	
Kabuli (4 cv)	RDS (%)	12.1 to 17.2	45.2 to 63.8	[22]
	SDS (%)	14.2 to 35.5	4 to 26.2	
	RS (%)	47.3 to 73.1	25.1 to 36.2	
Kabuli (2cv)	RDS (%)	11 to 15		[23]
	SDS (%)	50 to 60		
	RS (%)	24 to 37		
Kabuli	RDS (%)	19.83		[20]
	SDS (%)	46.7		
	RS (%)	33.4		
Desi (9cv)	RDS (%)	17.2 to 29.9	56.3 to 59.2	[25]
	SDS (%)	27.4 to 36.2	33.2 to 35.4	
	RS (%)	42.2 to 52.5	6.4 to 9.2	
Desi	RDS (%)	23.5		[20]
	SDS (%)	41.4		
	RS (%)	35.0		
Desi (2cv)	RDS (%)	10.95 to 11.34		[23]
	SDS (%)	48 to 50		
	RS (%)	37 to 40		
N.S.	RDS (%)	18	88.3	[21]
	SDS (%)	22.9	1.9	
	RS (%)	59.1	9.8	
N.S.	RDS (%)	51.3		[49]
	SDS (%)	16.9		
	RS (%)	15.2		

N.S. not specified. cv. Cultivar. RDS SDS RS.

Glycemic Index

The *in vitro* starch digestibility in starchy foods is closely related to the predicted glycemic response. Thus, this relative inexpensive test can save time and specialized facilities [18]. Table 3 shows predicted glycemic indexes of chickpea flours and isolated starches.

According to pGI values foods are classify as high (>69), medium (56-69) and low (<56) glycemic impact [51]. This is a relevant parameter to diabetes patients or with prevalence of metabolic syndrome. The pGI of raw starches and flours varied from 48 to 66% and 48 to 56%, correspondingly (Table 3). Therefore, these samples fall in the category of low or medium glycemic impact [52]. Interestingly, cooking increases pGI of chickpea flours to 78.47 to 84.88% and values are higher compared to corresponding pGI assayed in refined cooked starches (73 to 75%) [24, 25]. According to these pGI values, cooked chickpea flours can be classified as high glycemic impact foods [52]. Interestingly, chickpea samples processed by heat and rosting showed relatively low pGI values, which make them attractive for development of low glycemic impact foods. In this study the sample with the highest pGI was soaked before the thermal process [36].

Table 3. Predicted glycemic index of chickpea foods

Source	Type	Process	pGI	Ref
Isolated starch	Kabuli	Raw	65.6	[25]
	Kabuli	Raw	48.7	[20]
	Desi	Raw	53.1	[109]
	Desi	Raw	65.3 to 66.1	[25]
	Kabuli	Cooked	74.5	[25]
	Desi	Cooked	73.8 to 75.5	[25]
Chickpea flours	Desi	Raw	48.9	[19]
	Kabuli	Raw	54 to 56	[19]
	Kabuli	Cooked	84.88	[24]
	Desi	Cooked	78.4 to 83.8	[24]
	NS	Cooked	38 to 41	[36]
	NS	Roasted	29 to 72	[36]

pGI. Predicted glycemic index.

NON-DIGESTIBLE CARBOHYDRATES IN CHICKPEA

The non-digestible carbohydrates are formed by oligosaccharides, fiber, resistant starch and β-glucans. β-glucans is a homopolysaccharide formed by several glucose units linked together to form a large polymer chain. Barley (*Hordeum vulgare*), oats (*Avena sativa*), rye (*Secale cereale*) and fungi are good sources of this proven cholesterol lowering compound [53, 54]. Genç et al. (2001) quantified amounts of β-glucans in chickpea (0.9 g/100g) and concluded that amounts were similar to the amounts assayed in rye cultivars [54]. Moreover, Milán-Noris et al. (2019) reported β-glucans contents of 0.96 and from 1.02 to 2.42 g/100g in raw Kabuli and Desi chickpea flours, respectively. These values showed a strong correlation with the amounts of seed coats demonstrating that β-glucans are mainly associated to their fibrous rich cell walls [24]. Likewise, Wood et al. (2011) found that Desi chickpeas contain higher β-glucans in the outer palisade layer of the seed coats compared to Kabuli counterparts [55].

Dietary fiber is a food component that is not digested in the small intestine and is composed of poly/oligosaccharides, lignin and other substances [56]. The dietary fiber is classified into soluble and insoluble, where soluble is slowly digested in the colon and insoluble is metabolically inert and helps with bowel movements. Soluble fiber components are fermented in the colon helping the growth of the wide array of bacteria that comprise the microbiota [2]. The dietary fiber content of chickpea seed coats or testa is approximately 75% [2]. The Desi types have higher fiber content and insoluble fraction compared to the Kabuli counterparts. This is attributed to thicker seed coat in Desi (11.5% of total seed weight) compared to the Kabuli types (4.3-4.4% of total seed weight) [2, 57]. The soluble fiber contents of the Kabuli and Desi types are similar and also share same proportions of hemicelluloses, which constitute the largest part (~ 55%) of the total seed dietary fiber [58]. Xylose is the major constituent of soluble fiber fractions in chickpea whereas glucose is present in large amounts in seed coat and soluble fiber fractions. The hemicellulosic sugar arabinose/rhamnose is present in appreciable amounts in seed coat and insoluble fiber fractions of chickpea [59].

Table 4 depicts the dietary fiber contents of chickpea foods. The total dietary fiber content in raw chickpea varied according to the cultivars evaluated and methods of quantification employed and ranged from 9 to 29 g/100g. The insoluble fiber acounted more than 80% of total dietary fiber [24, 29, 34, 35, 60-64]. Chickpea cooking and soaking increased the dietary fiber content [34, 35, 61, 65, 66]. For instance, Wang et al. (2010) analyzed fiber content in Kabuli and Desi chickpeas subjected to cooking. Results of the study indicated that the raw Desi types were higher in dietary fiber components compared to the Kabuli type. The insoluble, soluble and total fiber of Desi raw chickpeas were 23.2, 1.38 and 24.6% whereas respective values for the Kabuli counterpart 14.1, 1.23 and 15.4%. Cooking significantly increases fiber fractions values in both types of chickpeas. The total dietary fiber in the cooked Desi and Kabuli chickpeas averaged 26.7 and 20.8% (dwb) respectively [35].

Pulses commonly had high amounts of oligosaccharides, in contrast to other seeds. Among these compounds, α-galactosides are the most abundant carbohydrate after sucrose which constitutes 62% of the total sugar content (mono, di and oligosaccharides). The two main groups of oligosaccharides in chickpeas belong to the raffinose family and galactosyl cyclitols. This family is comprised of the trisaccharide raffinose, the tetrasaccharide stachyose and the pentasaccharide stachyose also knwon as verbascose [2, 5]. The oligosaccharides cannot be absorbed or degraded due to absence of α-galactosidase in humans [2, 5]. According to literature [34, 35, 60, 61, 67-80], the total oligosaccharide content in raw chickpea ranged from 0.122 to 14.53 g/100 g db, being stachyose the most abundant (0.013 to 5.94 g/100 g) followed by raffinose (0.011 to 5.02 g/100 g), ciceritol (0.22 to 6.72 g/100 g) and verbascose (0.004 to 0.81 g/100 g) (Table

5). Others reports mention that after chickpea processing like soaking, cooking, extruding, and germination the oligosaccharide content is improved, and the range reported of total oligosaccharides varied from 0.20 to 17.60 g/100 g. Interestingly, ciceritol (0.25 to 7.22 g/100 g) is the most abundant form after chickpea processing [34, 35, 61, 68, 70, 71, 81].

Table 4. Dietary fiber contents of raw and processed chickpeas

Chickpea Type	Process	Fiber			Ref
		Soluble (g/100g)	Insoluble (g/100g)	Total (g/100g)	
Kabuli	Raw	5.5	5.02	10.52	[24]
Desi (9 cv)	Raw	5.3 to 5.9	4.0 to 4.8	9.7- to 10.4	[24]
N.S.	Raw	4.5	10.9	15.4	[110]
Desi	Raw	1.38	23.2	24.58	[35]
Kabuli	Raw	1.23	14.15	15.38	[35]
N.S.	Raw	0.96	20.48	21.44	[34]
N.S.	Raw	2.05	23.95	26	[29]
N.S.	Raw	3.6	12.6	16.2	[61]
N.S.	Raw	4.6	21.4	25.9	[60]
N.S.	Raw	6.55	18.62	25.13	[63]
Desi	Raw	1.003	28.29	29.29	[62]
Kabuli	Raw			17.8	[64]
N.S.	Cooked	2.91	6.50	9.41	[65]
N.S.	Cooked	0.0	15.8	15.8	[66]
Desi	Cooked	1.86	24.86	26.72	[35]
Kabuli	Cooked	1.41	19.55	20.75	[35]
N.S.	Cooked	4.4	11.6	15.9	[61]
N.S.	Soaked	1.28	19.8	21.04	[34]
N.S.	Soaked/Cooked	1.50	23.51	25.00	[34]
N.S.	S/Cooked/ D	1.96	25.30	27.25	[34]
Desi	Germinated	1.26	32.53	33.79	[62]

N.S. not specified. cv. Cultivar. S. Soaked Pr. Pressure. G. Germinated D. Dehydrated.

Table 5. Oligosaccharides of raw and processed chickpeas

Chickpea Type	Process	Oligosaccharides Raffinose (g/100g)	Stachyose (g/100g)	Verbascose (g/100g)	Ciceritol (g/100g)	Total (g/100g)	Ref
N.S. (3 cv)	Raw	0.57 to 0.86	0.74 to 1.17		2.51to 3.93	3.82 to 5.96	[69]
N.S.	Raw	0.32	1.77		2.76	4.85	[34]
Desi	Raw	0.52	1.67	0.004		2.194	[35]
Kabuli	Raw	0.72	2.13	0.026		2.876	[35]
N.S.	Raw	0.61	1.42		2.69	4.72	[71]
Desi (3 cv)	Raw	0.77 to 0.81	1.25 to 1.42			2.02 to 2.23	[68]
Kabuli (3 cv)	Raw	0.63 to 0.68	1.10 to 1.18			1.73 to 1.86	[68]
N.S.	Raw	0.46	1.68		2.70	4.84	[67]
N.S.	Raw	5.02	2.70		6.77	14.49	[70]
N.S.	Raw	0.673	1.61	0.81		3.09	[73]
N.S.	Raw	0.629	1.171		2.786	4.59	[72]
N.S.	Raw	0.50	2.28	0.22	3.05	6.05	[75]
N.S.	Raw	0.697	1.92	0.275	0.218	3.11	[61]
N.S. (17 cv)	Raw	0.75 to 0.99	1.70 to 1.87			2.45 to 2.86	[79]
N.S. (17 cv)	Raw	1.22 to 1.87	1.88 to 2.83	0.06 to 0.14		3.16 to 4.84	[78]
N.S. (17 cv)	Raw	0.087 to 1.41	1.82 to 5.94		2.62-7.18	4.53 to 14.53	[77]
N.S.	Raw	0.692	2.94	0.121		3.13	[80]
N.S.	Raw	0.516	1.824	0.082		2.49	[60]
Desi (171 cv)	Raw	0.015 to 0.196	0.017 to 0.30	0.09 to 0.12		0.122 to 0.62	[76]
Kabuli (171 cv)	Raw	0.011 to 0.184	0.013 to 0.14	0.10 to 0.18		0.124 to 0.50	[76]
N.S. (19 cv)	Raw	0.46 to 0.92	1.64 to 3.09	0.27 to 0.70	3.04 to 5.06	5.41 to 9.77	[74]
N.S. (3 cv)	Soaked	0.21 to 0.31	0.25 to 0.59		0.73 to 1.85	1.19 to 2.75	[69]
N.S.	Soaked	0.26	1.25		2.03	3.54	[34]
N.S.	Soaked	2.92	1.60		3.35	7.87	[70]
N.S.	Soaked	3.47	1.815			3.96	[111]

Table 5. (Continued)

Chickpea Type	Process	Oligosaccharides					Ref
		Raffinose (g/100g)	Stachyose (g/100g)	Verbascose (g/100g)	Ciceritol (g/100g)	Total (g/100g)	
Desi	Soaked	0.05	0.15			0.20	[112]
Kabuli	Soaked	0.26	0.79	0.05		1.10	[112]
N.S.	Soaked	0.17	0.68		1.82	2.67	[67]
N.S. (3 cv)	Cooked	0.097 to 0.12	0.27 to 0.55		0.47 to 1.09	0.84 to 1.76	[69]
Desi	Cooked	0.32	1.27	0.004		1.594	[35]
Kabuli	Cooked	0.42	1.38	0.030		1.830	[35]
Desi (3 cv)	Cooked	0.77 to 0.81	0.90 to 0.94			1.67 to 1.75	[68]
Kabuli (3 cv)	Cooked	0.63 to 0.68	0.80 to 0.81			1.43 to 1.49	[68]
N.S.	Cooked	4.09	6.29		7.22	17.60	[70]
N.S.	Cooked	0.75	2.13	0.27	0.24	3.40	[61]
N.S.	Cooked (Pr)	0.497 to 0.653	1.98-2.29		3.83-4.28	6.31 to 7.22	[81]
N.S.	Soaked/Cooked	0.14	0.98		0.98	2.10	[34]
N.S.	S/Cooked/D	0.13	0.88		1.09	2.10	[34]
Desi (3 cv)	Germinated	0.77 to 0.81	0.35 to 1.42			1.12 to 2.23	[68]
Kabuli (3 cv)	Germinated	0.63 to 0.68	1.10 to 1.18			1.73 to 1.86	[68]
N.S.	Extruded	0.75	1.89		2.88	5.52	[71]

N.S. not specified. cv. Cultivar. Pr. Pressure.

HEALTH EFFECTS OF CHICKPEA CARBOHYDRATES

Chickpea intake has been related to the reduction of the risk to chronic degenerative diseases mainly because chickpeas are good source of protein, carbohydrates and phytochemicals, which include phenolic compounds. The beneficial effects of its consumption may be synergistic. In this section, it summarizes the health effect information related to carbohydrates in chickpea. The content of fiber, resistant starch and oligosaccharides are known to dininish risks factors of cardiovascular diseases, metabolic syndrome, obesity, cancer and type 2 diabetes (T2D) [82-90]. Likewise, these compounds have demonstrated both prebiotic and anti-inflammatory effects [85, 86, 91].

Diabetes and Cardiovascular Diseases

Although there is a high recommendation of pulses in the diet for the prevention of diabetes or cardiovascular diseases, few studies have shown evidence of their protection against these diseases. For example, the daily intake of 16 to 35 g of pulses (6 to 9 g chickpea/day) in adults can prevent the development of type-2 diabetes [92]. The consumption of hummus by healthy subjects lowered approximately 4 times the glycemic index compared to white bread that contained similar amounts of carbohydrates. The two products did not exert alterations in insulin levels. Thus, the chickpea hummus showed potential to prevent type-2 diabetes, cancer or cardiovascular diseases [93].

Moreover, Hosseinpour-Niazi et al. (2015) studied the effect of legume-free and a non-soy legume-based diet with type-2 diabetic patients in a randomized cross-over clinical trial. The non-soy legume (pea, chickpea, beans) based diet showed good effects on inflammatory markers that are associated with type-2 diabetes. The intake of legumes decreased C-reactive protein, IL-6 and TNF-α in overweight diabetic patients. These beneficial effects could be attributed to the content of fiber and magnesium in legumes, which are known to exert anti-inflammatory effects. Previous studies have shown that a legume diet improved the biochemical parameters related to T2D including fasting blood glucose, fasting insulin and LDL-cholesterol, which also improved the anti-inflammatory status of the patients [85, 94]. In other clinical study, a similar pattern and conclusions were observed after the consumption of legumes (chickpeas, lentil) in a hypocaloric diet for 8 weeks. The diet intervention reduced pro-inflammatory markers and improved lipid profiles in obese subjects [86]. Similarly, the weekly intake of 5 cups of pulses (chickpea, peas, beans, lentils) for 8 weeks by overweight or obese patient with risk to metabolic syndrome, improved the blood glucose control, HDL-cholesterol and decreased insulin resistance and systolic blood pressure compared to control patients. The high protein and dietary fiber in pulses are the main responsible for the observed reduction of the risk of metabolic syndrome [88].

The consumption of chickpea meal lowered blood glucose after 30 and 60 min compared to a wheat meal. After 2 hours of ingestion, chickpea meal reduced insulin levels and its resistance compared to wheat meal. However, at the end of the study (6 weeks), these parameters did not show significant effects [83]. A similar study comparing chickpea versus wheat meal, did not find any significant effects in normoglycemic patients. Although chickpea meal ingestion reduced both serum total and LDL-cholesterol levels compared to individuals supplemented with wheat meals which provided similar fiber content [82, 84].

Furthermore, a long-time intake (8 months) of chickpea by rats showed that this pulse reduced glycaemia and insulianemia and also blood cholesterol and TAG both in liver and muscle [87]. Other interesting study in which sprouted chickpeas with and without vanadium were fed for 8 weeks to streptozotocin-induced diabetic rats showed interesting hypoglycemic and hypocholesterolemic effects [95]. Moreover, rats fed with extruded or boiled chickpea for 35 days showed lower blood total cholesterol than counterparts fed a control casein-based diet [96].

Cancer

For several decades, the consumption of fiber has been associated to prevention of colon-rectal cancer [97]. However, for the particular case of chickpeas, the prevention of cancer has been studied more relating to isoflavones or other bioactive compounds [7]. The intake of cooked chickpeas (2 or 10%) showed interesting results when tested with ICR mice induded with colon cancer. The daily consumption of cooked chickpeas reduced the protein, lipid and DNA oxidation, as well the expression of inflammatory enzymes (COX-2 and iNOS) and β-catenin, the last one is an important oncogenic protein related to colon cancer [98].

Prebiotic/Microbiota Modulation

The consumption of chickpea fiber and oligosaccharides stimulate the growth of colonic *Bifidobacterium* and exert beneficial effects on health [99]. Fernando et al. (2010) evaluated the daily supplementation of 5 g of purified raffinose from chickpea or 200 g of canned chickpea containing 3 to 5g of raffinose in healthy adults for 3 weeks. Both chickpea diets showed beneficial effects modulating the gut microbiota of patients with the increase of *Bifidobacterium spp* and a decreased in *Clostridium spp.* [100]. Additionally, the presence of *Clostridium histolyticum* and *Clostridum lituseburense,* pathogenic and putrefactive bacteria diminished in the group that consumed canned chickpea compared with gropus fed purified raffinose and control [100]. Similarly,

Queiroz-Monici et al. (2005) found bifidogenic effects of dietary fiber and resistant starch from cooked chickpea in rats fed isocaloric and isonitrogenous diets with cooked chickpeas for 28 days [66].

The α-galacto-oligosaccharides (α-GOS) from chickpea were supplemented for 6 weeks in a high fat diet (HFD) that induced the metabolic syndrome in male mice. The α-GOS purified from chickpea were fructose, sucrose, raffinose, ciceritol and stachyose. The feeding of HFD with α-GOS from chickpea to mice allowed to increase the short chain fatty acids (SCFA) production and beneficial *Bifidobacterium* and *Lactobacillus*. Besides α-GOS treated groups positively improve the biochemical parameters as blood glucose, TC, TG, LDL and HDL compared to HFD control [101]. In another study, C57BL/6 mice were fed with 20% of cooked chickpea and compared to counterparts fed an isocaloric control diet for 3 weeks. The cooked chickpea enhanced the epithelial barrier integrity, crypt mucus production and antimicrobial defenses. Furthermore, it modulated the cecal and fecal microbiota composition and production of SCFA. Chickpea supplementation in mice increased the abundance of *Provotella*, a genus that ferment carbohydrates and produce SCFA especially the particular type *Ruminococcus flavefaciens* which is difficult to find in obese and overweight patients. The authors associated those benefits effect to the fermentable carbohydrates (oligosaccharides, fiber, RS) and phenolics compounds present in chickpea [102].

Maccaferri et al. (2012) evaluated the *in vitro* fermentation of the pulse flour (50:50 chickpea/lentil) as natural source of prebiotics (fiber, oligosaccharides, β-glucans, fructo-oligosaccharides). The experiment was designed to emulate the fermentative human colon system in 3 validated stages. The pulses flour fermentation caused a modulation of colonic metabolome because it enhanced the growth of *Bacteriodes/Prevotella* species, which increased acetate and propionate concentration and reduced both *Clostridium spp* counts and butyrate concentration. The relative high concentration of propionate has been related to positive health effects on colon, lowering cholesterol levels and stimulating satiety [103]. In other study, the *in vitro* fermentation with human feces of chickpea purified ciceritol (>95%) improved the production of SCFA and the growth of *Bifidobacterium spp*, *Lactobacillus ssp* and *Enterococcus spp* [104].

Conclusion

The information in this chapter showed the importance of the determination of starch fraction and non-digestible carbohydrates (oligosaccharides, fiber and β-glucans) in processed samples, which are the main form of consumption. The non-digestible carbohydrates have shown several evidences as nutraceutical compounds that favor human health, tought more information is need it, as isolated source o the synergy effect of all bioactive compounds in chickpea seeds. Also, it is necessary to validate and unify methods of starch digestion, especially since SDS and RS have important nutraceutical potencial especially for diabetic patients.

REFERENCES

[1] FAOSTAT (2017). "Chickpea word production." Last Modified March, 2019, accessed March, 20, 2019. http://www.fao.org/faostat/en/#home.

[2] Jukanti, A. K., Gaur, P. M., Gowda, C. L. L., and Chibbar, R. N. (2012). "Nutritional quality and health benefits of chickpea (*Cicer arietinum* L.): a review." *British Journal of Nutrition* 108 (SupplementS1):S11-S26. doi: doi:10.1017/S0007114512000797.

[3] Rebello, C. J., Greenway, F. L., and Finley, J. W. (2014). "Whole grains and pulses: A comparison of the nutritional and health benefits." *Journal of Agricultural and Food Chemistry* 62 (29):7029-7049. doi: 10.1021/jf500932z.

[4] Kadam, S. S., Deshpande, S. S., and Jambahale, N. D. (1989). "Seed Structure and Composition." In *CRC Handbook of World Food Legumes: Nutritional Chemistry, Processing Technology and Utilization*. Boca Raton, FL. EUA. : CRC Press, Inc.

[5] Bar-El Dadon, S., Shahal, A., and Ram, R. (2017). "Leveraging traditional crops for better nutrition and health - The case of chickpea." *Trends in Food Science & Technology* 64:39-47. doi: http://dx.doi.org/10.1016/j.tifs.2017.04.002.

[6] Rachwa-Rosiak, D., Nebesny, E., and Budryn, G. 2015. "Chickpeas—Composition, nutritional value, health benefits, application to bread and snacks: A review." *Critical Reviews in Food Science and Nutrition* 55 (8):1137-1145. doi: 10.1080/10408398.2012.687418.

[7] Gupta, R. K., Gupta, K., Sharma, A., Das, M., Ansari, I. A, and Dwivedi, P. D. (2017). "Health risks and benefits of chickpea (*Cicer arietinum*) consumption." *Journal of Agricultural and Food Chemistry* 65 (1):6-22. doi: 10.1021/acs.ja fc.6b02629.

[8] Alvarez, M. D, Fuentes, R., Olivares, M. D, and Canet, W. (2014). "Effects of high hydrostatic pressure on rheological and thermal properties of chickpea (*Cicer arietinum* L.) flour slurry and heat-induced paste." *Innovative Food Science & Emerging Technologies* 21:12-23. doi: https://doi.org/10.1016/j.ifset.2013.11.005.

[9] Himmelsbach, D. S., Khalili, S., and Akin, D. E. (2002). "The use of FT-IR microspectroscopic mapping to study the effects of enzymatic retting of flax (*Linum usitatissimum* L) stems." *Journal of the Science of Food and Agriculture* 82 (7):685-696. doi: 10.1002/jsfa.1090.

[10] Krintiras, G. A., Göbel, J., Bouwman, W. G., van der Goot, A. J., and Stefanidis, G. D. (2014). "On characterization of anisotropic plant protein structures." *Food & Function* 5 (12):3233-3240. doi: 10.1039/C4FO00537F.

[11] Kaur, M., and Singh, N. (2005). "Studies on functional, thermal and pasting properties of flours from different chickpea (*Cicer arietinum* L.) cultivars." *Food Chemistry* 91 (3):403-411. doi: http://dx.doi.org/10.1016/j.foodchem.2004.06.015.

[12] Ma, Z., Boye, J. I., Simpson, B. K., Prasher, S. O., Monpetit, D., and Malcolmson, L. (2011). "Thermal processing effects on the functional properties and microstructure of lentil, chickpea, and pea flours." *Food Research International* 44 (8):2534-2544. doi: http://dx.doi.org/10.1016/j.foodres.2010.12.017.

[13] Luna-Vital, D. A., Mojica, L., González de Mejía, E., Mendoza, S., and Loarca-Piña, G. (2015). "Biological potential of protein hydrolysates and peptides from common bean (*Phaseolus vulgaris* L.): A review." *Food Research International* 76:39-50. doi: https://doi.org/10.1016/j.foodres.2014.11.024.

[14] Boye, J. I., Zare F., and Pletch A. (2010). "Pulse proteins: Processing, characterization, functional properties and applications in food and feed." *Food Research International* 43 (2):414-431. doi: https://doi.org/10.1016/j.foodres.2009.09.003.

[15] Boye, J. I., Aksay, S., Roufik, S., Ribéreau, S., Mondor, M., Farnworth, E., and Rajamohamed, S. H. (2010). "Comparison of the functional properties of pea, chickpea and lentil protein concentrates processed using ultrafiltration and isoelectric precipitation techniques." *Food Research International* 43 (2):537-546. doi: https://doi.org/10.1016/j.foodres.2009.07.021.

[16] Sun, Y., Ye, H., Hu, B., Wang, W., Lei, S., Wang, X., Zhou, L., and Zeng, X. (2015). "Changes in crystal structure of chickpea starch samples during processing treatments: An X-ray diffraction and starch moisture analysis study." *Carbohydrate Polymers* 121 (0):169-174. doi: http://dx.doi.org/10.1016/j.carbpol.2014.12.048.

[17] Chung, H. J., Liu, Q., Donner, E., Hoover, R., Warkentin, T. D., and Vandenberg, B. (2008). "Composition, molecular structure, properties, and *in vitro* digestibility of starches from newly released Canadian pulse cultivars." *Cereal Chemistry* 85 (4):471-479. doi: 10.1094/cchem-85-4-0471.

[18] Jeong, D., Han, J. A., Liu, Q., and Chung, H. J. (2019). "Effect of processing, storage, and modification on *in vitro* starch digestion characteristics of food legumes: A review." *Food Hydrocolloids* 90:367-376. doi: https://doi.org/10.1016/j.foodhyd.2018.12.039.

[19] Chung, H. J., Liu, Q., Hoover, R., Warkentin, T. D., and Vandenberg, B. (2008). "*In vitro* starch digestibility, expected glycemic index, and thermal and pasting properties of flours from pea, lentil and chickpea cultivars." *Food Chemistry* 111 (2):316-321. doi: http://dx.doi.org/10.1016/j.foodchem.2008.03.062.

[20] Miao, M., Zhang, T. and Jiang, B. (2009). "Characterisations of kabuli and desi chickpea starches cultivated in China." *Food Chemistry* 113 (4):1025-1032. doi: https://doi.org/10.1016/j.foodchem.2008.08.056.

[21] Ma, M., Wang, Y., Wang, M., Jane, J. L., and Du, S. K. (2017). "Physicochemical properties and *in vitro* digestibility of legume starches." *Food Hydrocolloids* 63:249-255. doi: https://doi.org/10.1016/j.foodhyd.2016.09.004.

[22] Yniestra Marure, L. M., Núñez-Santiago, M. C., Agama-Acevedo, E., and Bello-Perez, L. A. (2019). "Starch characterization of improved chickpea varieties grown in Mexico." *Starch - Stärke* 71 (3-4):1800139. doi: 10.1002/star.201800139.

[23] Hughes, T., Hoover, R., Liu, Q., Donner, E., Chibbar, R., and Jaiswal, S. (2009). "Composition, morphology, molecular structure, and physicochemical properties of starches from newly released chickpea (*Cicer arietinum* L.) cultivars grown in Canada." *Food Research International* 42 (5):627-635. doi: https://doi.org/10.1016/j.foodres.2009.01.008.

[24] Milán-Noris, A. K., De la Rosa-Millan, J., and Serna-Saldivar, S. O. (2019). "Comparative analysis of techno-functional properties, starch digestion and protein quality of pigmented chickpea flours." *International Journal of Food Science & Technology* 0 (0). doi: 10.1111/ijfs.14144.

[25] Milán-Noris, A. K., de la Rosa-Millán, J., Reyes-Moreno, C., and Serna-Saldivar, S. O. (2017). "Physicochemical, functional properties, and digestion of isolated starches from pigmented chickpea (*Cicer arietinum* L.) cultivars." *Starch - Stärke* 69 (5-6). doi: doi:10.1002/star.201600152.

[26] Englyst, H. N., Kingman, S. M., and Cummings, J. H. (1992). "Classification and measurement of nutritionally important starch fractions." *European Journal of Clinical Nutrition* 46:S33-S50.

[27] Sreerama, Y. N., Sashikala, V. B., Pratape, V. M., and Singh, V. (2012). "Nutrients and antinutrients in cowpea and horse gram flours in comparison to chickpea flour: Evaluation of their flour functionality." *Food Chemistry* 131 (2):462-468. doi: https://doi.org/10.1016/j.foodchem.2011.09.008.

[28] Aguilera, Y., Benítez, V., Mollá, E., Esteban, R. M., and Martín-Cabrejas, M. A. (2011). "Influence of dehydration process in Castellano chickpea: Changes in bioactive carbohydrates and functional properties." *Plant Foods for Human Nutrition* 66 (4):391-400. doi: 10.1007/s11130-011-0259-8.

[29] Bravo, L., Siddhuraju, P., and Saura-Calixto, F. (1999). "Composition of underexploited Indian pulses. Comparison with common legumes." *Food Chemistry* 64 (2):185-192. doi: https://doi.org/10.1016/S0308-8146(98)00140-X.

[30] Periago, M. J., Ros, G. and Casas, J. L. (1997). "Non-starch polysaccharides and *in vitro* starch digestibility of raw and cooked chick peas." *Journal of Food Science* 62 (1):93-96. doi: 10.1111/j.1365-2621.1997.tb04375.x.

[31] Bravo, L. (1999). "Effect of processing on the non-starch polysaccharides and *in vitro* starch digestibility of legumes/Efecto del procesado en el contenido de polisacáridos no amiláceos y la digestibilidad *in vitro* del almidón de legumbres." *Food Science and Technology International* 5 (5):415-423. doi: 10.1177/108 201329900500507.

[32] Ghavidel, R. A., and Prakash, J. (2007). "The impact of germination and dehulling on nutrients, antinutrients, *in vitro* iron and calcium bioavailability and *in vitro* starch and protein digestibility of some legume seeds." *Lwt-Food Science and Technology* 40 (7):1292-1299. doi: 10.1016/j.lwt.2006.08.002.

[33] Xu, Y., Thomas, M., and Bhardwaj, H.L. (2014). "Chemical composition, functional properties and microstructural characteristics of three kabuli chickpea (*Cicer arietinum* L.) as affected by different cooking methods." *International Journal of Food Science & Technology* 49 (4):1215-1223. doi: 10.1111/ijfs.12419.

[34] Aguilera, Y., Martín-Cabrejas, M. A., Benítez, V., Mollá, E., López-Andréu, F.J., and Esteban, R. M. (2009). "Changes in carbohydrate fraction during dehydration process of common legumes." *Journal of Food Composition and Analysis* 22 (7–8):678-683. doi: http://dx.doi.org/10.1016/j.jfca.2009.02.012.

[35] Wang, N., Hatcher, D. W., Tyler, R. T., Toews, R., and Gawalko, E. J. (2010). "Effect of cooking on the composition of beans (*Phaseolus vulgaris* L.) and chickpeas (*Cicer arietinum* L.)." *Food Research International* 43 (2):589-594. doi: https://doi.org/10.1016/j.foodres.2009.07.012.

[36] Simsek, S., Herken, E. N., and Ovando-Martinez, M. (2016). "Chemical composition, nutritional value and *in vitro* starch digestibility of roasted chickpeas." *Journal of the Science of Food and Agriculture* 96 (8):2896-2905. doi: 10.1002/jsfa.7461.

[37] Fares, C., and Menga, V. (2012). "Effects of toasting on the carbohydrate profile and antioxidant properties of chickpea (*Cicer arietinum* L.) flour added to durum wheat pasta." *Food Chemistry* 131 (4):1140-1148. doi: https://doi.org/10.1016/j.foodchem.2011.09.080.

[38] Khatoon, N. and Prakash, J. (2006). "Nutrient retention in microwave cooked germinated legumes." *Food Chemistry* 97 (1):115-121. doi: https://doi.org/10.1016/j.foodchem.2005.03.007.

[39] Rehman, Z-r. 2007. "Domestic processing effects on available carbohydrate content and starch digestibility of black grams (*Vigna mungo*) and chick peas (*Cicer arietium*)." *Food Chemistry* 100 (2):764-767. doi: https://doi.org/10.1016/j.foodchem.2005.10.041.

[40] Rehman, Z. U. and Shah, W. H. (2005). "Thermal heat processing effects on antinutrients, protein and starch digestibility of food legumes." *Food Chemistry* 91 (2):327-331. doi: https://doi.org/10.1016/j.foodchem.2004.06.019.

[41] Zhu, F. (2015). "Interactions between starch and phenolic compound." *Trends in Food Science & Technology* 43 (2):129-143. doi: https://doi.org/10.1016/j.tifs.2015.02.003.

[42] Hoover, R., Hughes, T., Chung, H. J., and Liu, Q. (2010). "Composition, molecular structure, properties, and modification of pulse starches: A review." *Food Research International* 43 (2):399-413. doi: http://dx.doi.org/10.1016/j.foodres.2009.09.001.

[43] Granfeldt, Y., Björck, I., Drews, A., Tovar, J. (1992). "An *in vitro* procedure based on chewing to predict metabolic response to starch in cereal and legume products." *European Journal of Clinical Nutrition* 46 (9):649-660.

[44] Xu, Y., Sismour, E. N., Narina, S. S., Dean, D., Bhardwaj, H. L., and Li, Z. (2013). "Composition and properties of starches from Virginia-grown kabuli chickpea (*Cicer arietinum* L.) cultivars." *International Journal of Food Science & Technology* 48 (3):539-547. doi: 10.1111/j.1365-2621.2012.03217.x.

[45] Huang, J., Schols, H. A., van Soest, J. J. G., Jin, Z., Sulmann, E., and Voragen, A. G. J. (2007). "Physicochemical properties and amylopectin chain profiles of cowpea, chickpea and yellow pea starches." *Food Chemistry* 101 (4):1338-1345. doi: http://dx.doi.org/10.1016/j.foodchem.2006.03.039.

[46] Singh, N., Singh S. K., and Kaur, M. (2004). "Characterization of starches separated from Indian chickpea (*Cicer arietinum* L.) cultivars." *Journal of Food Engineering* 63 (4):441-449. doi: http://dx.doi.org/10.1016/j.jfoodeng.2003.09.003.

[47] Chung, H. J., Liu, Q., and Hoover, R. (2009). "Impact of annealing and heat-moisture treatment on rapidly digestible, slowly digestible and resistant starch levels in native and gelatinized corn, pea and lentil starches." *Carbohydrate Polymers* 75 (3):436-447. doi: http://dx.doi.org/10.1016/j.carbpol.2008.08.006.

[48] Zhang, H., Yin, L., Zheng, Y., and Shen, J. (2016). "Rheological, textural, and enzymatic hydrolysis properties of chickpea starch from a Chinese cultivar." *Food Hydrocolloids* 54:23-29. doi: 10.1016/j.foodhyd.2015.09.018.

[49] Demirkesen-Bicak, H., Tacer-Caba, Z., and Nilufer-Erdil, D. (2018). "Pullulanase treatments to increase resistant starch content of black chickpea (*Cicer arietinum* L.) starch and the effects on starch properties." *International Journal of Biological Macromolecules* 111:505-513. doi: https://doi.org/10.1016/j.ijbiomac.2018.01.026.

[50] Huang, S., Ma, Y., Zhang, C., Cai, S., and Pang, M. (2017). "Bioaccessibility and antioxidant activity of phenolics in native and fermented *Prinsepia utilis* Royle seed during a simulated gastrointestinal digestion *in vitro*." *Journal of Functional Foods* 37:354-362. doi: https://doi.org/10.1016/j.jff.2017.08.004.

[51] Brand-Miller, J., Wolever, T. M. S., Colagiuri, S., Foster-Powell, K. (1999). *The glucose revolution.* New York: Marlowe & Company.

[52] Jenkins, D. J., Wolever, T. M., Taylor, R. H., Barker, H., Fielden, H., Baldwin, J. M., Bowling, A. C., Newman, H. C., Jenkins, A. L., and Goff, D. V. (1981).

"Glycemic index of foods: a physiological basis for carbohydrate exchange." *The American Journal of Clinical Nutrition* 34 (3):362-6.

[53] Inglett, G. E. (1993). "Amylodextrins containing β-glucan from oat flours and bran." *Food Chemistry* 47 (2):133-136. doi: https://doi.org/10.1016/0308-8146(93)90234-7.

[54] Genç, H., Özdemir, M., and. Demirbaş, A. (2001). "Analysis of mixed-linked (1→3), (1→4)-β-d-glucans in cereal grains from Turkey." *Food Chemistry* 73 (2):221-224. doi: https://doi.org/10.1016/S0308-8146(00)00290-9.

[55] Wood, J. A., Knights, E. J., and Choct, M. (2011). "Morphology of chickpea seeds (*Cicer arietinum* L.): Comparison of desi and kabuli types." *International Journal of Plant Sciences* 172 (5):632-643. doi: 10.1086/659456.

[56] DeVries, J. W., and American Association of Cereal Chemists United States of America. (2001). "The definition of dietary fiber." *Cereal Foods World* 46 (3):112-114.

[57] Rincón, F., Martínez, B., and Ibáñez, M. V. (1998). "Proximate composition and antinutritive substances in chickpea (*cicer arietinum* L) as affected by the biotype factor." *Journal of the Science of Food and Agriculture* 78 (3):382-388. doi: 10.1002/(SICI)1097-0010(199811)78:3<382::AID-JSFA128>3.0.CO;2-J.

[58] Singh, U. (1984). "Dietary fibre and its constituents in desi and kabuli chickpea (*Cicer arietinum* L.) cultivars." *Nutrition Reports International* 29:419-426.

[59] Dalgetty, D. D., and Baik, B. K. (2003). "Isolation and characterization of cotyledon fibers from peas, lentils, and chickpeas." *Cereal Chemistry* 80 (3):310-315. doi: 10.1094/CCHEM.2003.80.3.310.

[60] Njoumi, S., Amiot, M. J., Rochette, I., Bellagha, S., and Mouquet-Rivier, C. (2019). "Soaking and cooking modify the alpha-galacto-oligosaccharide and dietary fibre content in five Mediterranean legumes." *International Journal of Food Sciences and Nutrition* 70 (5):551-561. doi: 10.1080/09637486.2018.1544229.

[61] Felker, F. C., Kenar, J. A., Byars, J. A., Singh, M., and Liu S. X. (2018). "Comparison of properties of raw pulse flours with those of jet-cooked, drum-dried flours." *LWT* 96:648-656. doi: https://doi.org/10.1016/j.lwt.2018.06.022.

[62] Domínguez-Arispuro, D. M., Cuevas-Rodríguez, E.O., Milán-Carrillo, J., León-López, L., Gutiérrez-Dorado, R., and Reyes-Moreno, C. (2018). "Optimal germination condition impacts on the antioxidant activity and phenolic acids profile in pigmented desi chickpea (*Cicer arietinum* L.) seeds." *Journal of Food Science and Technology* 55 (2):638-647. doi: 10.1007/s13197-017-2973-1.

[63] Collar, C., and Armero, E. (2018). "Value-added of heat moisture treated mixed flours in wheat-based matrices: a functional and nutritional approach." *Food and Bioprocess Technology* 11 (8):1536-1551. doi: 10.1007/s11947-018-2125-2.

[64] Santos, F. G., Fratelli, C., Muniz, D. G., and Capriles, V .D. (2018). "Mixture design applied to the development of chickpea-based gluten-free bread with attractive technological, sensory, and nutritional quality." *Journal of Food Science* 83 (1):188-197. doi: 10.1111/1750-3841.14009.

[65] Goñi, I., and Valentín-Gamazo, C. (2003). "Chickpea flour ingredient slows glycemic response to pasta in healthy volunteers." *Food Chemistry* 81 (4):511-515. doi: https://doi.org/10.1016/S0308-8146(02)00480-6.

[66] Queiroz-Monici, K. da. S., Costa, G. E. A., da Silva N., Reis, S. M. P. M., and de Oliveira, A. C. (2005). "Bifidogenic effect of dietary fiber and resistant starch from leguminous on the intestinal microbiota of rats." *Nutrition* 21 (5):602-608. doi: https://doi.org/10.1016/j.nut.2004.09.019.

[67] Frias, J., Vidal-Valverde, C., Sotomayor, C., Diaz-Pollan, C., and Urbano, G. (2000). "Influence of processing on available carbohydrate content and antinutritional factors of chickpeas." *European Food Research and Technology* 210 (5):340-345. doi: 10.1007/s002170050560.

[68] Mahmood, T., Hameed, T., Hasnain, S., Ali, S., Qayyum, A., Mehmood, A., Liaquat, M., Khan, S. U., Saeed, M., and Khan, A. (2018). "Assessment of flatulence causing agents in Chickpea (*Cicer arietinum* L.) and their possible removal." *Food Science and Technology* 38:120-125.

[69] Sánchez-Mata, M. C., Cámara Hurtado, M. M., and Díez-Marqués, C. (1999). "Effect of domestic processes and water hardness on soluble sugars content of chickpeas (*Cicer arietinum* L.)." *Food Chemistry* 65 (3):331-338. doi: https://doi.org/10.1016/S0308-8146(98)00219-2.

[70] Han, I. H., and Baik, B. K. (2006). "Oligosaccharide content and composition of legumes and their reduction by soaking, cooking, ultrasound, and high hydrostatic pressure." *Cereal Chemistry* 83 (4):428-433. doi: 10.1094/CC-83-0428.

[71] Berrios, J. D., Morales, P., Camara, M., and Sanchez-Mata, M. C. (2010). "Carbohydrate composition of raw and extruded pulse flours." *Food Research International* 43 (2):531-536. doi: 10.1016/j.foodres.2009.09.035.

[72] Sánchez-Mata, M. C., Peñuela-Teruel, M. J., Cámara-Hurtado, M., Díez-Marqués, C., and Torija-Isasa, M. E. (1998). "Determination of mono-, di-, and oligosaccharides in legumes by high-performance liquid chromatography using an amino-bonded silica column." *Journal of Agricultural and Food Chemistry* 46 (9):3648-3652. doi: 10.1021/jf980127w.

[73] Sreerama, Y. N., Neelam, D. A., Sashikala, V. B., and Pratape, V. M. (2010). "Distribution of nutrients and antinutrients in milled fractions of chickpea and horse gram: Seed coat phenolics and their distinct modes of enzyme inhibition." *Journal of Agricultural and Food Chemistry* 58 (7):4322-4330. doi: 10.1021/jf903101k.

[74] Xiaoli, X., Yan, L., Hua, Shuang., Li, W., Sun, Y., Ma, H., Zhang, J., and Zeng, X. (2008). "Determination of oligosaccharide contents in 19 cultivars of chickpea (*Cicer arietinum* L) seeds by high performance liquid chromatography." *Food Chemistry* 111 (1):215-219. doi: http://dx.doi.org/10.1016/j.foodchem.2008.03.039.

[75] Rubio, L. A., Pedrosa, M. M., Pérez, A., Cuadrado, C., Burbano, C., and Muzquiz, M. (2005). "Ileal digestibility of defatted soybean, lupin and chickpea seed meals in cannulated Iberian pigs: II. Fatty acids and carbohydrates." *Journal of the Science of Food and Agriculture* 85 (8):1322-1328. doi: 10.1002/jsfa.1964.

[76] Wang, R., Gangola, M. P., Jaiswal, S., Båga, M., Gaur, P. M., and Chibbar, R. N. (2017). "Variation in seed-quality traits of chickpea and their correlation to raffinose family oligosaccharides concentrations." *Crop Science* 57 (3):1594-1602. doi: 10.2135/cropsci2016.08.0710.

[77] Raja, R. B., Balraj, R., Agasimani, S., Dinakaran, E., Thiruvengadam, V., Bapu, J. R. K., and Ram, S. G. (2015). "Determination of oligosaccharide fraction in a worldwide germplasm collection of chickpea ('*Cicer arietinum*' L.) using high performance liquid chromatography " *Australian Journal of Crop Science* 9 (7):605-613.

[78] Gangola, M. P., Jaiswal, S., Khedikar, Y. P., and Chibbar, R. (2014). "A reliable and rapid method for soluble sugars and RFO analysis in chickpea using HPAEC–PAD and its comparison with HPLC–RI." *Food Chemistry* 154:127-133. doi: https://doi.org/10.1016/j.foodchem.2013.12.085.

[79] Gangola, M. P., Jaiswal, S., Kannan, U, Gaur, P. M., Båga, M., and Chibbar, R. N. (2016). "Galactinol synthase enzyme activity influences raffinose family oligosaccharides (RFO) accumulation in developing chickpea (*Cicer arietinum* L.) seeds." *Phytochemistry* 125:88-98. doi: https://doi.org/10.1016/j.phytochem.2016.02.009.

[80] Moussou, N., Corzo-Martínez M., Sanz, M. L., Zaidi, F., Montilla, A., and Villamiel, M. (2017). "Assessment of Maillard reaction evolution, prebiotic carbohydrates, antioxidant activity and α-amylase inhibition in pulse flours." *Journal of Food Science and Technology* 54 (4):890-900. doi: 10.1007/s13197-016-2298-5.

[81] Pedrosa, M. M., Cuadrado, C. Burbano, C., Allaf, K., Haddad, J., Gelencsér, E., Takács, K., Guillamón, E, and Muzquiz, M. (2012). "Effect of instant controlled pressure drop on the oligosaccharides, inositol phosphates, trypsin inhibitors and lectins contents of different legumes." *Food Chemistry* 131 (3):862-868. doi: https://doi.org/10.1016/j.foodchem.2011.09.061.

[82] Udahogora, M. (2012). "Health benefits and bioactive compounds in field peas, faba beans, and chickpeas." In *Cereals and Pulses*, 199-215. Wiley-Blackwell.

[83] Nestel, P., Cehun, M., and Chronopoulos, A. (2004). "Effects of long-term consumption and single meals of chickpeas on plasma glucose, insulin, and triacylglycerol concentrations." *American Journal of Clinical Nutrition* 79 (3):390-395.

[84] Pittaway, J. K., Robertson, A. K., and Ball, M. J. (2008). "Chickpeas may influence fatty acid and fiber intake in an ad libitum diet, leading to small improvements in serum lipid profile and glycemic control." *Journal of the American Dietetic Association* 108 (6):1009-1013. doi: 10.1016/j.jada.2008.03.009.

[85] Hosseinpour-Niazi, S., Mirmiran, P., Fallah-Ghohroudi, A., and Azizi, F. (2015). "Non-soya legume-based therapeutic lifestyle change diet reduces inflammatory status in diabetic patients: a randomised cross-over clinical trial." *British Journal of Nutrition* 114 (2):213-219. doi: 10.1017/s0007114515001725.

[86] Hermsdorff, H. H., Zulet, M. A., Abete, I., and Martinez, J. A. (2011). "A legume-based hypocaloric diet reduces proinflammatory status and improves metabolic features in overweight/obese subjects." *European Journal of Nutrition* 50 (1):61-69. doi: 10.1007/s00394-010-0115-x.

[87] Yang, Y., Zhou, L., Gu, Y., Zhang, Y., Tang, J., Li, F., Shang, W., Jiang, B., Yue, X., and Chen, M. (2007). "Dietary chickpeas reverse visceral adiposity, dyslipidaemia and insulin resistance in rats induced by a chronic high-fat diet." *British Journal of Nutrition* 98 (4):720-726. doi: 10.1017/s0007114507750870.

[88] Mollard, R. C., Luhovyy, B. L., Panahi, S., Nunez, M., Hanley, A., and Anderson, G. H. (2012). "Regular consumption of pulses for 8 weeks reduces metabolic syndrome risk factors in overweight and obese adults." *British Journal of Nutrition* 108:S111-S122. doi: 10.1017/s0007114512000712.

[89] Chen, Z-Y., Jiao, R., and Ma, K. Y. (2008). "Cholesterol-lowering nutraceuticals and functional foods." *Journal of Agricultural and Food Chemistry* 56 (19):8761-8773. doi: 10.1021/jf801566r.

[90] Threapleton, D. E., Greenwood, D. C., Evans, C. E. L., Cleghorn, C. L., Nykjaer C., Woodhead, C., Cade, J. E., Gale, C. P., and Burley, V. J. (2013). "Dietary fibre intake and risk of cardiovascular disease: systematic review and meta-analysis." *Bmj-British Medical Journal* 347. doi: 10.1136/bmj.f6879.

[91] Al-Sheraji, S. H., Ismail, A., Manap, M. Y., Mustafa, S., Yusof, R. M., and Hassan, F. A. (2013). "Prebiotics as functional foods: A review." *Journal of Functional Foods* 5 (4):1542-1553. doi: http://dx.doi.org/10.1016/j.jff.2013.08.009.

[92] Becerra-Tomás, N., Díaz-López, A., Rosique-Esteban, N., Ros, E., Buil-Cosiales, P., Corella, D., Estruch, R., Fitó, M., Serra-Majem, L., Arós, F., Lamuela-Raventós, R. M., Fiol, M., Santos-Lozano, J. M., Díez-Espino, J., Portoles, O., Salas-Salvadó, J. and PREDIMED Study Investigators. (2018). "Legume consumption is inversely associated with type 2 diabetes incidence in adults: A

prospective assessment from the PREDIMED study." *Clinical Nutrition* 37 (3):906-913. doi: 10.1016/j.clnu.2017.03.015.

[93] Augustin, L. S. A., Chiavaroli, L., Campbell, J., Ezatagha, A., Jenkins, A L., Esfahani, A., and. Kendall, C. W. C. (2016). "Post-prandial glucose and insulin responses of hummus alone or combined with a carbohydrate food: a dose–response study." *Nutrition Journal* 15 (1):13. doi: 10.1186/s12937-016-0129-1.

[94] Hosseinpour-Niazi, S., Mirmiran, P., Hedayati, M., and Azizi, F. (2014). "Substitution of red meat with legumes in the therapeutic lifestyle change diet based on dietary advice improves cardiometabolic risk factors in overweight type 2 diabetes patients: a cross-over randomized clinical trial." *European Journal of Clinical Nutrition* 69:592. doi: 10.1038/ejcn.2014.228.

[95] Mao, X., Zhang, L., Xia, Q., Sun, Z., Zhao, X., Cai, H., Yang, X., Xia, Z., and Tang, Y. (2008). "Vanadium-enriched chickpea sprout ameliorated hyperglycemia and impaired memory in streptozotocin-induced diabetes rats." *BioMetals* 21 (5):563. doi: 10.1007/s10534-008-9142-y.

[96] Wang, Y. H. A., and McIntosh, G. H. (1996). "Extrusion and boiling improve rat body weight gain and plasma cholesterol lowering ability of peas and chickpeas." *The Journal of Nutrition* 126 (12):3054-3062. doi: 10.1093/jn/126.12.3054.

[97] Harris, P. J., and Ferguson, L. R. (1993). "Dietary fiber-Its composition and role in protection against colorectal-cancer." *Mutation Research* 290 (1):97-110. doi: 10.1016/0027-5107(93)90037-g.

[98] Chino, X., Sánchez, M., Jiménez Martínez, C., Vásquez Garzón, V. R., Álvarez González, I., Villa Treviño, S., Madrigal Bujaidar, E., Dávila Ortiz, G., and Baltiérrez Hoyos, R. (2017). "Cooked chickpea consumption inhibits colon carcinogenesis in mice induced with azoxymethane and dextran sulfate sodium." *Journal of the American College of Nutrition* 36 (5):391-398. doi: 10.1080/07315724.2017.1297744.

[99] Dwivedi, S., Sahrawat, K., Puppala, N., and Ortiz, R. (2014). "Plant prebiotics and human health: Biotechnology to breed prebiotic-rich nutritious food crops." *Electronic Journal of Biotechnology* 17 (5):238-245. doi: 10.1016/j.ejbt. 2014.07.004.

[100] Fernando, W. M. U., Hill, J. E., Zello, G. A., Tyler, R. T., Dahl, W J., and Van Kessel, A. G. (2010). "Diets supplemented with chickpea or its main oligosaccharide component raffinose modify faecal microbial composition in healthy adults." *Beneficial Microbes* 1 (2):197-207. doi: 10.3920/bm2009.0027.

[101] Dai, Z., Wanyong Lyu, Minhao Xie, Qingxia Yuan, Hong Ye, Bing Hu, Li Zhou, and Xiaoxiong Zeng. 2017. "Effects of α-galactooligosaccharides from chickpeas on high-fat-diet-induced metabolic syndrome in mice." *Journal of Agricultural and Food Chemistry* 65 (15):3160-3166. doi: 10.1021/acs.jafc.7b00489.

[102] Monk, J. M., Lepp, D., Wu, W., Graf, D., McGillis, L. H., Hussain, A., Carey, C., Robinson, L. E., Liu, R., Tsao, R., Brummer, Y., Tosh, S. M., and Power, K. A. (2017). "Chickpea-supplemented diet alters the gut microbiome and enhances gut barrier integrity in C57Bl/6 male mice." *Journal of Functional Foods* 38:663-674. doi: https://doi.org/10.1016/j.jff.2017.02.002.

[103] Maccaferri, S., Klinder, A., Cacciatore, S., Chitarrari, R., Honda, H., Luchinat, C., Bertini, I., Carnevali, P., Gibson, G. R., Brigidi, P., and Costabile, A. (2012). "*In vitro* fermentation of potential prebiotic flours from natural sources: Impact on the human colonic microbiota and metabolome." *Molecular Nutrition & Food Research* 56 (8):1342-1352. doi: 10.1002/mnfr.201200046.

[104] Zhang, Y., Su, D., He, J., Dai, Z., Asad, R., Ou, S., and Zeng, X. (2017). "Effects of ciceritol from chickpeas on human colonic microflora and the production of short chain fatty acids by *in vitro* fermentation." *LWT - Food Science and Technology* 79:294-299. doi: https://doi.org/10.1016/j.lwt.2017.01.040.

[105] Uppal, V., and Bains, K. (2012). "Effect of germination periods and hydrothermal treatments on *in vitro* protein and starch digestibility of germinated legumes." *Journal of Food Science and Technology* 49 (2):184-191. doi: 10.1007/s13197-011-0273-8.

[106] Goñi, I., Garcia-Alonso, A., and Saura-Calixto, F. (1997). "A starch hydrolysis procedure to estimate glycemic index." *Nutrition Research* 17 (3):427-437. doi: http://dx.doi.org/10.1016/S0271-5317(97)00010-9.

[107] de Almeida Costa, G. E., Queiroz-Monici, K. da. S., Reis, S. M., and de Oliveira, A. C. (2006). "Chemical composition, dietary fibre and resistant starch contents of raw and cooked pea, common bean, chickpea and lentil legumes." *Food Chemistry* 94 (3):327-330. doi: Doi 10.1016/J.Foodchem.2004.11.020.

[108] Angulo-Bejarano, P. I., Verdugo-Montoya, N. M., Cuevas-Rodriguez, E. O., Milan-Carrillo, J., Mora-Escobedo, R., Lopez-Valenzuela, J. A., Garzon-Tiznado, J. A., and Reyes-Moreno, C. (2008). "Tempeh flour from chickpea (*Cicer arietinum* L.) nutritional and physicochemical properties." *Food Chemistry* 106 (1):106-112. doi: doi 10.1016/J.Foodchem.2007.05.049.

[109] Abd El-Hady, E. A., and Habiba, R. A. (2003). "Effect of soaking and extrusion conditions on antinutrients and protein digestibility of legume seeds." *LWT - Food Science and Technology* 36 (3):285-293. doi: http://dx.doi.org/10.1016/S0023-6438(02)00217-7.

[110] Perez-Hidalgo, M. A., Guerra-Hernández, E., and García-Villanova, B. (1997). "Dietary fiber in three raw legumes and processing effect on chick peas by an enzymatic-gravimetric method." *Journal of Food Composition and Analysis* 10 (1):66-72. doi: https://doi.org/10.1006/jfca.1997.0522.

[111] Kliger, E., Fischer, L., Lutz-Wahl, S., and Saguy, I. S. (2011). "Controlled exogenous enzyme imbibition and activation in whole chickpea seed enzyme

reactor (SER)." *Journal of Food Science* 76 (4):E377-E382. doi: 10.1111/j.1750-3841.2011.02117.x.

[112] Brummer, Y., Kaviani, M., and Tosh, S. M. (2015). "Structural and functional characteristics of dietary fibre in beans, lentils, peas and chickpeas." *Food Research International* 67:117-125. doi: https://doi.org/10.1016/j.foodres.2014.11.009.

BIOGRAPHICAL SKETCHES

Ada Keila Milán-Noris

Affiliation: Universidad Autónoma de Sinaloa

Education:

PhD in Food Biotechnology (2017), Tec de Monterrey.
M.Sc. in Biotechnology (2011), Tec de Monterrey.
B.S. in Biochemical Engineering (2006), Universidad Autónoma de Sinaloa.

Business Address: Laboratorio de Nutracéuticos (18), Facultad de Ciencias Químico Biológicas, Universidad Autónoma de Sinaloa, Ciudad Universitaria, Blv. de las Américas y Josefa Ortiz de Domínguez, S/N, Culiacán, Sinaloa, México.

Research and Professional Experience:

Effect of cereal and pulses processing on nutritional, nutraceuticals and technological characteristics. Also, elucidation of mechanism effect of bioactive compounds.

Teaching experience (undergraduate): Food Processing, Nutraceuticals and Pharma and Food Development in Biochemical Engineering; and Advanced Chemical Analysis in Genomic Biotechnology.

Professional Appointments:

2018-present. Assistant Professor, Universidad Autónoma de Sinaloa.
2012-2013. Research assistant, Universidad Autónoma de Sinaloa.
2011-2012. Research assistant, Tec de Monterrey.

Honors:
Recognized by the National Researchers System of Mexico (SNI):
Candidate to National Researcher (January 2019-December 2021)

Publications from the Last 3 Years:

1. Milán-Noris, Ada Keila, Julián De la Rosa-Millan, and Sergio Othón Serna-Saldivar. 2019. "Comparative analysis of techno-functional properties, starch digestion and protein quality of pigmented chickpea flours." *International Journal of Food Science & Technology* 0 (0). doi: 10.1111/ijfs.14144.
2. Mora-Rochin, Saraid., Milán-Noris, Ada Keila., Milán-Carrillo, Jorge. 2019. "Maize: Processing, Product Development, and Nutritional Aspects." In *Whole Grains*, edited by Annamalai Manickavasagan Shabir Ahmad Mir, Manzoor Ahmad Shah. CRC Press
3. Milán-Noris, Ada Keila, Janet Alejandra Gutiérrez-Uribe, Arlette Santacruz, Sergio O. Serna-Saldívar, and Cristina Martínez-Villaluenga. 2018. "Peptides and isoflavones in gastrointestinal digests contribute to the anti-inflammatory potential of cooked or germinated desi and kabuli chickpea (Cicer arietinum L.)." *Food Chemistry* 268:66-76. doi: https://doi.org/10.1016/j.foodchem.2018.06.068.
4. Milán-Noris, Ada Keila, Julián de la Rosa-Millán, Cuauhutémoc Reyes-Moreno, and Sergio O. Serna-Saldivar. 2017. "Physicochemical, functional properties, and digestion of isolated starches from pigmented chickpea (Cicer arietinum L.) cultivars." *Starch - Stärke* 69 (5-6). doi: doi:10.1002/star.201600152.

Alvaro Montoya-Rodriguez

Affiliation: Universidad Autónoma de Sinaloa.

Education:
Ph.D. in Food Biotechnology (2015), Universidad Autónoma de Sinaloa.
M.Sc. in Food Science (2011), Universidad Autónoma de Sinaloa.
B.Sc in Biochemical Engineering (2009), Universidad Autónoma de Sinaloa.

Business Address: Laboratorio de Nutracéuticos (18), Facultad de Ciencias Químico Biológicas, Universidad Autónoma de Sinaloa, Ciudad Universitaria, Blv. de las Américas y Josefa Ortiz de Domínguez, S/N, Culiacán Sinaloa México

Research and Professional Experience:

His research is to understand the effect of bioactive peptides that are present in food after being processed, (*ie*. Extrusion, germination, nixtamalization) and also, its potential use as a preventive compound to control chronic diseases related to metabolic syndrome. Dr. Montoya Rodriguez teaches biochemistry classes to undergraduate students.
He has 9 peer-reviewed publications and book chapters (*h-index: 4*)

Professional Appointments:

2015-present. Associate Professor, Facultad de Ciencias Químico Biológicas, Universidad Autónoma de Sinaloa.

Honors:

Recognized by the National Researchers System of Mexico (SNI):
Candidate to National Researcher (January 2016-December 2018)
National Researcher Level 1 (January 2019-December 2021)
Recognized by the Sinaloense System of Researchers and Technologists:
Honorary Member since 2015- present.

Publications from the Last 3 Years:

1. Fitch-Vargas P. R., Aguilar-Palazuelos E., Vega-García M. O., Zazueta- Morales J. J., Calderón-Castro A., Montoya-Rodríguez A., Delgado-Nieblas C I., Camacho-Hernández I. L., 2018. Effect of a corn starch coating obtained by the combination of extrusion process and casting technique on the postharvest quality of tomato. *Revista Mexicana de Ingeniería Química*, Vol. 18, Pag. 789-801.
2. Gaxiola-Cuevas, N., Mora-Rochín, S., Cuevas- Rodriguez E. O., León-López, L., Reyes-Moreno, C., Montoya-Rodríguez, A., and Milán-Carrillo. J. (2017). Phenolic acids profiles and cellular antioxidant activity in tortillas produced from mexican maize landrace processed by nixtamalization and lime extrusion cooking, Plant Foods for Human Nutrition, Vol. 72, Pag. 314–320.

Julián de la Rosa Millán

Affiliation: Tecnologico de Monterrey. Campus Queretaro, Queretaro, Mexico.

Education: PHD in Biotic Products Development. Instituto Politécnico Nacional. México

Business Address: Av. Epigmenio Gonzalez 500, San Pablo, Queretaro, Mexico.

Research and Professional Experience:
Full time professor at Tecnologico de Monterrey. Campus Queretaro
Researcher in carbohydrates and proteins.

Professional Appointments:
AACC member, IFT member, ACS member.

Publications from the Last 3 Years:

1. Rivera–González, G., Amaya–Guerra, C. A., & de la Rosa–Millán, J. (2019). Physicochemical characterisation and *in vitro* starch digestion of avocado seed flour (persea americana V. hass) and its starch and fibrous fractions. *International Journal of Food Science and Technology*, doi:10.1111/ijfs.14160.
2. Chavez-Murillo, C. E., Orona-Padilla, J. L., de la Rosa Millan, J. (2019) Physicochemical, functional properties and ATR-FTIR digestion analysis of thermally treated starches isolated from black and bayo beans. *Starch/Starke*, 71, 3-4.
3. de la Rosa-Millán, J., Heredia-Olea, E., Perez-Carrillo, E., Guajardo-Flores, D., Serna-Saldívar, S. R. O. (2019). Effect of decortication, germination and extrusion on physicochemical and *in vitro* protein and starch digestion characteristics of black beans (Phaseolus vulgaris L.), *LWT-Food Science and Technology*. 102, 330-337.
4. Hernández-Reyes, K. E., Montemayor-Mora, G., Pérez-Carrillo, E., De la Rosa-Millán, J., García-Villanueva, C., Serna-Saldívar, S. O. (2019). Effect of soybean bagasse addition on texture, sensory properties, and protein quality of maize tortillas. *Cereal Chemistry*, 96, 283-291.
5. Milán-Noris, A. K., De la Rosa-Millan, J., Serna-Saldivar, S. O. (2019). Comparative analysis of techno-functional properties, starch digestion and protein quality of pigmented chickpea flours. *International Journal of Food Science and Technology. Early View.*
6. Buitimea-Cantúa, N. E., de la Rosa-Millán, J. (2018). Physicochemical, textural, and *in vitro* starch digestion properties of nixtamalized maize flour and tortillas enriched with sorghum (Sorghum bicolor (L.) Moench) bran. *Cereal Chemistry*, 95, 829-837.
7. de la Rosa-Millán, J., Orona-Padilla, J. L., Flores-Moreno, V. M., Serna-Saldívar, S. O. (2018). Physicochemical, functional and ATR-FTIR molecular analysis of protein extracts derived from starchy pulses. *International Journal of Food Science and Technology*. 53, 1414-1424.
8. Chávez-Murillo, C. E., Veyna-Torres, J. I., Cavazos-Tamez, L. M., de la Rosa-Millán, J., Serna-Saldívar, S. O. (2018). Physicochemical characteristics, ATR-FTIR molecular interactions and *in vitro* starch and protein digestion of thermally-treated whole pulse flours. *Food Research International*. 105, 371-383.

9. De La Rosa-Millán, J. (2017). Physicochemical, molecular, and digestion characteristics of annealed and heat-moisture treated starches under acidic, neutral, or alkaline pH. *Cereal Chemistry*, 94(4), 770-779. doi:10.1094/CCHEM-10-16-0250-R
10. Milán-Noris, A. K., de la Rosa-Millán, J., Reyes-Moreno, C., Serna-Saldivar, S. O. *Physicochemical, functional properties, and digestion of isolated starches from pigmented chickpea (Cicer arietinum L.) cultivars (2017) 69 (5-6),* art. no. 1600152. doi: 10.1002/star.201600152.
11. De La Rosa-Millan, J., Orona-Padilla, J.L., Flores-Moreno, V. M., *Serna-Saldivar, S. O. Physicochemical and digestion characteristics of starch and fiber-rich subfractions from four pulse bagasses (2017) 94 (3),* pp. 524-531. DOI: 10.1094/CCHEM-08-16-0208-R.
12. de la Rosa-Millán, J., Pérez-Carrillo, E., Guajardo-Flores, S. Effect of germinated black bean cotyledons (Phaseolus vulgaris L.) as an extruded flour ingredient on physicochemical characteristics, *in vitro* digestibility starch, and protein of nixtamalized blue maize cookies (2017) 69 (3-4), art. no. 1600085. DOI: 10.1002/star.201600085.
13. Chew-Guevara, A. A., Pérez-Carrillo, E., Othon Serna-Saldívar, S. R., De La Rosa-Millán, J. *Effect of decortication and protease treatment on physicochemical and functional characteristics of red sorghum (Sorghum bicolor) and yellow maize (Zea maiz) starches (2016) 68 (1-2),* pp. 1-8. doi: 10.1002/star.201500151.

In: Handbook of Chickpeas
Editors: Albert T. Lund and Noah D. Schultz
ISBN: 978-1-53616-374-2

Chapter 10

INTEGRATED PEST MANAGEMENT (IPM) AND AREA-WIDE MANAGEMENT (AWM) APPROACHES FOR INSECT PESTS OF CHICKPEA *CICER ARIETINUM* (*L.*) WALP CROP

Muhammad Sarwar*
National Institute for Biotechnology and Genetic Engineering (NIBGE), Faisalabad, Pakistan

ABSTRACT

Chickpea (*Cicer arietinum* L.), is one of the most important grain legumes, major source of protein and forms an integral part of diet. However, major constraints to chickpea during different stages of crop growth include pod borer *Helicoverpa armigera* (Hubner) (Lepidoptera: Noctuidae), cutworm (*Agrotis ipsilon* Hufnagel), leaf feeding caterpillar (*Spodoptere exigua* Hubner), leaf miner (*Liriomyza cicerina* Rondani) and aphid (*Aphis craccivora* Koch). Small larvae of *H. armigera* are primarily foliage feeders, while large caterpillars feed on pods and foliage, but have not any preference for flowers which show no serious yield loss incurred if pod borers are tolerated during flowering stage. During pods formation, the result is a sharp decline in chickpea production due to increasing losses of even up to 100%. In rare situations, control may be warranted during the vegetative and flowering stages, when pest pressure is extreme. If using controlling products that are only effective against small larvae, it may be necessary to apply a spray during flowering to control a population of small larvae to prevent them causing damage as large larvae during pod-set and pod-fill. A variety of predatory and parasitic insects, spiders, birds, bats, rodents and diseases attack pod borer

* Corresponding Author's Email: drmsarwar64@yahoo.com.

at different stages of its lifecycle. Traditionally, preferred control options are carbamates and pyrethroids to spray as soon as the threshold has been reached as insect grows rapidly and a few days delay in spraying can result in major crop damage and increased difficulty in control. Check crop regularly from flowering onwards to identify when crop is most susceptible, familiar with the economic thresholds by using them as a basis for control decision and time to control pest most effectively. Conserve populations of predatory and parasitic insects by using selective insecticides to reduce infestations below economic threshold if natural enemies are not disrupted by broad-spectrum sprays. Manage the local population with trap cropping, cultivating to destroy overwintering pupae, destroying weed hosts within the crop and surrounding areas and adopting effective chemical control strategy. Following the implementation of integrated pest management, pod borer incidence is significantly reduced, a two to six fold increase in grain yield obtained and there is also a two to five fold increase in net income.

Keywords: gram, *Cicer arietinum*, chickpea, major and minor pests, pest control, sucking and chewing pests

1. Introduction

Chickpea *Cicer arietinum* L., commonly known as gram in Fabaceae or Legume family, is a leading annual grain crop cultivated by humans. Chickpea is the principal source of dietary protein and rich in essential amino acids, carbohydrates, minerals, and vitamins A and C. The crop also fixes nitrogen in soil (40 kg N/ha) and reduces fertilizer costs in subsequent cereal crops such as rice and maize. However, the major pest problems in chickpea are pod borers (*Helicoverpa armigera* Hubner and *Helicoverpa punctigera* Wallengren), leafminer *Liriomyza cicerina* Rondani, black cutworm *Agrotis ipsilon* (Hufnagel), aphid (*Aphis craccivora* Koch) and semilooper (*Autographa nigrisigna* Walker). Among these pests, the pod borer, *Helicoverpa* (*Heliothis*) *armigera* is a leading and most serious one in many chickpea growing areas of the country wherein the small caterpillars are primarily foliage feeders while large larvae feed on pods and foliage. Other insects are rarely a problem in chickpea because it secretes an organic acid (malic acid) from hairs on their leaves, stems and pods, making the crop unattractive. Seedling insect pests, such as cutworm, can attack chickpeas, but are seldom an economic problem. Caterpillars of *Helicoverpa* sp., do most damage during pod set through to maturity, and can reduce both grain yield and quality (Mari et al., 2013; Sarwar, 2017).

Knowing when chickpea crops are susceptible to pest attack is the first step in good pest management. Controlling of pod borer through management options (cultural, biological and chemical controls) results in a significant increase in yield. Low to moderate levels of resistance have been identified in the germplasm, and a few improved cultivars with resistance to pod borer and high grain yield have been developed. Germplasm accessions of the wild relatives of chickpea (*Cicer bijugum*, *C. judaicum* and *C. reticulatum*) can be used to increase the levels and diversify the bases of resistance to

H. armigera. Efforts are also underway to utilize molecular techniques to increase the levels of resistance to pod borer. Synthetic insecticides, agronomic practices, nuclear polyhedrosis virus, entomopathogenic fungi, bacteria and natural plant products have been evaluated as components of pest management in chickpea (Minja et al., 2000; Sharma 2007). Herein *chapter*, something will be *examined* in greater detail by paying special attention to the elements of integrated pest management and area-wide management tactics for insect pests of chickpea crop.

2. Major Insect Pests of Chickpea

Major pests in a general term are more common and important that are harmful to crop comprising pod borers, which cause a huge damage to plant.

2.1. Pod Borers *Helicoverpa* Sp., (Lepidoptera: Noctuidae)

There are two species of pod borers, *Helicoverpa armigera* (Hubner) (corn earworm) (Figure 1) and *Helicoverpa punctigera* (Wallengren) (native budworm) (Figure 2). Species composition of pod borers in the crop can be influenced by the time of year. In temperate regions, the majority of the *H. armigera* populations over-winter from mid-March to onwards and emerge during September/October. The *H. punctigera* is usually the dominant species through September, but seasonal variation can lead to *H. armigera* dominance early in some years. Both these species of genus *Helicoverpa*, are serious pests of field crops (particularly grain legumes, summer grains and cotton) in the certain regions. Identification of species which is present, or knowing which species predominates in an area, helps to avoid products that may not give good pest control. Adult m*oths* are *dull light brown* with *dark markings* and *35 mm long* with a 30-45 mm wingspan. Borer *H. armigera has a small light or pale patch in the dark section of the hindwing, while the dark section is uniform* in *H. punctigera*. Forewings are brown in the female and creamy in the male. Commonly confused with other species include armyworms and loopers, but *Helicoverpa* larvae have a group of four pairs of legs in the back half of the body while loopers can have a group of two, three or four pairs of legs at the rear and loop when walking. Armyworms also have 4 pairs of prolegs, but are smoother and fatter, with more colored bands than *Helicoverpa*. The *H armigera* is generally regarded as the more serious pest because of its greater capacity to develop resistance to insecticides, broader host range and persistence in cropping areas from year to year. The *H. punctigera* numbers in cropping regions fluctuate based on conditions in inland breeding areas. Both species look the same at the egg and small larvae stages. Eggs are 0.5 mm in diameter and change from white to brown to a black head stage before hatching. Newly hatched larvae are light in color with tiny dark spots and dark heads. As larvae develop they become darker and the darker spots become more obvious. Medium larvae develop lines and bands running the length of the body and are variable in color. The *H. armigera* have a saddle of darker pigment on the fourth segment and at the back of the head and dark-colored legs. The *H. punctigera* have no saddle and light-colored legs. Large larvae of *H. armigera* have white hairs around the head, while *H. punctigera* have black hairs around the head. Pupae are found in soil underneath the crop, but *H. armigera* pupal tail spines are more widely spaced than those of *H. punctigera***.** Both *H. armigera* and *H. punctigera* take about 4-6 weeks to develop from egg to adult in summer, and 8-12 weeks in spring or autumn. Moths live for around 10 days during which females lay 1000 eggs most commonly laid on the top third of the plant and growing points (Miles, 2013).

The *Helicoverpa* larvae attack on all field crops can be present in crops from the vegetative stage to onwards. Larvae feed on leaves, but are most damaging when feeding on growing terminals, buds or squares, flowers, pods, and seeds or fruits. This includes direct losses through fruit shedding and reduced quality of produce. Recent research has determined the consumption rate of *Helicoverpa* larvae from hatchling to pupation to be 2.0 g of grain/larva. The crop needs to be closely monitored from emergence to maturity for the preence of eggs and larvae, however, most field crops only need to be monitored closely from budding or flowering through to maturity. The integrated control strategies incorporating chemical, cultural and biological methods are aimed to restrict populations to below damaging levels. For good pest management, it is essential that eggs, larval and adult stages of *Helicoverpa* should be easily identified. It is also important to be able to identify the different larval instars (very small, small, small-medium, medium-large, large) of *Helicoverpa* (Sarwar et al., 2009; 2011).

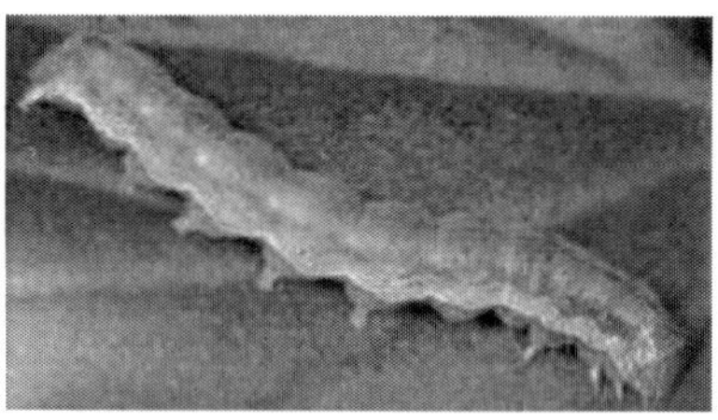

Figure 1. *Helicoverpa armigera* (Larva).

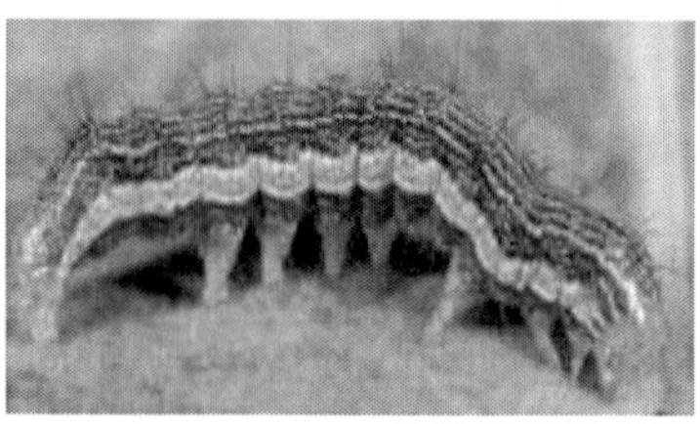

Figure 2. *Helicoverpa punctigera* (Larva).

Pheromone traps and computer modelling are two tools used routinely to monitor the arrival of the first brood of *H. punctigera* moths, and the emergence from diapause of local *H. armigera*. Pheromone traps attract male moths using the sex pheromone (a chemical attractant) that female moths emit to attract mates. As *H. armigera* and *H. punctigera* pheromones are different, so each species can be monitored separately. Pheromone traps, baited with synthetic pheromone lures, are used to monitor the arrival of *H. punctigera* moths over spring. The computer model uses current local temperature to calculate the rate of development of pupae in the soil and predicts the timing of emergence of moths (Sarwar, 2012; Ahmad and Sarwar, 2013).

Controlling of pod borer at flowering stage does not result in a significant increase in yield or quality over delaying of control until pod-set (expanded pods). If the level of *H. punctigera* is high, any registered product can control the larvae, but larvae can be best targeted when smaller than 7 mm. If *H. armigera* is the dominant species, then spray failures with carbamates or pyrethroids may occur due to resistance. The biopesticides *Helicoverpa* nucleopolyhedrovirus (HNPV) and *Bacillus thuringiensis* (*Bt*) currently have no known resistance problems. All stages of the *Helicoverpa* lifecycle are attacked by a wide range of predators, parasitoids and pathogens, and conserving of these in the crop through the avoidance of broad-spectrum insecticides can help to minimize the need for insecticide treatments (Knights and Siddique, 2002; Khan et al., 2010; Sarwar and Salman, 2016).

2.2. Population of Pod Borers during Flowering and Podding

Commencing from vegetative to early flowering stages, high populations of pod borers have generally no impact on yield or quality. In rare situations, control may be warranted during the vegetative and flowering stages if pest pressure is extreme. From mid-flowering to early podding, the impact is based on the value of crop loss that can be calculated by the following equation:-

Yield loss ($/ha) = (number of larvae/square meters × 2* × chickpea price ($/t) divided by 100 [*2.0 g grain/larva consumed]

In practice this means that in calculating how many larvae/m^2 are in a field, the following equation should be used:-

Number/square meter = [(S × 0.7) + M + L) divided by row spacing (meters)]

Where S = small, M = medium and L= large larvae, the number of larvae is based on the number/meter of row, assessed with a beat sheet, 0.7 multiplier = 70% surviving population, dividing by the row spacing (in meters) adjusts the density for different row spacing.

In practice, calculating of the potential yield loss requires information on pest numbers, size or age classes of larvae, crop row spacing, stage of crop development and expected crop value. This information is used to make a decision about whether control is warranted. In a field, there is an average of 4.2 larvae/m^2 (adjusted for mortality of small larvae). Assuming a chickpea price of \$400/t, the potential yield loss shows the cost of not controlling to be \$32/ha. In this example, if the cost of control is less than \$32/ha then it is economical to spray. The relationship between the number of pods damaged/m^2 and yield loss is: 1 damaged pod/m^2 (based on 1 m row spacing) = 1.7 kg/ha yield loss. For deciding whether to spray, there are additional factors that may influence the decision, timing and product choice. These include age structure of the larval population may need to be considered in relation to time to desiccation or harvest. For example, a late egg lay is unlikely to result in economic damage if the crop is 7-10 days away from harvest (Yassine and Omrane, 2012; Sarwar and Sattar, 2016).

3. Minor Insect Pests of Chickpea

Minor inects are those that are not causing huge damage to crop such less frequent *pests* include cutworms, locusts, aphids, false wire worms, thrips, loopers, green vegetable bugs and blue oat mites.

3.1. Black Cutworm *Agrotis ipsilon* (Lepidoptera: Noctuidae)

The common name of cutworm is derived from the larval habit of cutting the stems of young seedlings at or near ground level, causing the collapse of the plant. Several species of cutworms, including *Agrostis munda* Walker (brown cutworm), *Agrostis infusa* (Boisduval) (Bogong moth), *Agrostis ipsilon* (Hufnagel) (black cutworm) (Figure 3) and *Agrostis prophyricollis* (variable cutworm) can attack a wide range of crops including chickpea. The moths of black cutworm *A. ipsilon* are grey-brown or dull brown-black in color with a wingspan of 40 to 50 mm. Forewings are light brown, patterned with an ipsilon shape, and the hindwings are creamy white with brown edges. Moths prefer to lay their eggs in soil in lightly vegetated (e.g., a weedy fallow) or bare areas. Early autumn egg-laying results in most damage to young cereals. Larvae are up to 50 mm long, hairless with dark heads and usually darkish colored bodies, often with longitudinal lines or dark spots. Larvae curl up and remain still if picked up. All field crops are at most risk during seedling and early vegetative stages. Young caterpillars climb to plants and skeletonize the leaves or eat small holes. The older larvae may also climb to browse or cut off leaves, but commonly cut through stems at ground level and feed on the top growth of felled plants. Caterpillars that are almost fully grown often remain underground and chew into plants at or below ground level. They usually feed in the late afternoon or at night. By day time, they hide under debris or in the soil. Inspect crop twice weekly in seedling and early vegetative stage. Larvae usually feed late afternoons and evenings. Larvae hatch and feed on host plants right through emergence to maturity and mature larvae pupate in the soil. Under favorable conditions, the duration from egg-lay to adult emergence is 8-11 weeks, depending on the species. Usually there is a single generation during early vegetative stages.

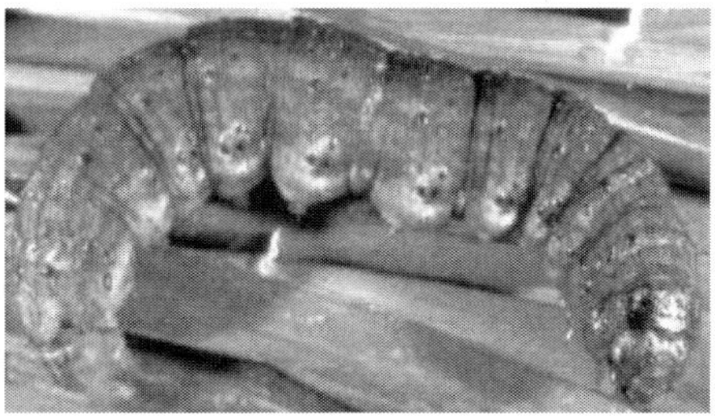

Figure 3. *Agrotis ipsilon* (Larva).

Chemical control is warranted when there is a rapidly increasing area or proportion of crop damage. Insecticide application is cost-effective and the whole crop may not need to be sprayed if pest distribution is patchy, and spot spraying may be sufficient. Control weeds 3-4 weeks prior to crop sowing and cutworms are attacked by a number of predators, parasites and diseases (Agegnehu and Medhin, 2014).

3.2. Pea Aphid *Aphis craccivora* Koch (Hemiptera: Aphididae)

Aphid *Aphis craccivora* Koch (Figure 4) variously known as the cowpea aphid, groundnut aphid or black legume aphid, is the most important insect pest of cowpea and also causes significant yield losses in other legume crops including chickpea. Aphid *A. craccivora* females have six-segmented antennae, apterae 1.2-2.3 mm long, shiny black or dark brown body with prominent cauda and brown to yellow legs. Immatures are slightly dusted with wax, adults without wax, and overwinter as eggs often at the base of young plants, but also on perennial weeds. The eggs hatch in early spring and the first larvae are known as fundatrix (stem mothers), and these aphids are all females and reproduce by parthenogenesis, producing nymphs, which molt four times over the course of eight to twelve days. A female aphid lives for 9 to 25 days and can produce from 25 to 125 young during its life, and there may be up to twenty generations in the year. Aphids attack flowering stage, seedling stage, vegetative growing stage or the pods of adult plants are attacked causing significant yield losses. Aphids cause direct damage to plants by stunting and distorting growth, but the honeydew produced is deposited on the plants and encourages the growth of sooty molds, which restrict photosynthesis. The aphid is the vector of a number of plant viruses including groundnut rosette virus, peanut mottle virus and stunt virus, and bean, cucumber and alfalfa mosaic virus (Dhingra, 1994).

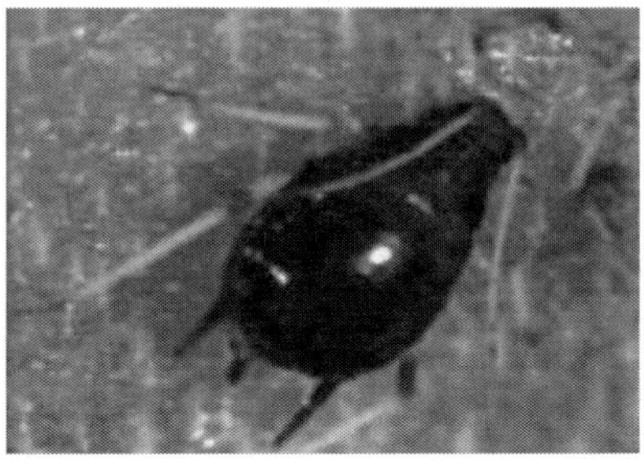

Figure 4. *Aphis craccivora*.

Particularly important parasites of *A. craccivora* are *Trioxys* (*Binodoxys*) *indicus* Subba Rao & Sharma (*Hymenoptera*: *Aphidiidae*), *Lysiphlebus fabarum* (Marshall) and *Lysiphlebus testaceipes* (Cresson) (Hymenoptera: Braconidae: Aphidiinae). *Alloxysta pleuralis* (Cameron) (Hymenoptera: *Cynipoidea*: Figitidae) is an important hyperparasite of *T. indicus and* similarly *Pachyneuron aphidis* (Bouche) (Hymenoptera: Pteromalidae) on *L. testaceipes*. Important predators include coccinellid beetles (Coleoptera: Coccinellidae) e.g., *Cheilomenes sexmaculata* (Fabricius) and *Coccinella septempuncta* Linnaeus; hover fly larvae (Diptera: Syrphidae) e.g., *Ischiodon scutellaris* (Fabricius); neuropteran larvae e.g., *Micromus timidus* Hagen (Hemerobliidae: Neuroptera); and a predatory dipteran e.g., *Aphidoletes aphidimyza* (Rondani) (Diptera: Cecidomyiidae). Control with natural predation by *Menochiles sexmaculatus* (Fabricius), is better when chickpea is intercropped with barley or linseed. Spiders may also be important predators in some areas. Recorded fungal pathogens include *Fusarium pallidoroseum, which* shows that an increase in spore concentration can result in a corresponding decrease in aphid population. *A. craccivora* can be controlled within integrated control systems practiced on numerous crops by combinations of selective insecticides, natural enemies, cultural methods and resistant crop varieties. Monitoring of pest populations to ensure effective insecticide spray application is combined with the use of mulching, which is found to be effective in delaying virus spread during the critical early growth and fruit-setting periods. The selected combinations of agronomic, chemical and cultural control measures, particularly when combined with early planting, provide better control of pests including *A. craccivora* than a single control strategy. A strategy that combines early planting, close planting (30 × 20 cm) and three insecticide applications, once each at the budding, flowering and podding stages, results in the highest yield gain over the farmer's traditional practices. The infestation of *A. craccivora* is lower at higher plant spacings (30 cm compared with 10 and 20 cm), low nitrogen fertilization levels (50 and 70 kg N/feddan) and early first irrigation (14, 21 and 28 days). Also transgenic chickpeas expressing the *Allium sativum* leaf agglutinin (ASAL) gene results in a significant reduction in survival and fecundity of *A. craccivora* (Kamphuis et al., 2012; Sarwar, 2009; 2013a; 2016 a).

3.3. Leaf Miners *Liriomyza cicerina* (Diptera: Agromyzidae)

The genus *Liriomyza* contains more than three hundred species, which are widely distributed in the new and old worlds. The agromyzid leafminers, *Liriomyza sativae* Blanchard and *Liriomyza trifolii* (Burgess) are major pests that attack and cause direct and indirect damage to chickpea. The damage caused to their host plants is very similar for both species and the larval stages feed within the leaves of the host plants, and at high densities severely reduce yields or kill the plants. In those areas where chickpea is cultivated, *Liriomyza cicerina* (Rondani) (Figure 5) must be considered as one of the most serious pests. The infestation of the plants is often severe and can strongly affect the vitality of the plants, and reduce the amount and quality of the yield. Its wing length is 1.3 - 6 mm and body coloration highly variable. Legs are normally brownish striated (marbled), but several specimens have yellowish legs. Frontorbital bristles are three or four and the anterior one can be much shorter than the others. Larval posterior spiracles each are with 7-9 pores, arranged in a semicircle and both spiracles widely separated. Puparium are pale brownish-yellow and pupation occurs in the soil. The number of generations seems to be variable (2-4 generations) depending on the availability of host plants and temperature (Asadi et al., 2006).

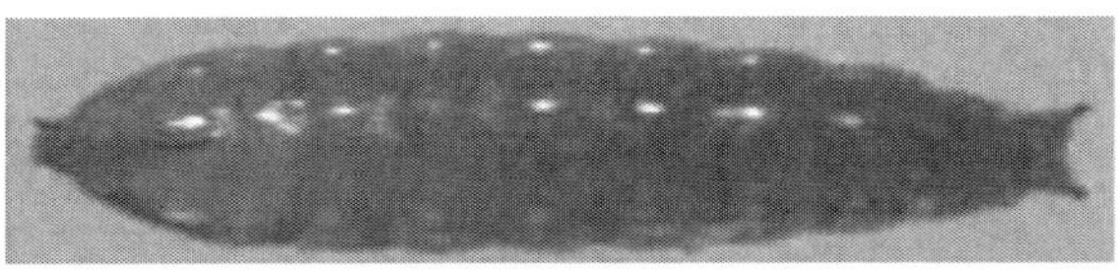

Figure 5. *Liriomyza cicerina* (Larva).

The previously prevailing attempts as control measures have been screening for resistance against leaf-miners and these efforts are partly successful. The majority of flies and larval parasitoids, which are *Diglyphus isaea* (Walker) (*Hymenoptera*: Eulophidae), *Hemiptarsenus zilahisebessi* (Mercet) (Hymenoptera: Eulophidae), *Chrysonotomyia* sp., *Pnigalio* sp., and *Cirrospilus* sp., emerge during the active season. Synchronization is found between the time of host emergence and abundance of the larval parasitoid *D. isaea* in the active season, but not in the diapause season. Asynchrony is observed between the larval-pupal parasitoid *Opius* sp., and flies, also for *Halticoptera circulus* (Walker) (*Hymenoptera*: Pteromalidae) and *Chrysocharis* sp., in both seasons. The population growth rates of larval parasitoids are lower than those of flies, which retar the biological control, particularly at the beginning of the season. To conserve and promote parasitoids, use of pesticides must be entirely avoided. The bio-effectiveness of the parasitoids is indicated by the decrease in emergence of flies from diapausing pupae. It is noteworthy to suggest that some agricultural practices i.e., spreading the harvested plants on plastic sheets to facilitate collection of the accumulated pupae, deep ploughing and applying kerosene can be used as safe control methods against flies pupae (El-Serwy, 2003).

3.4. Army Worm *Spodoptera exigua* (Lepidoptera: Noctuidae)

Army worm *Spodoptera exigua* (Hubner) (Figure 6), is cosmopolitan and polyphagous pest attacking rice, tobacco, tomato, maize, citrus, castor, groundnut, legumes and many other vegetables incliding chickpea. It forewings are brown in color with wavy white markings. Hind wing is white in color with a brown patch along the margin. Moths lay eggs on the underside of leaves in masses of 100 to 300. Egg masses appear golden brown, are round, white and covered with small hairs. Incubation period is 4-5 days, larvae are pale greenish with dark markings, gregarious in the early stages and migratory skeletonizer, and larval period lasts for 14-21 days. Pupa is brown and 1.9 cm long, pupation takes place in the soil and pupal period is 12-14 days. Longevity of adults is 9-10 days, total life cycle is 35-50 days and there are 6-8 generations per year. The larvae feed on the tender leaves and these scrape out the green matter in the leaf leaving the epidermis alone. The second and third instar larvae feed by making small holes, and late instars feed on entire lamina, petiole, even the twigs on the terminal shoots of plants. Later, the larvae spread in the field causing defoliation and larvae also feed on the developing pods. During fruiting stage, the larva scrapes the pods and feeds on it. Damage symptoms include leaf skeletonization, large irregular holes on leaves, defoliation in severe cases and deseeded pods in the field.

It is difficult to compare genotypic resistance to insects across seasons and locations because of the variation in the onset and severity of insect infestation. The results suggest that detached leaf assay could be used for large-scale screening of chickpea genotypes for resistance to *S. exigua*, while the diet incorporation assay could be used to gain additional information on the antibiosis mechanism of resistance to this insect. The genotypes suffering lower damage and with high grain yield potential can be used in chickpea improvement for resistance to *S. exigua.* Cultural control of *S. exigua* in chickpea includes plough of the soil to expose and kill pupae. Mechanical device takes account of collect and destroy the egg masses, gregarious larvae and grown up caterpillars (Shankar et al., 2013).

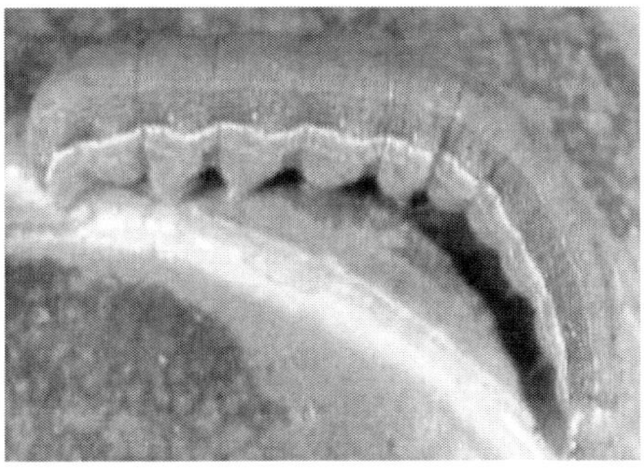

Figure 6. *Spodoptera exigua* (Larva).

3.5. Locusts and Grasshoppers (Orthoptera: Acrididae)

Locust and grasshopper have quite different behaviors, and several species of *grasshoppers* swarm as *locusts* in *different* parts of the world. Locusts may change from being individuals to becoming gregarious when populations of nymphs are dense. This change in locusts usually indicates large-scale migrations from breeding areas. Grasshoppers generally remain solitary and usually do not move far from their breeding area. Locusts and several species of grasshoppers breed in cropping areas where they may damage crops by chewing leaves and stems. The migratory locust *Locusta migratoria* (Linnaeus) (Figure 7) is large, heavily built, green or brown in the solitary form. Grows to about 14-60 mm in length and is straw colored when gregarious (swarming). Hind wings have no markings, but may be faintly greenish-yellow, mandibles (jaws) are dark-purple to black and hopper bands can be a striking black and tan color. Adult's flight is strong and steady, with the slight green wing tinge visible. Nymphs and adults can be found all year round, but damaging populations are restricted to the warmer months. Females lay eggs at least three times, but under favorable conditions may lay more than six times. Eggs are laid in the soil in pods at about 2 cm deep. Eggs pods are 1.5-2.0 cm long, tear-shaped and usually contain 10 to 16 eggs, each 4.5 mm long. In normal summer temperature, minimum life cycle is egg 11 days, hopper 30 days and laying adult 14 days. A wide range of field crops including legumes, native and introduced forbs, horticultural crops and trees are divested by locusts and grasshoppers (Hertel et al., 2013).

Figure 7. *Locusta migratoria*.

Landholders are required to control locusts on their land wherein cheap and effective control methods for isolated small areas include ground control for small hoppers (nymph stage) with boom sprays, misting machines or knapsack sprayers. Once locusts are on the wing, aerial application with higher rates is the only efficient method of control. Appropriate application methods for locusts vary with the locust's life stage and behavior. Bands of small hoppers (nymphs) are the best targets for ground spraying using agricultural chemicals. Large hopper bands can be sprayed with boom sprays. Isolated and small areas can be sprayed using misting machines or knapsack sprayers. Hoppers at densities below 30 square meters move very little and do not form bands. Spraying in the late afternoon is most effective, as hoppers tend to spread out during the day. Strong wind may cause hoppers to shelter in cracks in the ground or behind windbreaks. Once locusts have fledged or are flying, aerial spraying of agricultural chemicals is the only efficient method of control. A commercial product containing spores of the fungus *Metarhizium anisopliae* var. *acridum* has been developed for use against this species. Up to 30% of the eggs may be parasitized by *Scelio improcerus* Dodd and *Scelio parvicornis* Dodd (Hymenoptera: Scelionidae). Nymphs and adults are parasitized by several dipterans e.g., *Trichopsidea oestracea* Westwood (Diptera: Nemestrinidae) (Swaminathan et al., 2007).

3.6. False Wireworms (Coleoptera: Tenebrionidae)

Similar species of wireworm are *Pterohelaeus* sp., eastern false wireworm, striate false wireworm; and *Gonocephalum* sp., small false wireworm, southern false wireworm, northern false wireworm. Adult false wireworms (Figure 8) emerge from the soil during spring and early summer. Eggs are laid singly in moist soil, usually under trash or low-growing weeds. Larvae feed on decaying vegetable matter and crop residues in the soil, as well as on newly germinated seed (e.g., cereals, cotton, soybeans and sunflowers). Both the seeds and growing points are damaged, resulting in patchy stands. Larvae are up to 30 mm long, shiny and creamy, yellow or tan with three pairs of legs just behind the head. They are hard-bodied, cylindrical and segmented with a rounded head. Adult beetles of *Pterohelaeus* sp., are 20 mm long and dark grey-black with a distinctive pie-dish shape formed by flanges around the outline of the beetle. Adult beetles of *Gonocephalum* sp., are 9 mm long, dark grey-black and often covered in soil. There are flanges around the outline of the thorax (behind the head). Several species of false wireworms may occur in any particular crop, depending on locality, soil type, organic matter and tillage practices. They usually have a single generation per year. Larvae cause most damage to germinating crops during spring while adults are most troublesome in summer. Larvae feed on germinating seed and chew on seedling roots and shoots, resulting in patchy stands. Adults chew on seedlings at or above ground level, ring-barking or completely cutting the stem. For monitoring, use germinating seed baits or digging and sieving to detect larvae prior to planting. Treatment is required if more than 25 wireworm larvae are found in 20 seed baits. For adults, apply bait to the whole field if the soil insect rating is more than six. For chemical control of larvae, use seed dressings or in-furrow sprays, and for adults, use cracked grain baits. The false wireworm is mainly a pest in areas having heavy, black soils and it prefers cultivated soils rather than zero tillage. Use press wheels at sowing, which are set at 2-4 kg per cm width after planting grain or 4-8 kg per cm in dry soil (Teshale, 2015).

Figure 8. False Wireworm.

3.7. Thrips (Thysanoptera: Thripidae)

For many field crops (apart from beans), thrips are only a significant pest at the seedling stage. These infest the underside of cotyledons, young leaves and growing points. Adults and nymphs pierce the leaf and suck out sap. Affected areas are silvery-white and younger leaves become distorted in shape and growing points can die. Damage is more significant if seedlings are not actively growing. Flower thrips feed inside flowers causing flower abortion and pod distortion. Deformed pods may be difficult to thresh, resulting in further yield losses. Any thrips species can only be determined by observing microscopically.

3.7.1. Onion, Tobacco or Cotton Seedling Thrip (Thrips Tabaci)

Adult *Thrips tabaci* Lindeman is 2 mm long and dark, cigar-shaped and have narrow wings folded along their back (Figure 9). Nymphs are smaller, lack wings or with fixed *wing* pads and are pale. Adult thrips can infest a seedling's growing point as soon as it emerges from the ground. In cracking soils, seedlings may be infested before they emerge. Populations typically peak within four weeks of plant emergence. For many field crops (apart from beans), thrips are only a significant pest at the seedling stage. They infest the underside of cotyledons, young leaves and growing points. Adults and nymphs pierce the leaf and suck out cell sap. Affected areas are silvery-white and younger leaves become distorted in shape, and growing points can die. Damage is more significant if seedlings are not actively growing and thrips can feed inside flowers. This results in pods being curled, sometimes with external brown marks and these pods can be difficult to thresh. Open and microscopically examine the plant's growing points and can also be dunked in alcohol to dislodge thrips. If a decision is made to control thrips, apply a narrow band spray over the seedlings to preserve predators such as spiders in the inter-row. Seedling thrips damage can be greatly reduced in crop, if a systemic pesticide is applied soon after emergence within 3-4 days (Chaney, 2004).

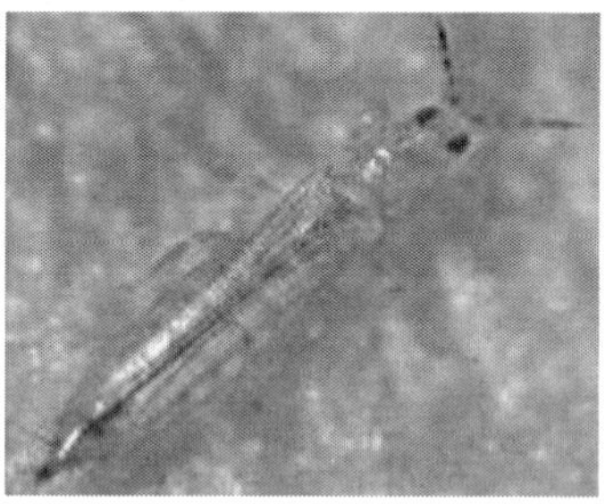

Figure 9. *Thrips tabaci*.

3.7.2. Western Flower Thrips (Frankliniella Occidentalis)

Adults *Frankliniella occidentalis* (Pergande) are 2 mm long, cigar-shaped and range from yellow-orange to grey-black (Figure 10). These have tiny, narrow wings carried over the back. Nymphs are similar in shape, pale yellow to orange-yellow, wingless and smaller. Thrips damage is distinguished from spider mite damage by the appearance of liquid fecal deposits, which cause dark green speckling, whereas spider mites produce black granules. Eggs are laid in slits made in leaves and growing points. There are two larval stages as well as a pre-pupal and pupal stage. Nymphs and adults feed in growing points and inside flowers. Pupation occurs in the soil and the life cycle can be completed in as little as 10 days at 20°C. Its populations decline when temperature exceeds 30°C. Crops can be infested during the early vegetative stage, but severe symptoms may not occur until much later in the crop's development. The earlier the thrips attack, the greater the risk of damage can occur. Symptoms of the western flower thrips feeding include discoloration and indentation of the leaf surface. Damage to buds and flowers can result in deformed pods. Check seedling and monitor flowering crop by peeling open the flower to look inside for thrips. Very few western flower thrips are required to spread tomato spotted wilt virus, which are resistant to most insecticides. Late plantings may help to avoid western flower thrips damage. Using soft pesticides against other pests may conserve other thrips species, which by competing with western flower thrips, may slow the latter's population growth (Herron and James, 2005).

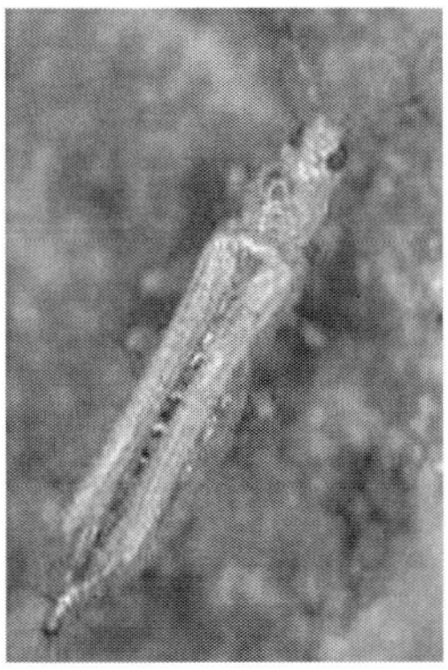

Figure 10. *Frankliniella occidentalis*.

3.8. Loopers (Lepidoptera: Noctuidae)

Loopers can be distinguished from *Helicoverpa* by their looping action when walking by their body, which tapers to the head and they have only two pairs of hind legs, as opposed to four for pod borer. The *Thysanoplusia* (soybean looper) moths are brown with large golden markings on their forewings. The *Chrysodeixis* sp., (tobacco and vegetable loopers) moths are brown with small silver markings on the forewings. The eggs are similar in shape to *Helicoverpa* eggs, but are more flattened and pale yellow-green in color. Larvae are green with white or dark stripes and can reach 50 mm in length. Soybean looper larvae are more prominently striped than the vegetable and tobacco loopers (Figure 11). The loopers can occur anytime throughout the crop growing season, they normally feed almost exclusively on leaves and do not usually warrant control and under some conditions, their populations can flare to high damaging levels. Control should be considered if leaf loss is likely to exceed 33% in the vegetative stage, or 20% from flowering to onwards. Loopers appear susceptible to all insecticides used for *Helicoverpa* control, with the exception of viruses (NPV), which only act against *Helicoverpa*, but thorough coverage is required for best results (Allen et al., 1996).

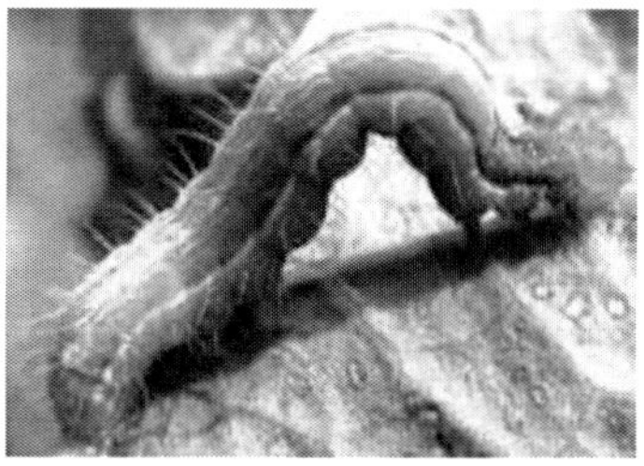

Figure 11. Looper (Larva).

3.9. Green Vegetable Bug *Nezara viridula* Linneaus (Hemiptera: Pentatomidae)

The green vegetable bug *Nezara viridula* (Linnaeus) (Figure 12), is a key pest of many crops. There are several other related pests that are similar in appearance to the green vegetable bug. This pest is bright green, shield-shaped, 13-15 mm long with three small white spots between their shoulders. Over-wintering adults are purple-brown and adults emit a foul smell when disturbed to deter predators. Eggs are laid in rafts (50-100 eggs per raft), newly-laid eggs are creamy, but turn bright orange prior to hatching and take 6 days to hatch at 25°C. Newly hatched nymphs (1.5 mm long) are orange and brown (sometimes black) in color. Later instars are either green or black, with white, cream, orange and red markings. Younger nymphs are round or oval rather than shield shaped and usually aggregate in large clusters, while older nymphs are more widely dispersed. There are five nymphal instars with a total development time of 30 days and has three to four generations during the summer seaon. This species is the most damaging pod sucking bug in pulses, by virtue of its abundance, widespread distribution, and rates of damage and reproduction. Adult bugs typically invade legumes at flowering, but is primarily a pod feeder with a preference for pods with well-developed seeds until pods are too hard to damage very near to harvest. Adults use their long thin mouthpart to suck nutrients from the seed, and damage to young pods produces deformed and shriveled seeds, thus reducing yield. Some pulse cultivars can compensate for early damage, but seeds damaged in older pods are blemished and difficult to grade out, and thereby resulting in reducing harvested seed quality, particularly that destined for human consumption.

Crops should be inspected for bug twice weekly from budding until close to harvest and can be sampled in the early to mid-morning when bugs bask at the top of the crop canopy. As a result, the threshold in typical edible crops is as low as 0.3 adult per square meter. Bugs should be controlled during early pod filling before nymphs reach a damaging size. Pesticides are best applied in the early to mid-morning to contact bugs basking at the top of the canopy. Where possible, avoid sequential plantings of legumes as bug populations can move progressively from earlier to later plantings, eventually building to very high levels. Eggs are frequently parasitized by a tiny introduced wasp *Trissolcus basalis* (Wollaston) (green vegetable bug egg parasite) (Hymenoptera: Scelionidae) and parasitized eggs are easily recognized as they turn black, while, nymphs are attacked by ants, spiders and predatory bugs, and final (5th) instar and adult are parasitized by tachinid fly *Trichopoda giacomellii* (Blanchard) (Diptera: Tachinidae) (Musolin, 2012; Sarwar and Saqib, 2010).

Figure 12. *Nezara viridula*.

3.10. Blue Oat Mite *Penthaleus major* Koch (Acarina: Penthaleidae)

Seedling pests, such as *mites* can attack *chickpea to* cause significant yield reduction in produce (Sarwar, 2019). Adults of blue oat mite *Penthaleus major* Koch, are 1 mm long, 8 legs, and both adults and nymphs have a purplish-blue rounded body with red legs (Figure 13). They move quickly when disturbed and the presence of a small red area on the back distinguishes it from the red-legged earth mite. Eggs laid in the soil hibernate over winter, allowing populations to build up over a number of years. Mainly, it is a pest of cereals and grasses, but can also feed on pasture legumes and many weeds. Blue oat mite is an important pest of seedlings, and when infestations are severe, the leaf tips wither and eventually the seedlings die. Adults and nymphs pierce and suck on leaves resulting in silvering of the leaf tips in cereals. When heavy infestations occur, the leaf tip withers and the seedling can die. In canola crop, leaves are mottled or whitened in appearance.

Check pest from planting to early vegetative stage, particularly in dry seasons. They are most easily seen in the late afternoon when begin feeding on the leaves. Foliar applications of insecticides may be cost-effective if applied within 2-3 weeks of emergence. The use of control tactics solely in spring may not prevent the carry-over of eggs into the following autumn. This can cause severe damage if crop rotation is not practiced. Natural enemies of this pest are predatory mites, thrips and ladybirds, otherwise the chemical control option can be followed (Sarwar et al., 2012; Sarwar, 2016 b).

Figure 13. *Penthaleus major*.

4. INTEGRATED INSECT PESTS MANAGEMENT

Integrated pest management (IPM) begins with understanding the pest situation in the region and field, and matching those conditions with gram varieties that are best suited for managing those stresses. A major emphasis is now underway to develop varieties with multiple pest resistance traits that when are incorporated into adapted germplasm, will simplify pest management (Romeis and Shanower, 1996). The following list summarizes those needs determined to be the most critical to gram crop pests management. Overall, top priorities of integrated insect pests management strategy are listed underneath:-

4.1. Resistant Plants

It is clear that most available chickpea cultivars are not resistant to most potential insect pests. Growers may not be so satisfied about this situation, but breeders can undoubtedly breed more tolerant germplasm and focus upon projects to select genotypes that are less susceptible to losses caused by insect pests, particularly *H. armigera*. Such insect-resistant cultivars will form the backbone of integrated pest management in future. The development and deployment of cultivars with resistance to insects would offer the advantage of allowing some degree of selection for specificity effects, so that pests, but not the beneficial organisms can be targeted. For pest management programs to be effective in future, there is a need for in-depth understanding of the population dynamics of insect pests to develop appropriate control strategies, combine resistance to insects along with resistance to important diseases and cold tolerance, utilization of wild relatives to diversify the genetic basis, and thus, increase the levels of resistance to the target insect pests, identification of quantitative trait loci (QTLs) associated with resistance to insects, develop insect-resistant varieties through genetic transformation using genes with diverse modes of action, and insecticide resistance management (Daniels, 2007).

4.2. Crop Monitoring

Regular monitoring of the crop for the presence of insects pests and damage is necessary in order to make timely decisions on control, especially when it is important to be targeting small and possibly insecticide resistant larvae. A first step in developing an IPM approach is to establish the economic threshold of the target pest. It is the number of insects per unit area or per plant above which a significant economic loss in crop yield will occur, with reference to timing in the crop season and stage of the insect life cycle. Sampling with a beat sheet is best practice for monitoring *Helicoverpa* in chickpeas. All crops should be scouted for insects at regular intervals, usually once per week prior to pod-set and 2-3 times a week from pod-set to onwards. Check at least 5×1 row meter sections at six widely spaced sites in the field. Count larvae that are causing damage by looking on leaves, stems and buds, as well as small white-cream or brown colored eggs. Also, observe moths activity as an indication of possible egg laying. Generally, 1 larva per meter row length is reported as the economic threshold and injury level of *H. armigera* in chickpea. Control measures may be undertaken based on pest forecasting by pheromone or light trap catches (5 to 10 males per trap per day) (Ali, 2012).

4.3. Cultural Practices

Pest attacks can be modified by a variety of cultural practices. If it is known that *Helicoverpa* attacks are likely to be severe at a particular time, then it may be possible to adjust the sowing date or to utilize a cultivar of appropriate flowering and maturity timing to ensure that the flowering and podding stage does not coincide with the peak *Helicoverpa* attack period. There is usually a pool of *Helicoverpa* in any area that may be supplemented or depleted by migration. By synchronous sowing of the crop in any area, the available pest population will be diluted by dispersion across the whole crop area. Early sown fields will probably act as magnets for the pests and may act as multiplication sites for a subsequent dispersal to the main crop. Late-sown crops may bear the brunt of the pest dispersal from the maturing main crop. Poor plant stands are commonly said to be a major factor in the poor yields obtained from this crop by many farmers, but there have been inductions that close spacing harbors more *Helicoverpa* larvae per unit area, so increased yields may be obtained only if the closer-spaced crop is protected by pesticide use. Thus, optimum spacing probably varies not only according to the cultivar used and to edaphic and climatic factors, but also to the degree of pest control afforded (Saeed and Ali, 1993; Hossain et al., 2010).

4.4. Natural Enemies of *Helicoverpa*

The most common *Helicoverpa* predators in field crops are predatory bugs, predatory beetles, spiders, lacewings and ants. Some predators are relatively permanent residents in fields (e.g., ants), but others migrate from nearby fields or other vegetation or even further away. Some parasitoids wasps and flies attack *Helicoverpa* eggs, larvae and pupae. Parasitoids kill their *Helicoverpa* host to complete their development. The parasitoids most active in field crops include smaller wasp species such as *Trichogramma*, *Telenomus* and *Microplitis,* relatively large parasitoid wasps (*Netelia*, *Heteropelma*, *Ichneumon*), and flies (*Carcelia* and *Chaetopthalmus*). Pathogens are viruses, fungi or bacteria that infect insects. Many naturally occurring diseases infect and kill *Helicoverpa*. The commonest pathogens infect larvae, including the nucleopolyhedrovirus (NPV) and fungal pathogens (*Metarhizium*, *Nomurea* and *Beauvaria*). Another disease ascovirus, stunts larval development and is spread by wasp parasitoids. Two pathogens of *Helicoverpa* are available commercially to control larvae, such as NPV and the bacterium *Bacillus thuringiensis* (commonly called *Bt*) (Sarwar and Salman, 2015; Shah et al., 2015).

4.5. Bio and Plant Origin Pesticides

The increasing concern for environmental and awareness of pesticide hazards have evoked a worldwide interest in the use of pest control agents of bio and plant origin. These bio control agents and botanical pesticides are safer to be used in pest control programs and may prevent several adverse effects caused by synthetic insecticidal application. Efficacy of some botanicals against pod borer *H. armigera* damage in chickpea resulted in a significant reduction in its density. Baculovirus pesticides are increasingly being used as effective biological control agents against *Helicoverpa* caterpillars worldwide. The bio-control agent, HNPV, showed equally the best performance like synthetic insecticides and also exhibited higher efficacy than neem (azadiractin) based insecticides (Hossain, 2007; Nguyen et al., 2011; Sarwar, 2015 a; 2015 b).

4.6. Chemical Control

Pod borer *H. punctigera* is easily killed by all registered products, including products to which *H. armigera* is resistant (e.g., synthetic pyrethroids). Application of endosulfan, indoxacarb, methidathion and deltamethrin at flowering or early pod setting is quite effective. Good timing of spray applications against *Helicoverpa* larvae occurs when sprays target very small (1-3 mm) to small (4-7 mm) larvae (that require a lower dose to kill), larvae that are feeding or moving in the open field and therefore more easily contacted by spray droplets, and larvae before they move into protected feeding locations (e.g., flowers, cobs, pods or bolls) (Rajput et al., 2003 a; 2003 b).

4.7. Control Deprived of Insecticides

Insecticides are not the only options for controlling and managing *Helicoverpa*, but pupae busting remains an important non-chemical option to reduce the size of overwintering populations. Pupae busting also reduces the carryover of insecticide resistant individuals from season to season. Weed management in and around crops can prevent the build-up of *Helicoverpa* and other insect pests. To avoid planting into high populations of wireworm, potential production fields may be bioassayed for wireworms before planting. Non-crop border habitat or refuge areas may be maintained to enhance populations of aphid natural enemies. An experiment has been conducted to develop an IPM approach for the management of pod borer, in chickpea field. Out of seven modules studied, module consisting of sequential first spray with *Helicoverpa* nuclear polyhedrosis viruses (HNPV) @ 500 LE/ha and second spray after seven days interval with Cypermethrin @ 1 ml/liter gave the best protection with the lowest pod borer damage, provided the highest yield and maximum net return (Ahmed and Awan, 2013).

Research and develop broadleaf weed controls, especially post-emergent, targeting kochia, nightshades, mayweed chamomile, prickly lettuce and sowthistle. Develop varieties with resistance to diseases and insects, and continue of additional research on biology of pathogens and insects. Further investigate transgenic breeding for pest management. Research on pests management through crop rotation and resistant varieties. Find an alternative to systemic insecticide dimethoate. Develop an effective aphid management program and associated tools (e.g., seed treatments that persist in systemic manner). Improve crop quality and breeding for competitiveness and desirable characteristics such as bleach resistance (Sarwar, 2013 b).

All farmers should sow synchronously at the optimum time and spacing, and use cultivars that are less susceptible to the problematic pests. If nonpolluting pesticides are known to be of undoubted economic value, then they should be applied as efficiently and as timely as possible, according to counts of eggs and young larvae. The crop should be harvested as soon as it ripens and crop residues should either be removed or plowed. If feasible, there should be a closed season during which the crop and the alternative hosts of the damaging pests are not grown in the area (Mahmudunnabi et al., 2013).

Recent strategies such as area-wide management (AWM) and integrated pest management aim to restrict the build-up of *Helicoverpa* populations to below damaging levels. Some tactics that may be considered in the context of an area-wide management approach and that relate to *Helicoverpa* management in chickpea are reducing of the spring *H. armigera* generation in commercial chickpea crops, minimizing the carryover of moths from chickpea to susceptible summer crops and insecticide resistance management. Successful IPM strategies (which combine biological, cultural and chemical control options) require a more sophisticated understanding of pod borer's lifecycle and biology than relying on insecticides alone (Sarwar, 2013 c).

Figure 14. A healthy chickpea

CONCLUSION

Chickpea, used for both, human consumption and as forage plant for livestock, is an important plant especially in this region Pakistan and elsewhere. The major insect pest of chickpeas common to the region is *Helicoverpa*, while other less frequent pests include early-season and late season cutworms, aphids, leaf miner, false wireworms, locusts or grasshoppers and blue oat mites. Leafminers feed on dryland chickpea and the adult of this insect is a small fly that along with its maggot-like larvae, feeds under the leaf surface, creating wandering mines on the leaf. Early season cutworms (army cutworm, pale western cutworm) could be a problem if chickpea is planted into a winter cereal cover crop or around field margins because these insects can move out from neighboring grasslands into chickpea fields. The western bean cutworm may be a late-season problem, but the extent to which this insect can feed on chickpea is unknown. Pea aphid is commonly found on alfalfa in the region and can also feed on chickpea. Normally, this insect is not overly damaging; however, if populations build at flowering or shortly thereafter, seed yields can be significantly reduced. Grasshoppers often seek plants with good nutrition and chickpea being a legume with high nitrogen content, would likely be a preferred target. In years when grasshoppers are at high levels, it is important to watch infestations around chickpea fields. Ideally, integrated pest management is unlikely to be a real success if applied only to an individual field. There is a much greater chance of success if all farmers of the crop in an area coordinate in united action. Recent strategies such as area-wide management (AWM) and integrated pest management (IPM) aim to restrict the build-up of *Helicoverpa* populations to below damaging levels. Successful AWM and IPM strategies (which combine biological, cultural and chemical control options) consisted of an improved tolerant chickpea variety, seed treatment with insecticide, wider row spacing, and need and weather-based application of insecticide for the management of pests and pod borers. Promotion of AWM and IPM technology necessarily involves the participatory approach like farmers participation on-farm research and promotion conducted in collaboration between the agricultural research council, non-government organizations, farmers self-help groups, international agricultural research centers, advanced research institutes and universities. Through improved pest management farmers can increase yields and decrease expenditure on pesticides usages.

REFERENCES

Agegnehu, E. & Medhin, Z. G. (2014). Demonstration of Apron Star 42 WS Seed Dressing Chemical against Cutworm (*Agrotis segtum*) on Chickpea using Farmer Research Group Approach at Dembia Woreda, North Gondar, Ethiopia. *Open Access Library Journal*, *1*, 1-5.

Ahmad, N. & Sarwar, M. (2013). The Cotton Bollworms: Their Survey, Detection and Management through Pheromones: A Review. *Journal of Agriculture and Allied Sciences, 2* (3), 5-8.

Ahmed, K. & Awan, M. S. (2013). Integrated Management of Insect Pests of Chickpea *Cicer arietinum* (L. Walp) in South Asian Countries: Present Status and Future Strategies - A Review. *Pakistan Journal of Zoology*, *45* (4), 1125-1145.

Ali, R. B. (2012). New Perspectives in Plant Protection. *InTech*, 246 p.

Allen, D. J., Ampofo, J. K. O. & Wortmann, C. S. (1996). Pests, Diseases, and Nutritiona Disorders of the Common Bean in Africa. A Field Guide. CIAT, Colombia. Publication No. 260, 132 *p.*

Asadi, R., Talebi, A. A., Fathipour, Y., Moharramipour, S. & Rakhshani, E. (2006). Identification of Parasitoids and Seasonal Parasitism of the Agromyzid Leafminers Genus *Liriomyza* (Dip.: Agromyzidae) in Varamin, *Iran. J. Agric. Sci. Technol.*, *8*, 293-303.

Chaney, W. E. (2004). Insectary plants for vegetable crops. California Conference on Biological Control IV, Berkeley, California, USA, 13-15 July, 2004, p. 53-54.

Daniels, C. H. (2007). Pest Management Strategic Plan for Pulse Crops (Chickpeas, Lentils and Dry Peas) in the United States. Summary of a workshop held on February 27-28, 2006, Spokane, Washington. p. 80.

Dhingra, S. (1994). Development of resistance in the bean aphid, *Aphis craccivora* Koch to various insecticides used for nearly a quarter century. *Journal of Entomological Research, 18*, 105-108,

El-Serwy, S. A. (2003). Studies on the leafminers; *Liriomyza cicerina* (Rondani) and *Liriomyza bryoniae* (Kaltenbach) (Diptera: Agromyzidae) and their parasitoids in faba bean in Egypt. *Egyptian Journal of Agricultural Research, 81* (4), 1581-1593.

Herron, G. A. & James, T. M. (2005). Monitoring insecticide resistance in Australian *Frankliniella occidentalis.* Pergande (Thysanoptera: Thripidae) detects fipronil and spinosad resistance. *Australian J. Entomol., 44*, 299-303.

Hertel, K., Roberts, K. & Bowden, P. (2013). Insect and mite control in field crops. *New South Wales Department of Primary Industries.* p. 104.

Hossain, A. (2007). Efficacy of Some Synthetic and Biopesticides Against Pod Borer, *Helicoverpa armigera* (Hubner) in Chickpea, *Tropical Agricultural Research & Extension, 10*, 74-78,

Hossain, M. A., Azizul-Haque, M., Ahmad, M. & Prodhan, M. Z. H. (2010). Development of an Integrated Management Approach for Pod Borer, *Helicoverpa armigera* (Hubner) On Chickpea. *Bangladesh J. Agril. Res., 35* (2), 201-206.

Kamphuis, L. G., Gao, L. & Singh, K. B. (2012). Identification and characterization of resistance to cowpea aphid (*Aphis craccivora* Koch) in *Medicago truncatula. BMC Plant Biology, 12*, 101.

Khan M. H., Sarwar, M., Farid, A. & Syed, F. (2010). Compatibility of pyrethroid and different concentrations of neem seed extract on parasitoid *Trichogramma chilonis* (Ishii) (Hymenoptera: Trichogrammatidae) under laboratory conditions. *The Nucleus, 47* (4), 327-331.

Knights, E. J. & Siddique, K. H. M. (2002). Chickpea status and production constraints in Australia. In: Bakr, M.A., Siddique, K.H.M. and Johansen, C. (eds.). Integrated Management of Botrytis Grey Mould of Chickpea in Bangladesh and Australia. *Australian Centre for International Agricultural Research*, Canberra., p. 33-41.

Mahmudunnabi, M., Dutta, N. K., Rahman, A. K. M. Z. & Alam, S. N. (2013). Development of biorational-based integrated pest management package against pod borer, *Helicoverpa armigera* Hubner infesting chickpea. *Journal* of *Biopesticides, 6* (2), 108-111.

Mari, J. M., Chachar, S. D., Chachar, Q. I. & Kallar, S. A. (2013). Insect diversity in chickpea ecosystem. *Journal of Agricultural Technology, 9* (7), 1809-1819.

Miles, M. (2013). Chickpea insect pest management. Northern grains region. *Queensland Department of Agriculture, Fisheries and Forestry.*, p. 16.

Minja, E. M., Shanower, T. G., Silim, S. N. & Karuru, O. (2000). Efficacy of different insecticides for pigeonpea pest management in Kenya. International Chickpea and Pigeonpea. *News letter, 7*, 53-55.

Musolin, D. (2012). Surviving winter: diapause syndrome in the southern green stink bug *Nezara viridula*. *Physiological Entomology*, *37* (4), 309-322.

Nguyen, Q., Qi, Y. M., Wu, Y., Chan, L. C., Nielsen, L. K. & Reid, S. (2011). *In vitro* production of *Helicoverpa* baculovirus biopesticides- automated selection of insect cell clones for manufacturing and systems biology studies. *J. Virol. Methods*, *175* (2), 197-205.

Rajput, A. A., Sarwar, M., Bux, M. & Tofique, M. (2003a). Evaluation of synthetic and some plant origin insecticides against *Helicoverpa armigera* (Hubner) on chickpea. *Pakistan Journal of Biological Sciences*, *6* (5), 496-499.

Rajput, A. A., Sarwar, M., Ahmad, N., Siddiqui, Q. H. & Tofique, M. (2003b). Evaluation for resistance in some local and exotic chickpea genotypes against *Helicoverpa armigera* (Hubner). *Pakistan Journal of Biological Sciences*, *6* (18), 1612-1615.

Romeis, J. & Shanower, T. G. (1996). Arthropod natural enemies of *Helicoverpa armigera* (Hubner) (Lepidoptera: Noctuidae) in India. *Biocontrol Science and Technology*, *6*, 481-508.

Saeed, S. & Ali, H. (1993). Dates of sowing cum varietal impact on the larval prevalence of *Heliathis armigera* (Hb.) on gram crop. *Pak. Entomologist*, *15* (1-2), 123-124.

Sarwar, M. (2009). Populations' synchronization of aphids (Homoptera: Aphididae) and ladybird beetles (Coleoptera: Coccinellidae) and exploitation of food attractants for predator. *Biological Diversity and Conservation*, *2* (2), 85-89.

Sarwar, M. (2012). Competency of Natural and Synthetic Chemicals in Controlling Gram Pod Borer, *Helicoverpa armigera* (Hubner) on Chickpea Crop. *International Journal of Agricultural Sciences*, *2* (4), 132-135.

Sarwar, M. (2013a). Exploration on resource of resistance in chickpea (*Cicer arietinum* L.) genotypes to gram pod borer *Helicoverpa armigera* (Hubner) (Lepidoptera). *African Journal of Agricultural Research*, *8* (26), 3431-3435.

Sarwar, M. (2013b). Survey on screening resistance resources in some chickpea (*Cicer arietinum* L.) genotypes against gram pod borer *Helicoverpa armigera* (Hubner) (Lepidoptera: Noctuidae) pest. *International Journal of Agricultural Sciences*, *3* (3), 455-458.

Sarwar, M. (2013c). Integrated Pest Management (IPM) - A Constructive Utensil to Manage Plant Fatalities. *Journal of Agriculture and Allied Sciences*, *2* (3), 1-4.

Sarwar, M. (2015a). Biopesticides: An Effective and Environmental Friendly Insect-Pests Inhibitor Line of Action. *International Journal of Engineering and Advanced Research Technology*, *1* (2), 10-15.

Sarwar, M. (2015b). Usage of Biorational Pesticides with Novel Modes of Action, Mechanism and Application in Crop Protection. *International Journal of Materials Chemistry and Physics*, *1* (2), 156-162.

Sarwar, M. (2016a). Recognition of some lady beetles (Coleoptera: Coccinellidae) deadly sighted for insect and mite pests in agroecosystems. *International Journal of Entomology Research*, *1* (2), 29-34.

Sarwar, M. (2016b). Comparative life history characteristics of the mite predator *Neoseiulus cucumeris* (Oudemans) (Acari: Phytoseiidae) on mite and pollen diets. *International Journal of Pest Management*, *62* (2), 140-148.

Sarwar, M. (2017). Integrated Control of Insect Pests on Canola and Other Brassica Oilseed Crops in Pakistan. In: *Integrated Management of Insect Pests on Canola and Other Brassica Oilseed Crops* (Ed. Reddy, G. V. P.). CABI, Wallingford, UK. p. 193-221.

Sarwar, M. (2019). Biology and Ecology of Some Predaceous and Herbivorous Mites Important from the Agricultural Perception. In: *Acarology, Mite Biology and Ecology* (Ed. Hufnagel, L.). IntechOpen Ltd., London, UK. p. 29.

Sarwar, M. & Saqib, S. M. (2010). Rearing of Predatory Seven Spotted Ladybird Beetle *Coccinella septempunctata* L. (Coccinellidae) on Natural and Artificial Diets under Laboratory Conditions. *Pakistan Journal of Zoology*, *42* (1), 47-51.

Sarwar, M. & Salman, M. (2015). Biological insecticide *Trichogramma* spp. (Hymenoptera: Trichogrammatidae) strikes for caterpillar control. *International Journal of Entomology Research*, *1* (1), 31-36.

Sarwar, M. & Sattar, M. (2016). An Analysis of Comparative Efficacies of Various Insecticides on the Densities of Important Insect Pests and the Natural Enemies of Cotton, *Gossypium hirsutum* L. *Pakistan Journal of Zoology*, *48* (1), 131-136.

Sarwar, M. & Salman, M. (2016). From Production to Field Application Methodology of Generalist Predator Green Lacewing, *Chrysoperla carnea* [Stephens] (Neuroptera: Chrysopidae). *International Journal of Zoology Studies, 1* (1), 35-40.

Sarwar, M., Ahmad, N. & Tofique, M. (2009). Host plant resistance relationships in chickpea (*Cicer arietinum* L.) against gram pod borer (*Helicoverpa armigera*) (Hubner). *Pakistan Journal of Botany*, *41* (6), 3047-3052.

Sarwar, M., Ahmad, N. & Tofique, M. (2011). Identification of susceptible and tolerant gram (*Cicer arietinum* L.) genotypes against gram pod borer (*Helicoverpa armigera*) (Hubner). *Pakistan Journal of Botany*, *43* (2), 1265-1270.

Sarwar, M., Xuenong, X. & Kongming, W. (2012). Suitability of webworm *Loxostege sticticalis* L. (Lepidoptera: Crambidae) eggs for consumption by immature and adults of the predatory mite *Neoseiulus pseudolongispinosus* (Xin, Liang and Ke) (Acarina: Phytoseiidae). *Spanish Journal of Agricultural Research*, *10* (3), 786-793.

Shah, M. S., Khan, I. A., Salman, M., Ahmad, J., Akbar, I., Shah, J. A., Khan, G. Z. & Sarwar, M. (2015). Study on different life parameters of *Trichogramma chilonis* (Ishii) on eggs of *Sitotroga cerealella* (Olivier) fed on old and new varieties of wheat, maize and sorghum. *Journal of Entomology and Zoology Studies*, *3* (6), 397-400.

Shankar, M., Sharma, H. C., Ramesh, B. T. & Sridevi, D. (2013). Evaluation of Chickpea Genotypes for Resistance to Beet Armyworm, *Spodoptera exigua*. *Ind. Journal of Plant Protection*, *41* (4), 275-281.

Sharma, H. C., Gowda, C. L. L., Stevenson, P. C., Ridsdill-Smith, T. J., Clement, S. L., Rao, G. V. R., Romeis, J., Miles, M. & El-Bouhssini, M. (2007). Host plant resistance and insect pest management in chickpea. Yadav, S. S., Redden, R. J., Chen, W., Sharma, B. (Eds.). In: *Chickpea breeding and management.* CABI, Wallingford, UK. p. 520-537.

Swaminathan, R., Hussain, T. & Bhati, K. K. (2007). Influence of crop diversity on host preference by major insect pests of kharif pulses. *Ind. Journal of Applied Entomology*, *21*, 59-62.

Teshale, J. (2015). Assessment of chickpea losses at different pre-harvest and post-harvest chains in Dandi woreda and Ambo woreda of Oromia region. *International Journal of Advanced Agricultural Research*, *1* (1), 12-15.

Yassine, M. & Omrane, B. (2012). Integrated Pest Management in Chickpea, New Perspectives in Plant Protection, Ali, R. B. (Ed.), *InTech*, 38 p.

In: Handbook of Chickpeas
Editors: Albert T. Lund and Noah D. Schultz
ISBN: 978-1-53616-374-2

Chapter 11

MULTIPLE INSECT PESTS MANAGEMENT OPTIONS FOR CHICKPEA (*CICER ARIETINUM* L.) WALP GRAIN WITH REFERENCE TO STORAGE

Muhammad Sarwar*

National Institute for Biotechnology and Genetic Engineering (NIBGE), Faisalabad, Pakistan

ABSTRACT

Chickpea (*Cicer arietinum* L.), (Fabales: Fabaceae), is one of the most popular vegetables in several areas of the world having an important source of protein particularly for vegetarian population. The chickpea seeds are vulnerable, both in the field and storage conditions to attack by several insect pests of economic importance, whereby infested grains lose their quality and viability. Five insect pests such as *Callosobruchus chinensis* (L.), *Tribolium castaneum* (H.), *Trogoderma granarium* (E.), *Rhizopertha dominica* (F.), and *Sitophilus oryzae* (L.), are mainly observed in stored chickpea. The losses due to insects activity during storage are physical loss, losses in carbohydrates and proteins, nutritional losses, and contamination of products with uric acid, fragments and fecal matter. Bruchids or pulse beetles (Coleoptera: Bruchidae) belonging to the genus *Callosobruchus* (*C. chinensis* Linnaeus, *C. maculates* Fabricius and *C. analis* Fabricius) are major storage insect pests of chickpea and cause severe damage to stored grains. Their adults and grubs feed on the grain by making a small hole and white eggs are seen on damaged seed. Owing to the bad smell created by the attack of the pest, the damaged grains do not remain fit for humans as well as animals consumptions. Contrasting to cereal grains, chickpea cannot be merely treated with protectants to prevent insect infestations in storage. Therefore, basic steps in order to control insects in warehouse or

* Corresponding Author's Email: drmsarwar64@yahoo.com.

processing plants that must be followed are monitoring for infestation detection, identification of the problem and insect control measures. Insect control measures can be any one or a combination of subsequent methods like, physical and mechanical methods, chemical methods, sanitation, exclusion and Integrated Pest Management (IPM). Sun drying is the most common and preferred method not only to kill the insects, but it also reduces grain moisture to the limits of safe storage. Adoption of meticulous hygiene and aeration cooling to manage storage temperature and moisture are crucial to prevent insect damage and from downgrading of stored chickpea. Genetic resistance in grain is considered the most desirable control method since it is more cost effective and environment friendly than the use of chemicals. Most of the physical methods suggested when may not give total insects control then application of chemicals sometimes becomes essential for complete pests mortality and prevention of their growth.

Keywords: *Cicer arietinum,* gram, chickpea, grain, seed, integrated pest management

1. Introduction

Chickpea (*Cicer arietinum* L.), commonly known as gram (Fabales: Fabaceae), is one of the most popular pulse crop in several states of the world for vegie people. Chickpea seeds are susceptible, equally in the field and storage environments, to the invasion of quite a lot of insect pests of financial significance. This important leguminous crop is commonly cultivated in different parts of world, where it is often severely damaged in storage conditions (Sarwar et al., 2009; 2011). So, the main constraint for production of chickpea is post-harvest loss during storage situations. Five insect pests specifically *Callosobruchus chinensis* (L.), *Tribolium castaneum* (H.), *Trogoderma granarium* (E.), *Rhizopertha dominica* (F.), and *Sitophilus oryzae* (L.), have been commonly detected in stored chickpea. A high rate of reproduction and a short development period enable these insects to cause important damage by rapidly developing from a small number of individuals to a large mass. Beetles of the genus *Callosobruchus* (Coleoptera: Bruchinae), are the most important storing pests of chickpea and inflict significant monetary damages for stored grains. Chickpea pods are seldom infested in the field, but the pests attack nearly mature and dried pods during storage. Their adults and grubs feed on the grain by making a small hole. As a result, the round exit hole and the white eggs on the pod wall are clearly visible. Infested stored seed can be recognized by the presence of eggs on the seed surface and the round exit holes with the 'flap' of seed coat. Infested chickpeas lose their viability and damaged grains do not remain fit for humans along with animals feeding due to the bad smell created by the attack of the pest. In Africa, Asia and Oceania, bruchid beetles mainly belonging to the genus *Callosobruchus* (*C. chinensis* Linnaeus, *C. maculates* Fabricius and *C. analis* Fabricius) are considered as pests of economic importance for stored-legume and cause severe damage to chickpea grains. Their presence in certain chickpea growing areas is considered as a limiting factor to the expansion of the crop (Weigand, 1990).

Unlike cereal grains, chickpea cannot be repeatedly treated with protectants to avoid insect invasions. *The pest control methods adopted are chemical control, biological control, treatment with botanical pesticides and* integrated pest management *treatments (in which more than one of the above practices are included).* Therefore, meticulous hygiene and aeration cooling to manage storage temperature and moisture are crucial to prevent insects damage and molds from downgrading stored chickpea. Genetic resistance is considered the most desirable control method since it is more cost effective and environment friendly than the use of chemicals. Breeding host plants for resistance to a large range of abiotic and biotic stresses has proved to be very profitable. In particular, breeding for resistance to insect pests has been extremely successful in chickpea. Among the tested chickpea genotypes, tolerant materials can be used as a source of insect pests resistance in breeding programs that could then be grown in organic cultivation free from pesticides (Rajput et al., 2003). In this chapter, it is examined the developments in integrated management of insect pests in storage chickpea *ecosystem.*

2. Insect Pests of Stored Chickpea

Family Bruchidae *of order Coleoptera*, made up of nearly 1400 species of bean and pea beetles, is comprises of major seeds pests of both cultivated and wild leguminous plants throughout the world. Many of these species are well-known for the characteristic injury, which their larvae cause in seeds. Species of genus *Callosobruchus* are important pests of pulses globally. In tropical developing countries, where legume seeds are often the main source of protein in the human's diet, the losses caused by these insects are of major significance. Infestation may start in the pods before harvest and carry over into storage where substantial losses may occur (Taylor, 1981). Bruchids *Callosobruchus* multiplying rapidly in stored condition, give rise to new generation every month. Thus, delayed harvest can result in more damage by *Callosobruchus* insects causing in more seed damage and seed weight loss by adversely affecting germination (Ouedraogo et al., 1996).

There are multiple reports of damage caused by *Callosobruchus* pests, both in the field and on stored gram grains (Figure 1). In the early stages of attack the only symptoms are the presence of eggs cemented to the surface of the pulse. As development occurs entirely within the seed, the immature stages are not normally seen. The adults emerge through windows in the grain, leaving round holes that are the main evidence of damage. The study also found that seed germination, seedling length and vigor index of chickpea seeds are reduced due to Bruchidae attack. Biochemical assays of infested pulse showed that there is a gradual increase in moisture content, crude fiber and crude fat.

Crude protein initially increased after some days, but later showed a slight decrease. Total carbohydrate content is also decreased in chickpea grain (Parameshwarappa et al., 2012).

Observation on bruchids associated with seeds showed that there are differences in the susceptibility to attack of different varieties and in general, the larger seeded varieties tested being the better suitable as hosts. This is shown not to be due to larval competition for food within the seeds. But, there is a definite preference for oviposition on seeds having a smooth testa when there is a choice given for smooth and rough surfaces, however there seemed to be no color preference (Sarwar, 2015 a).

Chickpea seed is particularly susceptible to bruchid (*Callosobruchus* spp.), attack and precautions are necessary in storing of grains. Bruchids can infest a standing crop and so the seed should be cleaned and sun dried prior to storage. Effective storage can be achieved, even at village level, by an adequate sun drying, adding naphthalene balls to the seed, grain sealing in a polythene bag, placing the bags in an air-tight container and retaining the container in a dry place well above ground level. Furthermore, the effective control of bruchids in storage of chickpea seed can be obtained through following the subsequent procedure, for instance, drying of threshed seed in bright sunlight for 5-7 hr, inspection of seed for absence of bruchids immediately prior to packing for storage, placing seed in a polythene bag with 1 gm (= 1 ball) naphthalene kg^{-1} seed and sealing the bag, placing the bag in a vermin-proof container, and storing the container above ground level in a dry and airy location (Musa, 2000).

As chickpeas are damaged by a large number of insect species, both under field conditions and in storage, thus, it is wastage of labor, seed and other resources where the damage is significant if no protective measures are taken (Clement et al., 2000). Amongst the many insect pests damaging food legumes, the most important insects damaging pulses in field and storage are referred as bruchids or pulse beetles. The genus *Callosobruchus* has large number of species, representing *C. maculatus* (Fabricius), *C. chinensis* (Linnaeus), *C. analis* (Fabricius) and *C. phaseoli* (Gyllenhal) as more common in subtropical regions. However, *C. rhodesianus* (Pic) and *C. sunnotatus* (Pic) are present in tropical region. Beetle *C. theobromae* (Linnaeus) is found in pods of pigeonpea and *Acanthoscelides obtectus* (Say) has been originated as serious pest in kidney bean. Other insects pests which cause damage to stored legumes are *Trogoderma granarium* (Everts), *Rhyzopertha dominicia* (Fabricius), *Tribolium castaneum* (Herbst), *Ephestia cautella* (Walker), *Corcyra cephalonica* (Stainton), *Latheticus oryzae* (Waterhouse), *Lasioderma serricorne* (Fabicius), *Stegobium paniceum* (Linnaeus), *Oryzaephilus surinamensis* (Linnaeus), *Cryptolestes ferrugineus* (Steph) and few species of mites. Fungi from genera *Penicillium* and *Aspergillus* in association with these insects, enhance the rate of grains deterioration. Pulses are mostly consumed in form of dehusked split, which in absence of seed coat are more prone to moisture gain and fungal infestation, besides preferred by insects other than bruchids. The split pulses are attacked by *R. dominica*, *T. granarium*, *T.*

castaneum and *Cadra cautella* (Walker) under storage conditions. The losses due to these insects activities during storage are physical loss, loss in carbohydrates and proteins, nutritional losses and contamination of products with uric acid, fragments and fecal matter. Bruchid beetles belonging to the genus *Callosobruchus* (*C. chinensis*, *C. maculates*, *C. analis*) are the most devastating pests of chickpea in Asia, Africa, and Australia (Van Emden et al., 1988).

Figure 1. Bruchids damage.

Research and development (R & D) on pests biology, ecology, economic thresholds, sampling methods and effective control options have contributed to build a robust integrated management strategy for chickpea insects. The management endeavors of industry have been supported through the provision of diagnostics, responsive R & D, training and broad-based communication.

2.1. Bruchids or Pulse Beetles

Of the 20 bruchid species identified as pests of stored grain legumes, the most destructive and economically significant are members of the genera *Callosobruchus*, *Acanthoscelides* and *Zabrotes* (Credland, 1994).

Beetles, are members of *Callosobruchus* genus in the family Chrysomelidae (formerly the family Bruchidae), of the subfamily Bruchinae (bean weevils), within order Coleoptera and unlike true weevils, bruchids have no snout. Most species in the genus are native to Asia and have long been reported to destroy the seeds of leguminous plants. These occur in places outside of their native range as introduced species. There are at least 20 species in genus *Callosobruchus,* which are well known as economically important pests that infest stored foodstuffs (Tuda et al., 2006). These beetles specialize on legumes of the tribe Phaseoleae, which includes many types of beans used for food. Host plants include mung bean (*Vigna radiata* L.), adzuki bean (*Vigna angularis* L.), rice bean (*Vigna umbellata* L.), cowpea (*Vigna unguiculata* L.), Bambara groundnut (*Vigna subterranea* L.), pigeon pea (*Cajanus cajan* L.), lablab (*Lablab purpureus* L.) and common bean (*Phaseolus vulgaris* L.). These can also be found in peas, lentils, chickpeas and peanuts. Likewise, these feed on seeds and flowers of non-leguminous plants belonging to the families Compositae, Malvaceae, Convolvulaceae, Anacardiaceae, Rosaceae, Umbelliferae, Papavaraceae and Palmaceae (Tuda et al., 2005).

The adult bruchids of bean weevils are less injurious to the stored seeds since these do not feed on them: rather they are a nuisance, thereby compromising the purity of the seeds. The larvae of *Callosobruchus* spp., are destructive because they bore holes into the seeds in an attempt to feed on the germ layer of the seed (Jackai and Dauoust, 1986).

Some are considered pests because these invade stores of legume foods, they lay eggs on the seeds and the larvae consume them during development, and emerge from the seeds as adults. Among the several species of bruchids attacking edible legumes, the two most widespread species of beetles are *C. maculatus* and *C. chinensis*, which are distributed throughout the tropics and sub-tropics. Both these beetles are most destructive and attack almost all edible legumes, including chickpea. But, *C. maculatus* in particular is well adapted to live in dry beans and seeds because it does not require food or water to reproduce. It can invade a store of chickpea and infest 100% of the seeds when incursion is severe (Erler et al., 2009).

2.1.1. Cow Pea Weevil Callosobruchus analis (Fabricius)

Cow pea weevil *Callosobruchus analis* (F.), is considered an important pest in several countries in Africa, Asia and Oceania. It has been observed infesting seeds belonging to 15 Leguminosae genera, including peanut, bean, chickpea, pea, cowpea and soybean (Waterworth, 1986). Among different legume seeds, the gram is most preferred host for oviposition and development of the bruchid *C. analis* under common storage conditions. It has been reported that seed quality of chickpea during storage is significantly affected in the damaged seed. The pest attacks nearly matured and dried pods during storage (Sharmila and Roy, 1994).

Adults are oval in shape, chocolate or reddish brown in color with long and erected antennae. Both male and female are more or less similar, lateral black spots on elytra surrounded by white setae and have dark stripes on each side of female dorsal abdomen (Figure 2). The average length and breadth of male and female adults are 3.25 ± 0.23 mm, 3.60 ± 0.08 mm and 2.16 ± 0.05 mm, 2.02 ± 0.04 mm, respectively. Adults male C. *analis* have an average antenna length of 2.96 ± 0.08 mm, which is slightly longer than 2.42 ± 0.12 mm on average for female bruchids. This species have a moderately long tooth on the inner carina of the hind femur, rather than the small or minute tooth of normal strains, but is identical to the normal form of this species in all other respects (Deeba et al., 2006).

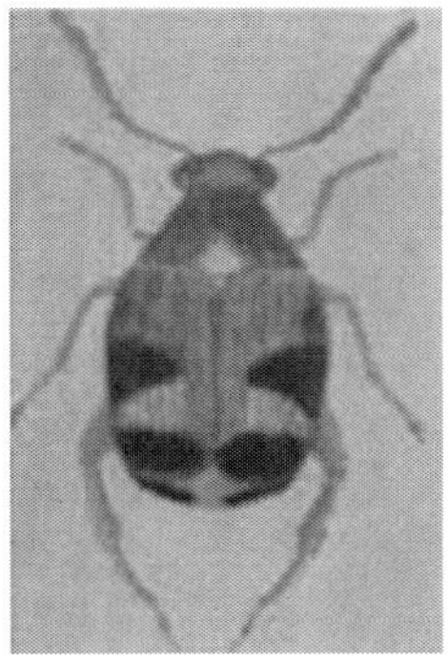

Figure 2. *Callosobruchus analis*.

The male bruchid mounts the female, but does not restrict their movement as both are found moving from one place to another. Eggs glued to the seed testa are visible to the competent eye sight. Eggs are dome-shaped, more-or-less oval in outline and colorless or creamy. The length of the eggs is on average 0.58 ± 0.00 mm, while their width 0.34 ± 0.00 mm on average. Seeds may bear dark 'windows', behind which lie maturing pupae. Seeds may also bear one to several holes left behind by emerging adults. The developmental stages of *C. analis* have been observed under laboratory conditions at 28 ± 2°C and 72 ± 5% relative humidity on cowpea (*V. unguiculata*). The results showed that female *C. analis* began to lay eggs within 24 hours and have an oviposition period of 8.50 ± 0.70 days and an average of 13.00 ± 0.50 eggs are laid by an individual throughout the lifetime. The mean developmental periods from egg to larva and larva to pupa are observed to be 8.50 ± 0.79 and 4.50 ± 0.70 days, respectively. The total developmental period of *C. analis* is observed to be 20.5 days on average. The unmated bruchids have been observed to live longer than the mated bruchids as the unmated bruchids lifespan is on average 10.50 ± 0.81 days, as opposed to 2.50 ± 0.75 days for mated females. The adults emerge by chewing and removing a circular piece of the seed coat to form a round hole (Sarwar et al., 2003; Olufemi-Salami et al., 2017).

Among natural enemies, three species of parasitoids are present in the cowpea (*V. unguiculata*) stock and subsequently in the storage systems. Parasitoid *Uscana lariophaga* Steffan (Hymenoptera: Trichogrammatidae) is an oophagous species (parasitizes the eggs), whereas two solitary ectoparasitoids (Hymenoptera), *Dinarmus basalis* (Rond) (Pteromalidae) and *Eupelmus vuilleti* (Craw) (Eupelmidae), develop on bruchid larvae and pupae inside the seeds. The use of such Hymenoptera parasitoids as biological control agents in storage has frequently been considered, but its efficacy has never been shown in a real storage situation. However, amongst one of its main natural enemies, controlling efficiency of parasitoid *D. basalis*, has already been demonstrated in several studies (Costa et al., 2007). In one test, the introduction of *D. basalis* adults into storage systems resulted in a great reduction in numbers of bruchids and ensured the maintenance of seeds in good quality during storage (Sanon et al., 1999).

2.1.2. Bean Weevil Callosobruchus chinensis (Linnaeus)

It is commonly known as the adzuki bean weevil, but its other common names include the pulse beetle, Chinese bruchid or cowpea bruchid, and is distributed across the tropical and subtropical regions of the world. Beetle *C. chinensis* is one of the most damaging pests to the stored legume industry due to its generalized legume diets and wide distribution. Some of their common host plants include green gram, lentil, cowpea, pigeon pea, chickpea and other pea species though they are known to live on many more legume hosts. The species most preferred habitat is in the tropics, on green gram or chickpeas. Pod infestation can start in the field before harvest and the pest then gains entrance into storage bins. It may cause substantial damage, coming to over 80% losses in weight and in germination rates. The adult and grub feed on the grain by making a small hole. Infested stored seed can be recognized by the white eggs on the seed surface and the round exit holes with the 'flap' of seed coat. Infested seeds are less nutritious and become unfit for humans *feeding (Neog, 2012).*

Adult beetle is 3-4 mm long, oval shaped, chocolate colored with black and grey patches over the body, and body is pointed towards front. The abdomen of the female is slightly longer than the elytra and it is white in color with two oval black spots on it, and elytra are red-brown with yellow makings. This species exhibits some sexual dimorphism where the female is larger and heavier than the male beetle, but female's tip of abdomen is exposed while in males it is covered by elytra. The antennae are pectinate in males while females bear the serrate antennae (Figure 3).

The Adzuki bean weevil females lay their many eggs (up to 70) directly on the surface of the legume singly and move on to either a different part of the legume or else to a different one depending on the bean density and competition among other females to lay more eggs. They can lay as many as 90 eggs after a single fertilization. Fecundity is relative to which legume is being used as a host and fitness of female. The eggs occur singly and have yellow colorings which become opaque when hatched. A female may lay 50-100 eggs on smooth legume pods or fecundity is about 100 eggs per female. The eggs usually hatch after 3-5 days and the new larvae will burrow into the bean for the rest of development. Eggs are whitish, elongated and stuck on the grains or on pods and sometimes on the surface of the container. Incubation period is 3-6 days and grubs are scarabeiform or eruciform, plump and with short legs (Deng *et al.*, 1999).

The larvae are yellowish-whitish in color with a brown head and reduced legs, and about 5 mm long. First instar larvae bear functional legs and a pair of thoracic plates to facilitate boring into the seeds. They feed on the inner contents of the grain and may damage several grains during development by feeding on the seed leaving only the seed coat. Larval period may vary between 12 and 20 days for completing four larval instars. The larvae chew tunnels through the bean until these are ready to pupate.

The pupae are dark brown in color and pupation occurs inside the legume grain with a development period of 5.67-7.33 days. Occasionally, pupation may take place outside the grain in a cocoon made of excretory matter. Completion of life cycle takes 4-5 weeks or 29.0 days in cowpea seeds and there may be 6-7 overlapping generations in a year (*Yanagi* et al., *2013*). Mature adults emerge from the bean, biting a neat circular exit from the pod as soon as 25 days after hatching. The mungbean is the most suitable host for *C. chinensis* and broad bean the least appropriate. The optimum development conditions for *C. chinensis* are around 32°C and 90% RH, and the minimum development period is 22-23 days *(Kyogoku and Nishida, 2013).*

The adults live up to two weeks after emerging from the pupa and these are active beetles and readily fly when disturbed. Beetle *C. chinensis* shows death feigning (playing dead) behavior as an anti-predatory technique. During this behavior, under certain stimulus that startles the insect, the beetle will roll onto its back and curl its legs up. This is likely to be used in order to dissuade parasitoid wasps from preying on the beetle (*Miyatake* et al., *2008*).

Intercropping with cereals and early harvesting of legumes before pest attack, good store hygiene including the removal of residues from last season's harvest and heating seeds to 50°C for one hour kills the eggs and larvae. Certain legume varieties with thick, hairy walls are resistant to beetle infestation. Periodical exposure of the grain to sun light helps to check infestations.

Dry the pods to optimum kernel moisture level of about 7%, store the pods in polythene-lined gunny bags and fill the top surface of the bag with a layer of 3 cm height sand. Dip the old gunny bags in boiled water for 15 minutes. Mouth of bags should not be stitched or closed to avoid germination loss, and care should be taken to avoid seeds breakage and broken grains should not be stored for long periods. Coat the seed with small quantities of vegetable oil or mix neem leaves in the stored grain as oviposition deterrents and for beetle control. Organophosphate compounds have been used in different states, so dip the gunny bags in 10% malathion solution. Apply E.D.B. ampulses @ 3 ml/q of seed in air tight storage structure and fumigation with aluminum phosphide protects the seed without affecting the viability.

Several parasitoids of the families Braconidae and Pteromalidae attack *C. chinensis* in various parts of the world (Islam and Kabir, 1995). The predatory mite *Pyemotes* also parasitizes the pest in bulk grain and food facilities (Sarwar, 2019).

Figure 3. *Callosobruchus chinensis*.

2.1.3. Southern Cowpea Weevil Callosobruchus maculatus (Fabricius)

Bean weevil *Callosobruchus maculatus* (Coleoptera: Chrysomelidae: Bruchinae), is a species of beetles commonly known as the cowpea weevil or cowpea seed beetle. Its other common names are oriental cowpea bruchid, mung bean bruchid, cowpea weevil (also applied to *C. maculatus*), cowpea beetle, adzuki bean weevil, adzuki bean seed beetle, pulse beetle, Chinese bean weevil, and Chinese bruchid. This common pest of stored legumes has a cosmopolitan distribution, occurring on every continent except Antarctica. The beetle is most likely originated in West Africa and moved around the globe with the trade of legumes and other crops. The larvae of this species feed and develop exclusively on the seed of legumes (Fabaceae) and are a major pest of cowpeas, green gram and lentils, in Africa and Asia that presently range throughout the tropical and subtropical world. It has been reported that the favorite hosts for *C. maculatus* are cowpea and chickpea (Desroches et al., 1997).

The level of quality deterioration of cowpea varieties caused to the grains by *C. maculatus* determined that total protein in the control experiment showed a progressive decrease with increase in storage period, and percentage ash in the control remained higher than the infested and it decreased with storage period. The percentage fat decreased with infestation and control has a higher percentage fat than the infested, but the fat decreased with storage period. The percentage fiber is higher in the infested grains than the control and it increased with infestation period. The percentage carbohydrate decreased with infestation and control samples have higher carbohydrate than the infested ones (Bambini et al., *2006)*.

Adults do not require food or water and spend their limited lifespan (1-2 weeks, usually no more than 12 days) in mating and laying eggs on beans. Adults emerge from seeds and emerging females soon assume a 'calling posture' and release a pheromone and continue for one week that attracts males. Male and female *C. maculatus* mate within an hour after emergence. A single mating of 3-8 min is usually sufficient to fertilize a female's eggs (Qi and Burkholder, 1982).

Bean weevil *C. maculatus* is more elongated in shape than other members of the leaf beetle family and adults are 2.0-3.5 mm long. It is reddish-brown overall with black and gray elytra marked with two central black spots. The last segment of the abdomen extends out from under the short elytra and also has two black spots, and antennae of both sexes are slightly serrate. Male and female bean beetles (of the sedentary morph) are easily distinguished from one another by general appearance. The most distinguishing characteristic is the coloration on the plate covering the end of the abdomen. In the female, the plate is enlarged and darkly colored on both sides, while in the male, the plate is smaller and lacks stripes. Generally, females are larger in size than males, but there is much variation. In some strains, females are black in coloration and males are brown, but in others both sexes are brown. Females often have strong markings on the elytra consisting of two large lateral dark patches mid-way along the elytra and smaller patches at the anterior and posterior ends, leaving a paler brown cross-shaped area covering the rest, while males are much less distinctly marked. In common with other species of *Callosobruchus, C. maculatus* has a pair of distinct ridges (inner and outer) on the ventral side of each hind femur, and each ridge bears a tooth near the apical end. The inner tooth is triangular and equal to or slightly longer than the outer tooth (Figure 4).

Just at once inseminated, adult females lay (oviposit) many eggs up to 115 on the external surface of seed that are cemented to the surface of pulses and are smooth, domed structures with oval, flat bases. Individual eggs are 0.75 mm long, oval or spindle shaped, clear, shiny and firmly glued to the bean surface. Eggs hatch within 5-6 days of oviposition and upon hatching, the larva bites through the base of the egg, through the testa of the seed and into the cotyledons. Detritus produced during this period is packed into the empty egg as the insect hatches, turning the egg white and making it clearly visible to the naked eye (Horng, 1997).

The larvae and pupae are normally only found in cells bored within the seeds of pulses. The larva that hatches from the egg burrows from the egg through the seed coat and into the bean endosperm without moving outside the protection of the egg. Once the larva burrows into the bean, the remaining egg (shell) becomes opaque white or mottled as it fills with frass (feces) from the larva. The larvae burrow and feed on the bean endosperm and embryo undergoes a series of molts, and burrow to a position just underneath the seed coat prior to pupation. Although the seed coat of the bean is still intact, a round 1-2 mm window is apparent at the location where the beetle is pupating. Pupation is the complete metamorphosis of the larval maggot to a winged adult. At an average temperature of 30 ± 10 °C, mean duration larval plus pupal period in different hosts ranged from 19.66 days (cowpea), 27.73 days (chickpea) to 38.29 days (soybean). The adult that results from pupation, chews through the seed coat and emerges from the bean. The adults are fully matured in 24 to 36 hours after emergence. Fertilization occurs as females lay eggs and glue them to the surface of a bean seed (Fabaceae). Early embryonic development occurs inside the transparent egg until the first instar larva (maggot) burrows through the seed coat into the seed endosperm directly from the egg. Bean beetles development is long-germ (Patel et al. 1994) similar to *Drosophila* in which body segment determination occurs by the end of the blastoderm formation. The empty egg shell is typically filled with white frass (fecal waste) as the larva feeds. There are four larval instars (Devereau et al., 2003) all feeding inside the endosperm of the seed on which the egg is laid. Pupation occurs inside the seed and an adult emerges by chewing and removing a circular piece of the seed coat to form a round exit hole.

At 25°C and 70% RH, the total development period of *C. maculatus* breeding on seeds of *V. unguiculata* is about 36 days (minimum about 21 days) and pupation taking place within the seed 26 days after oviposition. Infestation commonly begins in the field, where eggs are laid on maturing pods (Fitzner et al., 1985). As the pods dry, the pest's ability to infest them decreases. Thus, dry seeds stored in their pods are quite resistant to attack, whereas the threshed seeds are susceptible to attack throughout storage.

The elapsed time from newly laid eggs to the emergence of adult beetles varies between bean beetle strains and environmental conditions. Within a limited range, increasing temperature will decrease the generation time. In a study, generation times have observed as short as 3-4 weeks in a 30°C incubator (12:12 day: night light cycle) and ambient humidity (averaging 30% RH and ranging from 20%-40% RH). Cultures raised in a 25°C incubator (12:12 day: night light cycle) and ambient humidity (averaging 50% RH and ranging from 40%-60% RH) have generation times of 5-6 weeks. Cultures maintained on a laboratory bench at room temperature (22°C) with indirect outdoor window lighting and ambient humidity (averaging 50% RH and ranging from 40%-60% RH) have a generation time of a full 7 weeks. Simply, placing a tray of water in a temperature-controlled incubator may be all that is necessary to bring RH to the 40%-60% range and improve emergence success rates. Generation time also depends on the host species of bean, which have been chosen. Longer generation times have been found in adzuki beans compared to either mung or black-eyed peas. At 30°C, it takes 7 weeks for emergence from adzuki beans compared to 3-4 weeks from mungbeans (Howe and Currie, 1964).

Figure 4. *Callosobruchus maculatus.*

A study investigated the potential of light spectra in the control of the bean weevil; In the first trial, ten individual bean weevils have been directly irradiated with green, yellow, red and white and ambient (control) light generated by electric bulbs at 25 Watts. In the second trial, ten individual weevils have been housed with 700 bean seeds and exposed to the light spectra. Each light treatment has been replicated thrice and weevils have been exposed for two hours daily for 10 days. White, blue and red lights have been lethal to beans weevil by killing them and reducing their activities (Kehinde *et al., 2019).*

Bruchidae are attacked by parasitoids belonging to 10 families of Hymenoptera and 1 of Diptera. The parasitoids of *C. maculatus* include several parasitoid wasps, specifically wasps *Anisopteromalus calandrae* (Howard) as well as *Dinarmus basalis* (Rondani) are two important pteromalid hymenopteran and *Uscana mukerjii* (Mani) (Hymenoptera: Trichogrammatidae) targets *Callosobruchus* species. Parasitoid *D. basalis* parasitizes small larvae and halts their development. This limits the damage they can do to beans, but their presence still make the beans unfit for human consumption and usually create them unfit for sowing, as well. Parasitoid *U. mukerjii* is an egg parasite, which prevents the egg from hatching, thereby preventing damage to the legume (Kapila and Agarwal, 1995).

2.2. Red flour beetle *Tribolium castaneum* (Herbst)

Red flour beetle *Tribolium castaneum* (Herbst) (Coleoptera: Tenebrionidae), the darkling beetle, is of Indo-Australian origin and found in temperate areas. It can survive the winter in protected places, especially where there is heat centrally, within most tropical and subtropical countries of the world. Red flour beetles may be present in large numbers in infested grain, but generally are unable to attack sound or undamaged grain (Walter, 1990). The adults are attracted to light, but will go towards cover when disturbed. Typically, these beetles can be found not only inside infested grain products, but in cracks and crevices where grain may have spilled. Red flour beetles may fly and these are attracted to grain with high moisture content and can cause a grey tint to the grain they are infesting. The beetles give off a displeasing odor and their presence encourages mold growth in grain (Grunwald et al., 2013).

The red flour beetle is reddish-brown in color and approximately 3-4 mm long, its antennae end in a three-segmented club-shape and the sides of the thorax are slightly curved (Figure 5). The head and upper part of the thorax are covered with minute punctures and the wing cases (elytra) are ridged along their length. The antennae are enlarged at the tip (capitate) with the last three segments wider than preceding segments and eyes are reddish-black in colour. The front, middle and hind legs of the beetle show the 5-5-4 tarsal formula. Female red flour beetles are polyandrous in mating behavior and within a single copulation period, a single female will mate with multiple different males for obtaining a greater amount of sperm to increase fertility (Pai et al., 2005).

The eggs are microscopic and often have bits of flour stuck to their surface. Eggs are approximately 0.5 mm long, cylindrical and white or colourless. They are sticky which causes them to become covered in flour and stick to containers. Females lay between 300-400 eggs in their 5-8 months adult's lifespan. After the female lays the eggs they will hatch within nine to twelve days. The slender and cylindrical larva has pale-brown head, creamy yellow to light brown in body color and is covered with fine hairs. They have two dark pointed projections on the last body segment. The larval period lasts from 15 to more days and larvae seem to be visible after 13 ± 1 days. The larval development proceeds through five instars, divided depending upon the size and is completed in about 12-14 days from the day of appearance followed by the non-feeding pre-pupal stage, a stage at which the larva commits itself for metamorphosis to pupae. The pre-pupal stage extends over 1-2 days followed by the pupal stage, which lasts for about 4-5 days. The pupae are lighter in shade having white to yellowish shade, while becoming brown later on and pupa does not have a cocoon (Wade, 1984).

These beetles can breed throughout the year in warm areas. The life cycle takes from 40 to 90 days and the adult can live for three years. All forms of the life cycle may be found in infested grain products at the same time. Under optimum conditions of 35°C and 60-80% relative humidity, the development times for each stage are approximately 3 days for eggs, 16 days for larvae and 5 days for pupae (Bousquet, 1990).

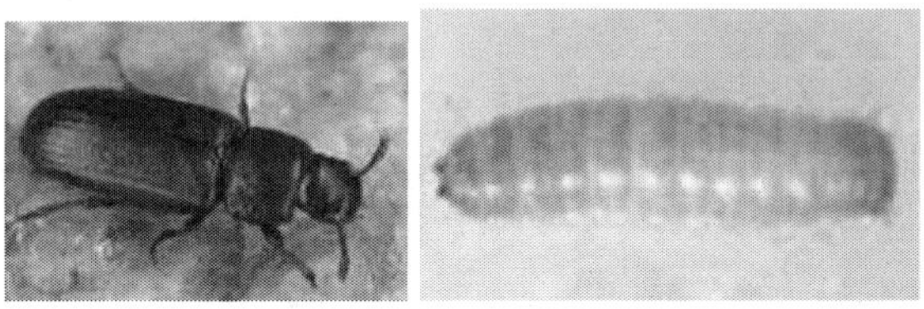

Adult Larva

Figure 5. *Tribolium castaneum*.

Both adult and larva stages cause significant damage to grains that are already damaged physically or by the action of other pests. Adults move quickly and rapidly to run for a cover when disturbed. They can be found at the surface or deep within grain stores. Adults disperse over short distances through flight. The adult is long-lived, sometimes living more than three years.

Flour beetles can feed and survive on even the smallest bits of grain, so, cleaning is a crucial part of controlling these pests. The infested material may also be placed into a freezer for four to five days. Red flour beetle *T. castaneum* generally attacks the stored grain products like nut, flour, meal pasta, spices, beans, cereals, seeds, chocolate, cake mix and also museum specimens (Via, 1999). This flour beetle has a very broad host range, and is a major pest of maize, groundnut, oats, Brazil nut, barley, walnuts, lentil, rice, beans, peas, almond, rye, sorghum and wheat. Red flour beetle also attacks a wide range of dried stored products. It is possible to confuse the red flour beetle (*T. castaneum*) with the confused flour beetle (*Tribolium confusum* Jacquelin du Val), the eyes of red flour beetle are larger and closer together than those of the confused flour beetle (Ali *et al., 2009; 2011).*

Red flour beetles can be detected by simple traps baited with food such as carob beans, ground nuts or cereals, and can help to detect individuals when their numbers are low. At higher densities of holes and tunnels in grains and seeds, the adults and larvae, as well as the dust created by their feeding activities, can be seen. The eggs that stick to storage containers can be detected and there may be a pungent odour at high densities. The severity of a red flour beetle infestation and population can be reduced by removal of adult insects and larvae from the grain by sieving, cleaning the store between harvests, removing and burning of infested residues, immersing grain sacks in boiling water, removing wood from stores, selection of only un-infested material for storage, and fumigating the store to eliminate residual infestations (Sarwar et al., 2004; Sarwar, 2009; 2015 b).

2.3. Khapra Beetle *Trogoderma granarium* Everts

The Khapra beetle *Trogoderma granarium* (Coleoptera: Dermestidae), also called cabinet beetle, which originated in South Asia, is one of the world's most destructive pests of grain products and seeds. These beetles are primarily spread by the transportation of infested materials such as bagged cereals, grains and seeds. Worldwide, there are over 120 described *Trogoderma* sp., and many genera within the Dermestidae family that are similar to *T. granarium*. Identification of the most common stored product pest *T. granarium* and separation of the local *Trogoderma* sp., from them is difficult as specimens collected in the infested commodity are in larval form and adult specimens are usually scarce and damaged, therefore, the genitals should be dissected and examined for identity (Banks, 1994).

The khapra beetle *T. granarium*, is a voracious feeder of stored products and is considered one of the most important quarantine pests globally. It is a serious pest of stored grains and stored products (groundnut, cotton, barley, rice, millet, sesame, sorghum, wheat, maize and cowpea etc.). Its primary seed and cereal grain hosts are *Avena sativa* L. (oat), *C. arietinum* (garbanzo bean), *Glycine max* L. (soybean), *Hordeum vulgare* L. (barley), *Lens culinaris* Medikus (lentil), *Oryza sativa* L. (rice), *Pisum sativum* L. (garden pea), *Sorghum bicolor* L. (grain sorghums), *Triticum aestivum* L. (wheat), *V. unguiculata* (cowpea), and *Zea mays* L. subsp. *mays* (corn).

The beetle does not bore into host material, but young larvae feed on damaged seed and older larvae nourish on whole grains. The larvae are voracious eaters of grains, grain products and seeds. They destroy valuable commodities by feeding on grains and seeds, often leaving only the husk behind. Larvae contaminate products by shedding their cast-off skins and hair, as well as also leaving behind excrement and urine. An infestation lowers product quality and affects the possibility of exportation. If the beetle is left undisturbed in stored grain, it can cause important weight loss and in case of seeds it may lead to significant reduction in seed viability. Severe infestation may cause unfavorable changes in chemical composition of grain or seed. Larvae typically attack the embryo point or a weak place in the pericarp of grain or seed, but will attack other parts as well during heavy infestations (Borzoui et al., 2015).

Besides weight losses, it also reduces grade of grain, renders unfit for consumption and may result in less profit for wholesalers. Apart from the destruction of grain dry products by khapra beetle, ingesting products contaminated with body parts, setae and cast larval skins can result in allergic reactions or gastrointestinal irritation. Handling or consuming contaminated grain and seed products can lead to other health issues such as skin irritation and digestive distress. Asthmatics and sensitized individuals are also at risk, as contaminants are highly allergenic and dermatitis in nature.

Larval skins cast by the beetles may cause skin irritation in people handling heavily infested grains. In rare cases, if swallowed, the barbed hairs of the larvae may be a serious health hazard and can cause digestive frustration (Sarwar and Sattar, 2007).

Adult beetles are brownish and reddish, oval-oblong, covered in dense hairs, head and pronotum dark reddish-brown, elytra reddish-brown usually with indistinct lighter reddish-brown fasciae, venter of thorax and abdomen reddish-brown, legs yellowish-brown and antennae are also yellowish-brown, 9, 10 or 11 segmented with a distinct 3-5 segmented club-shape (Figure 6). Males are 1.4-2.3 mm long, 0.75-1.1 mm wide and brown to black, while females 2.1-3.4 mm long, 1.7-1.9 mm wide, slightly larger with lighter color, and both sexes are ovate and densely hairy beetles. The setae on the dorsal surface are of two types; evenly distributed coarse, semi-erect, yellowish-brown ones; and few scattered, dark reddish brown setae, while color of setae follows the color of cuticles. The pronotum medially and laterally has indistinct patches of yellowish-white, ensiform setae, and elytra with two or three indistinct band of yellowish-white, ensiform (flattened) setae. The median ocellus on the front side is always distinct (Kavallieratos et al., 2019).

The eggs of the khapra beetle are cylindrical with one end more rounded and the other more pointed (pointy end has a number of spine-like projections), nearly 0.7 mm long, 0.25 mm broad and weighing about 0.02 mg. The eggs are initially a milky white, but over several hours turn to a pale yellowish color. In optimal conditions, females lay between 50 and 100 eggs in a year, which can create up to 10 generations of beetles in one year. The eggs, which are loosely scattered in host material, hatch nearly in 3-14 days (Athanassiou et al., 2016).

When hatched, larvae are approximately 1.6 to 1.88 mm long and yellowish-white with brown hairs covering the head and body. The early stage larvae are yellowish-white with a brown head and mature larvae are golden to reddish-brown. As they grow, the larvae's color changes to a golden or reddish-brown and become hairier with dense reddish-brown hair. Larval development usually takes 4-6 weeks during which time they moult 4-15 times. The larval stage however, can last from a month to a year, if it enters diapause (a dormant phase) and can continue for up to seven years in adverse conditions. The pupal stage lasts for 2-5 days and quiescent adult stage 1-2 days. All stages are very hairy and adults live for 12-25 days and produce between one and nine generations per year depending on factors such as the host species, temperature, light and moisture. Usually, the lifespan of adult khapra beetle is between five and ten days. The beetle prefers hot, dry conditions and can be found in areas where grain and other potential food is stored, such as pantries, malt houses, grain and fodder processing plants, and stores of used grain sacks or crates. Khapra beetle is a synanthrope, predominantly living in close association with humans (Athanassiou et al., 2019).

This species cannot always be controlled by insecticides and other nonchemical methods that are usually effective against other pests of stored products, particularly owing to its ability to enter diapause by late larval stage and due to its habit of hiding in cracks and crevices. It can rapidly develop at elevated temperatures and under dry conditions, which are not normally favorable for many major stored-product insects.

Visual inspection can be used to detect pest, but as the beetles crawl into crevices deep into grain stores, they can be difficult to become aware of until their numbers become quite large. Pheromone traps can be used to detect khapra beetles at lower densities. The severity of a khapra beetle infestation can be reduced by good store hygiene such as cleaning the store between harvests, removing and burning infested residues, immersing grain sacks in boiling water, filling in cracks and crevices or fumigating the store to eliminate residual infestations, and the selection of only uninfected material for storage. The removal of adults and larvae from the grain by sieving can reduce populations and addition of inert dusts such as ash and clay to the grain can reduce insect numbers by causing the pest to die from desiccation (Sarwar and Sattar, 2012).

Prolonged exposure to extreme cold and heat have demonstrated marginal impact, but most larvae are found to have survived in extreme conditions well beyond the threshold needed to kill adult beetles. Inspection at ports and entry points provides an effective way to restrict entry of this pest to export marketing. In storage facilities trapping proved to be a useful surveillance tool using pheromone and larval traps. Heat treatment has proved to be very effective and the treatment involves a 30-minute exposure at 60°C (140°F), which has given a 100% kill of all stages of the khapra beetle. Treatment with fast electrons, using a linear accelerator, could provide an efficient method of controlling khapra beetle in store grain. The most effective treatment is phosphine or methyl bromide fumigation and facilities that cannot be fumigated may be sanitized and treated with a surface application of insecticide. Malathion applied repeatedly is currently approved for control of khapra beetle infestations in structures and surrounding surface areas (Sarwar, 2008).

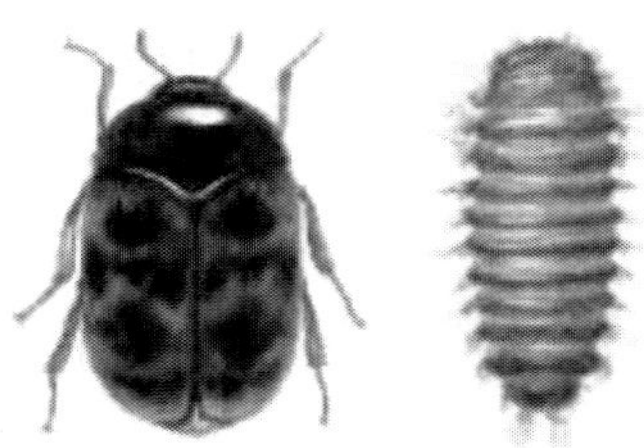

Figure 6. *Trogoderma granarium.*

2.4. Lesser Grain Borer *Rhizopertha dominica* (Fabricius)

Genus *Rhyzopertha* comprises a monotypic (species that does not include subspecies) beetle consisting of only *Rhyzopertha dominica* (Fabricius), the false powderpost beetle in the family Bostrichidae of order Coleoptera. The sole species *R. dominica* commonly referred to as auger beetle, is known generally as the lesser grain borer, wheat weevil or stored grain borer. Currently, *R. dominica* has a worldwide distribution, especially in warmer temperate climates zones and also in tropical countries, however the Indian subcontinent is its most probable native home. It is predominantly found in forested and grain storage environments. It is possible to confuse the lesser grain borer with other storage insect pests such as the large grain borer *Prostephanus truncatus*) (Horn) and the maize weevil *Sitophilus zeamais* (Motschulsky). However, the saw-toothed front edge of the pronotum in the lesser grain borer family (Bostrichidae) is very distinctive. In contrast, the large grain borer is larger (3-4.5 mm long), dark brown in colour, its body surface has many small wart-like outgrowths (tubercles) and the end of its body terminates in a straight edge. The end of the body of the maize weevil is more rounded than that of the lesser grain borer, its head is visible from above and mouthparts are 'beak-like' and the antennae elbowed shape (Sarwar, 2015 c).

There are various substrates that make up the resources and diet for the *R. dominica*. This includes grains, such as rice, wheat, sorghum, oat, pearl, millet and malt barley from the family Poaceae, and chickpeas, peanuts and beans from the family Leguminosae. Although most insect prefers foods such as trees and dried fruit, *R. dominica* seems to be pre-adapted for feeding on dry grains. It feeds on the whole grain in both larval and adult stages. Together with the deep movement into the grain mass and the cryptic feeding on the kernels, it can become difficult to detect initial *R. dominica* infestation. Large amount of frass is also produced from adults feeding activities, containing ovoid granules of undigested endosperm mixed with a finer flour, larvae exuvae, feces, fragments of immature insects and various by products affecting the overall quality of the grain. Both adult and larval stages of *R. dominica* damage the grain by boring, and feed on the germ and endosperm. The pest hollows previously undamaged grains, eating the starch and holes are characteristic with an even edge. Adults and larvae feed on the entire grain apart from the bran leaving behind empty husks and flour (Edde, 2012).

The average *R. dominica* is 2.1-3.0 mm in length, its body displays a reddish brown coloration with a slim cylindrical shape and having 11 antennae segments ending in a distinctive 3-segmented antennal club-shape. The head is not visible when viewed from above, pronotum is located near the base of the body with no depressions and in addition, the basal part of the pronotum has a wrinkled appearance. The front edge of the pronotum (plate-like covering of front segment of the thorax) has a saw-toothed appearance (Figure 7). Distinct tubercles on the *R. dominica* are found on the anterior margin, but appear to be slightly apart at the median. The elytra (wing cases), which cover membranous hind wings, have regular rows of coarse punctures (finer at sides) covered with curved setae (hairs). Moreover, it has clear elytral strioles that are angularly rounded at the apex and short, yellowish, bent setae (Park et al., 2015).

Externally, there are no major recognizable differences between male and female adults of *R. dominica*, however, the last ventral abdominal sternite of the female is seen as pale yellow as compared to the uniformly brown in males. The mating behavior of *R. dominica* follows within 24 hour after the individual hatches from the egg. Female attraction to the male occurs upon physical contact, whereby the close proximity allows for the olfactory senses to detect the male produced pheromones. Stimulation from the pheromones is characterized (in both male-to-male and male-to-female interaction) by an excited and rapid walking motion, and the head, thorax and antennae are extended forward and up. Copulation lasts for 2 hours and can occur multiple times in *R. dominica*, as females require more than one mating to fertilize effectively all the eggs produced during its lifetime (Velasquez et al., 2017).

The eggs are ovoid in shape, 0.6 mm in length, 0.2 mm in diameter, white when first laid and turn rose to brown before hatching. Within the grain mass, females deposit their eggs loosely and allow the first instar larva to enter the kernel. Lesser grain borer females lay between 200 and 500 eggs in their lifetime either singly or in clusters of up to about 20 on outside of grains. The larvae are white to cream colored, with biting mouthparts and three pairs of legs. The young larvae are mobile in grain bulks, but become immobile and gradually more C-shaped as they complete their development concealed within grain or flour. The larva after undergoing 4 larval instar developments will emerge from the kernel as an adult. The duration of development takes up to 35 days, with optimal conditions of 28°C and 50% humidity. Adults, which can live up to 240 days, emerge by chewing through the outer grain layers (Jia et al., 2008).

Figure 7. *Rhyzopertha dominica.*

Sieving a sample of grain is the simplest detection method and removing of spilt grain from around storage facilities can reduce lesser grain borer populations. The addition of inert dusts such as ash and clay to the grain can decrease insect numbers by affecting the insects to die from dehydration. Where suitable infrastructure exists, low oxygen and carbon dioxide-enriched atmospheres can be used to control stored product pests and heating for 24 hours. The fumigant phosphine can be used in sealed storage facilities. Pesticides are poisons, so it is essential to follow all safety precautions on labels (Lin, 1981).

There are a few parasites and predators that attack the lesser grain borer, for instance, *Cephalonomia rhizoperthae* Westwood (Hymenoptera: Bethylidae) parasitizes *R. dominica* (Xu et al., 2008), but fungus *Beauveria bassiana* can be used as a biological insecticide.

2.5. Rice Weevil *Sitophilus oryzae* (Linnaeus)

Insect pest *Sitophilus oryzae* (L.), is particularly called as rice weevil'(lesser grain), because its breeding habits and life cycle have been first of all studied in rice. It is an insect of family Curculionidae in order Coleoptera, cosmopolitan in distribution and has originated in Far East region. They can exist anyplace where physical conditions for growth are favorable and the grain is left undisturbed for some time. This is supposed to be a native of India, although few workers are of the opinion that some temperate country is its original home. Nowadays, it is believed to be most widely distributed among the known insect pests of the biosphere anywhere shipments of grains have reached all over the world. However, this pest is particularly common in warm countries of the world. It is a serious pest not only of paddy and rice, but almost of all cereals and their products. In fact, it is the commonest pest that one encounters in all kinds of stores usually found in grain storages and processing plants. They infest wheat, oats, rye, barley, rice and corn. Sometimes, these are also found infesting beans, sunflower seeds and dried corn. Weevils have chewing type of mouthparts and attacks from *S. oryzae* can start in the field, when the moisture content of grain is about 20% before harvest. Commonly, adults attack sound grain, and both adults and larvae feed inside them, leaving large cavities and emergence holes. The adults also attack damaged grains, and during large infestations, heat and moisture are produced, leading to colonization and secondary damage by insects and mites. Rice weevils pretend death by drawing their legs close to body and remaining silent even when disturbed. Field bean and bengal gram proved its best hosts, where the weevil showed highest fecundity, fertility, progeny production, adult longevity and short developmental period (Deepthi and Manjunatha, 2015).

The most significant identification feature of weevils is their snout, which is good-looking and long. The adult rice weevil is reddish-brown to black in color and 2-3 mm long. It has irregularly round shaped pits on its thorax and also four light spots are found on wing covers or yellowish spots on the elytra (wing covers). The adult weevil can fly and is attracted to lights. The maize weevil is very similar to the rice weevil, but larger in size having circular punctures on pronotal dorsum instead of longitudinally elliptical punctures on pronotal dorsum as in rice weevil (Figure 8). Males of *S. orzyae* produce an aggregation pheromone while females create a pheromone which attracts only males. Rice weevil lays on an average of 4 eggs per day in crevices of kernels or dust and female during its life span of four to five months produces 250-400 eggs. The female also uses strong mandibles to chew a hole into a grain kernel after which it deposits a single egg within the hole and sealing it with gelatinous fluid from its ovipositor. Females can be separated by having slender and longer rostrum than male bearing with thick-stout rostrum (Singh et al., 2018).

Each egg is tiny, white, and oval in shape measuring 0.7 mm long and 0.3 mm broad. The eggs hatch into a young larva in about 3 days, which migrates toward the center of the grain kernel and feeds inside for an average of 18 days. At this time inside the grain, the larva feeds, grows and develops into a pupa. The larva consumes food several times more than its own weight and periodically molts to increase in size. During larval stage, these are legless and white to creamy white in color, with a small tan head. The larva develops within the grain, hollowing it out while feeding and then pupates within the grain kernel. A fully matured grub makes a pupal cell inside the grain for pupation. Before pupation, it may pass through pre-pupal stage for 1-2 days (Povey and Holloway, 1992).

The pupa undergoes major internal and external changes inside the grain to develop into an adult. The pupal stage lasts for an average of 6 days with 5-16 days range. The new adult will remain in the seed for 3 to 4 days while its cuticle hardens and matures. The adult formed after pupation bores its way out of the grains and emerge out through the hole of emergence. A single generation can be completed in around 28 days. The emerged adults are ready to mate and begin their generation and the life span of the adult weevil is 4-5 months. The duration of life cycle of rice weevil and number of generations completed in a year depends upon the weather conditions, but 5-7 generations are generally completed in a year (Somnath et al., 2014).

Rice weevils are the most destructive pests of stored grain and they completely destroy the grain. Attack is evidenced by surface heating of grain, dampness which may even cause germination and presence of numerous adults. Control of this insect involves inspection and removal of infested food products, discarding the heavily infested material, repackaging material in new containers, and vacuuming kitchen cabinets. Products that need to be retained may be placed in the freezer for several weeks to kill adults and larvae.

Figure 8. *Sitophilus oryzae*.

3. Chickpea Grain Insect Pests Management

Insect pests in storage are generally not considered to be a major problem in stored chickpea seed. However, where a prior infestation exists in the storage structure (most commonly as a result of cereals residue), then the infestation can develop and spread in the chickpeas. Ensure all handling equipment and storages are cleaned of old cereal grains before these are used to handle chickpea. Good hygiene, combined with aeration cooling, can prevent infestations developing. If insect pests are found in stored chickpea, the only registered treatment is phosphine fumigation.

The most important insect damaging pulses in field and storage are referred as bruchids or pulse beetles. Additional insect pests, which cause damage to stored gram are *T. granarium*, *R. dominicia*, *T. castaneum* and *S. oryzae* resulting in physical losses and contaminations with undesirable excrements. The easiest way to avoid damages by insect pests is to prevent their occurrence and spread. Some basic steps for insects control in a storage warehouse or processing plants are their monitoring, identification and control as given below:-

3.1. Monitoring of Insects Infestation

Inspection and monitoring are excellent tools for early detection of insect infestations in stored chickpea. Monitoring is an inspection or surveillance program, which will yield prompt awareness of a possible problem (presence, level, source) before it occurs. Regular visual observations on the adult or larval populations and inspection should be carried out frequently and thoroughly to detect actual problem or the potential of it, and evaluate its seriousness. This is much important that it might be scheduled routinely and carried out by trained and qualified staff, which has an adequate knowledge about the stored-grain pests. Monitoring for insect infestations should be carried out as visual observations involving sampling, sifting and counting; detection of latent infestation through chemical, X-rays and sound amplification; use of chemical attractants such as synthetic pheromones and food attractants where insects are difficult to locate and their population is hard to assess; use of bait stations like attractive sticky food, especially for beetles; use of traps such as light traps that function as an early warning monitor system nets, or aspirators may be used; and combination of the above methods is the best option. Presence of either of the succeeding states in the sample of grain such as clustering of grains, decay or powder, foul smell, grain bag temperature noted higher than the ambient range, presence of white spots on seed coat, existence of eggs, and presence of larvae skin, are an indication of the occurrence of insects.

3.2. Identification of the Problem

In order to minimize the damage caused by insects infestation, determine the extent and nature of the possible problem (species, type, level and means of transmission). Identification of the insects involved is an important step in this direction. Different species of the insects behave differently and may not respond to one control measure in the same way. Identification of the species of the insect and its means of propagation is therefore helpful in devising a suitable control measure in stored chickpea.

3.3. Insect Pests Control Measures

Devise a plan for controlling the problem (integration of all possible means to achieve good, cheap and safe pest control) in stored chickpea. Plan for control is assessing the magnitude of the problem, selecting control measure based on potential risk of contamination, use of facilities available and cost involving. The efforts to protect grain against insects can take two arrangements, a preventive effort before storage of the grain, from the time it is received, even if no insects are visible, and a curative effort during or even before storage, if necessary. In both cases, the insects must be destroyed without altering the food quality of the grain. But, to accomplish this, some general hygienic measures and steps for treatment of the premises must be observed. Even though the chickpea's resistant varieties are less in availability, the pests can be controlled effectively through application of a combined pest management strategy. Scientists working in various institutions in this area have undertaken extensive studies and research to develop Integrated Pest Management (IPM) and Integrated Crop Management (ICM) in order to reduce chickpea grain yield losses. Insect control measures can be any one or a combination of following methods and some of strategic IPM practices are specified underneath:-

3.3.1. Harvesting and Threshing

Harvest crop when leaves start to senesce, pods turn yellow, plants are dry, and seeds feel hard and rattles within the pod. Harvesting and threshing both can be done by commercially available combine harvesters.

3.3.2. Seed Processing

The dried seeds should be graded and free from plant parts, soil particles, stones, weed seed, other crop seed, as well as shriveled, broken, or damaged seed. The cleaning and grading is achieved by winnowing and mechanical sieving of grain.

3.3.3. Seed Storage

Generally, seed moisture level is 10-12% for short-term grain storage (up to 8 months). After drying, the seed should be stored in polythene-lined gunny bags or in safe storage structures (metal bins or earthen containers) in air tight containers. Fumigate the seed bins or the seed store periodically with commercially available fumigants (ethylene dibromide and phosphine).

Seed moisture must be carefully watched during storage of pulse to prevent insect pests problems. Chickpea can safely be stored at 14-16 percent moisture. If moisture levels are too high, grain dryers can be used, but they must be used with extreme caution because these may cause mechanical and thermal damage to all pulse grains. Moisture is tested several times during first few weeks of storage to maintain proper levels and to prevent seed sweating. Aeration is used to cool and dry the seed and to avoid storage problems.

3.3.4. Physical Methods

In general, all pulse commodities need to be dried to safe moisture level, below 11-12%, prior to storage. Sun drying of grain in an open yard is a common practice in semi-arid tropical regions. Seeds can be kept under small transparent polyethylene bags. Solar heating can reduce *Callosobruchus* damage and during a week of observation solar heating raised the temperature in seed bags to 65°C each day. Solar heat-treated seeds remained free from bruchids damage even after 41 weeks of storage and no reduction in seed germination is observed in the solar-heated seeds (Chauhan and Ghaffar, 2002). Solar drying of legume grains can be done on cemented platform, mat, jute cloth or metal sheets. In order to prevent the damage caused by *Callosobruchus* or other insects in storage, farmers commonly expose the seeds and pods to the sun radiations for several days before their storing. The investigated study indicated that the different biological parameters are significantly affected when the insects are reared under sun radiations including longevity, fertility, infertility rate, eggs, development time, larval survival rate and emergence rate. In fact, it appears that sun exposure significantly inhibits egg laying, embryonic development and postembryonic development of pests resulting in complete inhibition of the emergence of the offspring (Moumouni et al., 2015). Thus, solar drying seems to be an effective method of preventing insects damage during storage.

3.3.5. Hermetic or Sealed Storage

In airtight conditions, reduced oxygen and increased carbon dioxide will eventually arrest insects development. It is reported that polythene bags are better storage containers than cloth bags against the protection of bruchids. Portable hermetic storage bags, with low oxygen permeability (less than 4 cc/ m^2/ day) also prevented *C. maculatus* and *R. dominica* infested pulse subjected to air tightness resulting into check in insect population build-up. Accumulation of carbon dioxide and depletion of oxygen levels adversely affect the insect growth. Incorporation of eucalyptus and mint oil in airtight conditions results in quick arrest of infestation and ensures better protection in small quantity of legumes (Garcia-Lara et al., 2013).

Various diatomaceous earth formulations are mostly effective against *C. maculatus*. Adults mortality increased progressively with increased doses of product and exposure time, and treatments are most effective at 50% relative humidity. Combined effect of *silk oak Grevillea robusta* A. Cunn. Ex R. Br., and diatomaceous earth gave 100% adult mortality of *C. chinensis* at 30°C and 65% RH (Wakil et al., 2011).

Adults pulse beetle, being very weak and having a short life, cannot move in grain mass and are restricted to top 15 cm layer. Even the adults emerging out of the infected material cannot move in the inter-granular space and could die before mating. Movement of adult pulse beetle can be prevented by placing a 7-10 cm layer of dry sand or dry activated clay at the top of grain mass. For preventing mixing of sand or clay with the material, a paper or polythene sheet can be placed on the top surface of grain and then sand or clay is placed.

3.3.6. Use of Irradiation

In addition to traditional storage bags and different storage systems, various novel technologies like ionizing radiation, radio frequency (RF) and micro waves are used to disinfest the pulses (Alice et al., 2014). One of the means of reducing storage insect pests losses is through the application of gamma irradiation. A major pest *C. maculatus* in seeds is killed within 8 days of irradiated with 0.25 KGy gamma rays. Exposure to 200, 300 and 400 W microwaves for 28 seconds to different life stages of *C. maculatus* in seeds caused complete mortality of all life stages (Darfour et al., 2012). Hence, saving chickpea from bruchids attack can certainly improve the nutrition security around the globe through radiation processing.

3.3.7. Insect Growth Regulators and Pheromones

Insect growth regulators (IGR) and pheromones are behavioral manipulation methods for management of insect pests. When chickpea seeds infested with *C. maculatus* are treated with the IGR cyromazine at 30°C, the reproductive rate of *C. maculatus* is drastically decreased. This treatment also decreased the consumption of stored food by *C. maculatus*. Cryomazine treatment also lowered the numbers of hatching eggs, pupation and adults emergence in mungbean and pea seeds (Al-Mekhlafi et al., 2012). A multi-attractant lure BFL-225, is used as trap bait for *C. maculatus* in commercial warehouse of chickpea (Ukeh et al., 2013).

3.3.8. Host Plant Resistance

Although chickpea grains are considered to be particularly susceptible to insects damage, large differences in susceptibility to the major pests have been detected in the germplasm and host resistance appears to be an attractive alternative to insecticides (Sarwar and Tofique, 2006; Sarwar et al., 2005). At certain locations, resistant varieties of chickpea have been developed and the resistant source can be utilized in further breeding programs. The host plant resistance is deemed to be the method of reducing storage insects attack in an environmentally friendly manner. The common improved chickpea cultivars show significantly non-preference type resistance for oviposition and feeding to have few eggs, high larval and pupal mortalities, low adult emergence and prolonged adult emergence period to show little damage or bruchid population development based on the growth indices of the insect. Reduced ovipositions, less seed damage, low adult emergence and prolonged development period have been recorded in the wild plants of chickpea. The cultivars with rough and thick seed coat are more resistant compared to those having smooth and thin seed coat. The minimum numbers of holes made by pulse beetle in grains, minimum eggs laid by adults, minimum grain damage, minimum grain weight loss, minimum adults attracted towards resistant cultivar have been found than on susceptible (Shaheen et al., 2006).

The tolerant chickpea genotypes exhibit hard and wrinkled seed coat, dark brown color and small size grain. These characteristics demonstrate a significant harmful effect to pest appearance and grain damage. The vulnerable genotypes have soft and smooth seed coat, white seed color and bigger grain size that cause vulnerability to *C. maculatus* (Sarwar, 2012). Feeding preference, weight lost after feeding, frass weight obtained, reproduction and development times of the bruchid pest, thickness of the outer seed coat, and percentage germination have been measured to determine the reaction of pulse genotypes. Relation between seed thickness and resistance level observed that the seed coat thickness has a vital role in the grain's resistance, thicker the seed coat, the greater resistance stability existed against *C. analis* (Sarwar, 2013 a).

Dry grain legume seeds possessing αAI-1, an α-amylase inhibitor from common bean (*Phaseolus vulgaris* L.), under the control of a cotyledon-specific promoter have been shown to be highly resistant to several important bruchid pest species. One transgenic chickpea line expressing αAI-1, has been assessed for resistance to the bruchids *Acanthoscelides obtectus* (Say), *C. chinensis* and *C. maculatus*. The transgenic line found highly resistant to both *Callosobruchus* species and their larvae suffered from significantly increased mortality rates inside transgenic seeds, while *A. obtectus* known to be tolerant to αAI-1, is able to develop in transgenic line. The αAI-1 present in seeds of transgenic chickpea line significantly increases the resistance to two important bruchid pest species (*C. chinensis* and *C. maculatus*) essentially due to immunity. To control αAI-1 tolerant bruchid species such as *A. obtectus* and to avoid the development of resistance to αAI-1, varieties carrying this transgene should be protected with additional control measures (Luthi et al., 2013).

For genetic improvement aimed at insects resistance, chickpea plants expressing Bt genes as well as pyramids have been developed. The next generation approach for crop protection is to knock down the crucial physiology-related genes of insect pests using transgenic plants, this is called Plant-mediated RNA interference (RNAi), which holds great promise for future development for chickpea protection (Khatodia et al., 2017).

3.3.9. Cultural Control and Sanitary Methods

It is crucially important to reduce the initial pest population and prevent development of any insect pests in the storage. The removal of infested residues from last season's harvest is essential, as a general hygiene. Before bringing a new grain into store, do not mix new grain with old, clean the storage structure, handle equipment and machinery carefully, and disinfect sacks and baskets by sunning or chemical treatment. Good storage hygiene plays an important role in limiting infestation by *Callosobruchus* as well as other pest species. When small lots are stored, sun-drying the chickpea periodically in a thin layer for periods of up to 4 hours can give substantial protection.

3.3.10. Atmospheric Regulation

Modified atmosphere based on lack of O_2 offers a safe and residue-free alternative to chemical fumigants for pest control in stored grains. Elevated CO_2 levels have been reported to suppress bruchid populations and 18% CO_2 in the atmosphere significantly increased the mortality of *C. maculatus* adults, later stages of eggs and 1st to 4th instars larva by hypoxia (Cheng et al., 2013). It is reported that adult bruchids exposed to CO_2 concentrations of 15% and 20% for 45 days, suffered 100% mortality. In an another study, on kabuli chickpea, 99% purity CO_2, under 20 bars and at an exposure time of 180 minutes, is sufficient to kill all the development stages of *C. maculatus* and 60 second exposure found sufficient to kill adult *C. maculatus*. The pupal stage is relatively more tolerant to high pressure CO_2 (Kutbay et al., 2011).

3.3.11. Eco-Friendly Management

Eco-friendly management of pulse beetles in stored chickpea includes several pre-storage treatments such as sun exposure to air, packing in gunny bags, storage of seeds in earthen mud pot often lined or mixed with ash, dried ash and leaves, smearing of vegetable oils on seeds, storage in triple lining of polyethylene sheets, and splitting or de-husking of pulses and roasting. These substances act as insect repellents, antifeedant and oviposition deterrents (Singh and Pandey, 2016).

Biopesticides are now experimented around the globe as an alternative to chemical pesticides. Among many plant extracts, powders, ashes and oils tested, these materials successfully showed their efficacies against pulse beetles infestation in gram storage and provided protection against pests for few months without adversely affecting the seed germination (Sarwar, 2010; Sarwar et al., 2012; Hina et al., 2015).

Entomopathogenic fungi, *Metarhizium anisopliae* and *B. bassiana* have been used as the bio-control agents against *C. chinensis* in stored chickpea grain and proved good alternatives to chemicals. These non-chemical control strategies might have significant contributions towards development of commercial formulations of bio-pesticides to be used against this grain beetle and other stored product insect pests (Iqbal et al., 2018).

3.3.12. Chemical Control

When the insects population is not controlled by non-chemical methods, then application of chemical sprays can be used as and when needed. Application of chemicals, therefore, sometimes becomes essential for complete insect mortality and prevention of insect growth. It features two broad types of treatment, for instance, treatment by contact insecticide and treatment by fumigation. Treatment of grain with contact insecticide consists of covering the grain with a film of insecticide that acts on contact with insects, with effects that vary in rapidity and persistence. These products come in various forms (powders for dusting, powders to be mixed with water, liquid concentrates) that dictate their techniques of application. For grain that is to be stored in bulk, the insecticide is incorporated directly into the grain by spraying before the silos are filled. For storage in bags, previously cleaned grain is mixed with powder or sprayed before bagging. In order to avoid re-infestation of grain stored in bags, further repeated dustings or sprayings are carried out while the bags are being stacked and during the storage period. The chemicals malathion, deltamethrin and cypermethrin are found most effective, resulting in 100% mortality of adult *C. chinensis* after 7 days of exposure. Seed treatment with spinosad is found to be most effective against *C. maculatus* and *C. chinensis* in seeds. Black gram seeds are well protected against *C. chinensis* for up to 5 months by the chemicals malathion and deltamethrin when the cotton bags used in storage are sprayed. Among new insecticides tested, lufenuron and emamectin benzoate are found to effectively control *C. chinensis* in chickpea, with 85% seed germination after nine months of storage.

The host grains, temperature and species of *Callosobruchus* can all alter the effectiveness of insecticides. Species of *Callosobruchus* can be controlled by fumigation treatment with phosphine, however admixture of approved grain insecticides, especially organophosphates, will protect against the pests and their populations are susceptible to deltamethrin.

Fumigation is a treatment that rids stored grain insects by means of a poisonous gas called a fumigant. This substance, produced and concentrated as a gas, is lethal for specific living species. Unlike contact powders, the fumigant penetrates to the interior of the grain mass and reaches the largely invisible incipient forms (eggs, larvae) developing there. Fumigants spread throughout the area where released, therefore, are used in totally sealed enclosure. Thus, when grains stored in bulk are fumigated, the bins must be perfectly airtight. For grain stored in bags, the usual method is to cover the bags with a tarpaulin whose edges are sealed to the ground or the walls. After fumigation, the grain storage should be aerated and thoroughly cleaned, and the collected trash should be removed away from the storage area (Sarwar, 2016).

4. Integrated Pest Management

Integrated pest management (IPM), is a sustainable approach to manage pests by combining biological, cultural, physical and chemical strategies in a way that minimizes economic, health and environmental risks. Strategies of IPM include biological controls (beneficial insects, beneficial pathogens), cultural controls (planting and harvest dates, crop rotation, sanitation), host plant resistance, mechanical and physical controls (weather events, trapping), and chemical control (pesticides). An IPM in chickpea promotes the judicious use of pesticides to maintain pest populations below economic levels, prevents pests from causing unacceptable losses to crop quality or yield, and reduces impacts on non-target organisms and the environment. For scouting and determining economic thresholds, insect population estimates are calculated by averaging counts taken from several separate areas of the field and storage. If insecticides are necessary when pulse crop is flowering, producers should apply applications during the evening to minimize any negative impacts on pollinators, including bees and natural enemies. The sustainable approaches to reduce the incidence of insects and achieve sustainability in chickpea production systems are through the adoption of an integrated approach to reduce the survival and damage of pests (Sarwar, 2004; 2013 b).

Conclusion

Chickpea storage plays an important role in preventing losses, which are caused mainly by insect pests. There are several sources of infestation in chickpea such as fields, carried-over commodities, waste and rejected produces, agricultural machineries, processing plants, farm grain stores, re-used sacks, means of transportation, and alternative hibernation sites and hosts. Examples of successful IPM approaches operating in farmers' fields have been sought, for their possible extrapolation to other situations. Admittedly, there are few examples of direct farmer involvement in evolution of IPM packages for food legume crops including chickpea. However, there is dire need to involve farmers themselves in the adoption and evaluation of IPM strategies, and it is hoped that this assembly of information relating to IPM of chickpea will facilitate increased in farmer-participatory IPM activities. It is concluded that an integrated approach is most effective for long-term sustainable management programs. An IPM is a sustainable approach to manage pests by combining biological, cultural, physical and chemical strategies in a way that curtails economic, health and environmental risks. The integrated pest management strategies would further require the use of not only novel Bt transgenics plant-mediated RNAi, but also the modern biotechnological tools like targeted CRISPR/ Cas-mediated plant genome editing for chickpea protection against insect pests.

Figure 9. Healthy chickpea grains

References

Ali A, Sarwar M, Khanzada S, Abro GH (2009). Reaction of Certain Wheat Varieties to the Action of Red Flour Beetle, *Tribolium castaneum* (Herbst) (Coleoptera) Under Insectary Conditions. *Pakistan Journal of Zoology,* 41 (1): 51-56.

Ali A, Sarwar M, Khanzada S, Abro GH Abro. (2011). Evaluating Resistance of Wheat Germplasms to Attack by Red Flour Beetle, *Tribolium castaneum* (Herbst) (Coleoptera). *Pakistan Journal of Zoology,* 43 (4): 793-797.

Alice J, Sujeetha RP, Kavitha, Abirami CV, Alagusundaram K (2014). *Biointensive safe storage methods for pulses: Review. Jiopest,* 7: 98-103.

Al-Mekhlafi FA, Mashaly AMA, Wadaan MA, Al-Mallah NM (2012). Effect of Different Applicable Conditions of the Insect Growth Regulator (Cyromazine) on the Southern Cowpea Weevils, *Callosobruchus maculatus* Reared on Peas. *Pakistan Journal of Zoology,* 44 (2): 481-488.

Athanassiou CG, Kavallieratos NG, Boukouvala MC (2016). Population Growth of the Khapra Beetle, *Trogoderma granarium* Everts (Coleoptera: Dermestidae) on Different Commodities. *Journal of Stored Products Research,* 69: 72-77.

Athanassiou CG, Phillips TW, Wakil W (2019). Biology and Control of the Khapra Beetle, *Trogoderma granarium*, a Major Quarantine Threat to Global Food Security. *Annual Review of Entomology,* 64: 131-148.

Bambini LJ, Onu I, Amatobi CI, Dike MC (2006). Effect of *Callosobruchus maculatus* infestation on nutritional loss on stored cowpea grains. *Archives of Phytopathology and Plant Protection,* 39 (2): 119-127.

Banks HJ (1994). Illustrated identification keys for *Trogoderma granarium, T. glabrum, T. inclusum* and *T. variabile* (Coleoptera: Dermestidae) and other *Trogoderma* associated with stored products. Division of Entomology, Commonwealth Scientific and Industrial Research Organisation (CSIRO), Canberra, Australia,

Borzoui E, Naseri B, Namin FR (2015). Different Diets Affecting Biology and Digestive Physiology of the Khapra Beetle, *Trogoderma granarium* Everts (Coleoptera: Dermestidae). *Journal of Stored Products Research,* 62: 1-7.

Bousquet Y (1990). Beetles associated with stored products in Canada. Canadian Government Publishing Centre, Ottawa. 189-192.

Chauhan YS, Ghaffar MA (2002). Solar heating of seeds - A low cost method to control bruchid (*Callosobruchus* spp.) attack during storage of pigeonpea. *Journal of Stored Products Research,* 38 (1): 87-91.

Cheng W, Lei J, Ahn JE, Wang Y (2013). CO_2 enhances effects of hypoxia on mortality, development, and gene expression in cowpea bruchid, *Callosobruchus maculatus. Journal of Insect Physiology,* 59 (11): 1160-1168.

Clement SL, Wightman JA, Hardie DC, Bailey P, Baker G, McDonald G (2000). Opportunities for integrated management of insect pests of grain legumes. In: Linking research and marketing opportunities for pulses in the 21st century, R. Knight, (ed.), 467-480. Kluwer Academic, Dordrecht, The Netherlands.

Costa VA, Guzzo EC, Lourençao AL, Garcia MA, Tavares CM, Vendramim JD (2007). Occurrence of *Dinarmus basalis* in *Callosobruchus analis* in stored soybean in Sao Paulo, Brazil. *Sci. Agric.,* (Piracicaba, Braz.), 64 (3): 301-302.

Credland PF (1994). Bioassays with bruchid beetles: problems and (some) solutions, pp. 509-516. In: E. Highley, E.J. Wright, H.J. Banks, and B.R. Champ [eds.], *Proceedings of the 6th International Working Conference on Stored-product Protection,* April 17-23, 1994.

Darfour B, Ocloo FCK, Wilson DD (2012). Effects of irradiation on the cowpea weevil (*Callosobruchus maculates* F.) and moisture sorption isotherm of cowpea seed (*Vigna unguiculata* L. Walp). *Arthropods,* 1 (1): 24-34.

Deeba F, Sarwar M, Khuhro RD (2006). Varietal Susceptibility of Mungbean Genotypes to Pulse Beetle, *Callosobruchus analis* (Fabricius) (Coleoptera: Bruchidae). *Pakistan Journal of Zoology,* 38 (4): 265- 268.

Deepthi N, Manjunatha M (2015). Comparative development of rice weevil, *Sitophilus oryzae* (L.) in stored split legumes. *Journal of Eco-friendly Agriculture,* 10 (2): 180-183.

Deng Y, Li L, Zhao Z (1999). Development and reproduction of *Callosobruchus chinensis* (Coleoptera: Bruchidae) on four legume plant seeds Proceedings of the 7th International Working Conference on Stored-Product Protection, 14-19 October 1998, Beijing, China. 1: 107-108.

Desroches P, Mandon N, Baehr JC, Huignard J (1997. Mediation of host-plant use by a glucoside in *Callosobruchus maculatus* F. (Coleoptera: Bruchidae). *Journal of Insect Physiology,* 43 (5): 439-446.

Devereau AD, Gudrups I, Appleby JH, Credland PF (2003). Automatic, rapid screening of seed resistance in cowpea, *Vigna unguiculata* (L.) Walpers, to the seed beetle *Callosobruchus maculatus* (F.) (Coleoptera: Bruchidae) using acoustic monitoring. *Journal of Stored Products Research,* 39: 117-129.

Edde PA (2012). A review of the biology and control of *Rhyzopertha dominica* (F.) the lesser grain borer. *Journal of Stored Products Research*, 48 (1): 1-18.

Erler F, Ceylan *F*, Erdemir T, Toker C (2009). Preliminary results on evaluation of chickpea, *Cicer arietinum*, genotypes for resistance to the pulse beetle, *Callosobruchus maculatus*. *Journal of Insect Science, 9 (58): 1-7.*

Fitzner MS, Hagstrum DW, Knauft DA, Buhr KL, McLaughlin JR (1985). Genotypic diversity in the suitability of cowpea (Rosales: Leguminosp) pods and seeds for cowpea weevil (Coleoptera: Bruchidae) oviposition and development. *Journal of Economic Entomology,* 78 (4): 806-810.

García-Lara S, Islas SO, Villers P (2013). Portable hermetic storage bag resistant to *Prostephanus truncatus, Rhyzopertha dominica,* and *Callosobruchus maculate*. *Journal of Stored Products Research,* 54: 23-25.

Grunwald S, Adam IV, Gurmai AM, Bauer L, Boll M, Wenzel U *(2013). The Red Flour Beetle Tribolium castaneum as a Model to Monitor Food Safety and Functionality. Advances in Biochemical Engineering/ Biotechnology, 135: 111-122.*

Hina HK, Sarwar M, Lohar MK (2015). Repellence Activity of Plant Oils against Red Flour Beetle *Tribolium castaneum* (Herbst) (Coleoptera: Tenebrionidae) in Wheat. *International Journal of Animal Biology,* 1 (3): 86-92.

Horng SB (1997). Larval competition and egg-laying decisions by the bean weevil, *Callosobruchus maculatus*. *Animal Behaviour,* 53 (1): 1-12.

Howe RW, Currie JE (1964). Some laboratory observations on the rates of development, mortality and oviposition of several species of Bruchidae breeding in stored pulses. *Bulletin of Entomological Research,* 55: 437-477.

Iqbal M, Shaheen FA, Naz F, Raja MU, Fiaz M, Nadeem M (2018). Management of Callosobruchus chinensis L. (Coleoptera: Bruchidae) in Stored Chickpea Grains by using Entomopathogenic Fungi. *Pakistan Journal of Agricultural Research*, 31 (4): 408-418.

Islam W, Kabir SMH (1995). Biological control potential of *Dinarmus basalis* (Rond.) (Hymenoptera: Pteromalidae), a larval-pupal ectoparasitoid of the pulse beetle, *Callosobruchus chinensis* (L.). *Crop Protection,* 14: 439-443.

Jackai LEN, Dauoust RA (1986). Insect pests of cowpea. *Annual Review of Entomology,* 31: 95-119.

Jia F, Toews MD, Campbell JF, Ramaswamy SB (2008). Survival and reproduction of lesser grain borer, *Rhyzopertha dominica* (F.) (Coleoptera: Bostrichidae) on flora associated with native habitats in Kansas. *Journal of Stored Products Research,* 44 (4): 366-372.

Kapila R, Agarwal HC (1995). Biology of an egg parasite of *Callosobruchus maculatus* (Fab.) (Coleoptera: Bruchidae). *Journal of Stored Products Research,* 31 (4): 335-341.

Kavallieratos N, Athanassiou C, Boukouvala M, Tsekos GT (2019). Influence of different non-grain commodities on the population growth of *Trogoderma granarium* Everts (Coleoptera: Dermestidae). *Journal of Stored Products Research,* 81: 31-39.

Kehinde FO, 1Dedeke GA, Popoola OI, Isibor PO (2019). Potential of Light Spectra as a Control of Cowpea Weevil, *Callosobruchus maculatus*, Activity. 8th International Biotechnology Conference, Exhibition and Workshop IOP Conf. Series: Earth and Environmental Science 210 0120081.

Khatodia S, Bhatotia K, Behl RK (2017). Helicoverpa Resistant Chickpea Plants: From Bt Toxins to Plant-Mediated RNAi. *Ekin Journal of Crop Breeding and Genetics,* 3 (1): 52-60.

Kutbay F, Varol I, Bayram M, Ozdemir A (2011). The effect of carbon dioxide at high pressure under different developmental stages of *Callosobruchus maculatus* (F.) hosting on chickpeas. *African Journal of Biotechnology,* 10 (11): 2053-2057.

Kyogoku D, Nishida T (2013). The mechanism of the fecundity reduction in Callosobruchus maculatus caused by Callosobruchus chinensis males. Population Ecology, 55 (1): 87-93.

Lin T (1981). Studies on the improved control measures of stored grain insects. *Journal of Agricultural Research of China,* 30 (1): 57-62.

Luthi C, Alvarez-Alfageme F, Ehlers JD, Higgins TJ, Romeis J (2013). Resistance of αAI-1 transgenic chickpea (*Cicer arietinum*) and cowpea (*Vigna unguiculata*) dry grains to bruchid beetles (Coleoptera: Chrysomelidae). *Bulletin of Entomological Research,* 103 (4): 373-381.

Miyatake T, Harano T, Okada K (2008). Negative relationship between ambient temperature and death-feigning intensity in adult Callosobruchus maculatus and Callosobruchus chinensis. Physiological Entomology, 33 (1): 83-88.

Moumouni D, Doumma A, Seyni IS (2015). Effect of solar drying on the biological parameters of the cowpea weevil, *Callosobruchus maculatus* Fab. (Coleoptera: Bruchinae), in Sahelian area. *Journal of Applied Biosciences, 84* (1): *7723*-7729.

Musa AM (2000). On-farm chickpea seed priming trials and demonstrations in the High Barind Tract of Bangladesh, 1999-2000. Research Report, People's Resource Oriented Voluntary Association, 204/ A Uposhahar, Sopura, Rajshahi, Bangladesh. 29 pp.

Neog P (2012). Studies on adult longevity of *Callosobruchus chinensis* (L.) developing in different pulses. *International Journal of Bio-resource and Stress Management*, 3 (3): 383-386.

Olufemi-Salami FK, Akinneye JO, Samson-Salami O (2017). Morphometric Analysis of the Developmental Stages and Insecticidal Efficacy of Three Botanical Oils against Adult *Callosobruchus analis. Journal of Horticultural Research,* 25 (2): 81-84.

Ouedraogo PA, Sou S, Sanon A, Monge JP, Huignard J, Tran B, Credland PF (1996). Influence of temperature and humidity on populations of *Callosobruchus maculatus* (Coleoptera: Bruchidae) and its parasitoid *Dinarmus basalis* (Pteromalidae) in two climatic zones of Burkina Faso. *Bulletin of Entomological Research,* 866: 695-702.

Parameshwarappa SG, Salimath PM, Upadhyaya HD, Patil SS, Kajjidoni ST (2012). Genetic variability studies in minicore collection of chickpea under different environments. *K. J. Agric. Sci.,* 25: 305-308.

Pai A, Bennett L, Yan G (2005). Female multiple mating for fertility assurance in red flour beetles (Tribolium castaneum). Canadian Journal of Zoology, 83 (7): 913-919.

Park S, Lee S, Hong KJ (2015). Review of the family Bostrichidae (Coleoptera) of Korea. *Journal of Asia-Pacific Biodiversity*, 8 (4): 298-304.

Patel NH, Condron BG, Zinn K (1994). Pair-rule expression patterns of *even-skipped* are found in both short- and long-germ beetles. *Nature,* 367: 429-434.

Povey SR, Holloway GJ (1992). The Effect of Energy Trade-Offs on Life History and Fitness in the Rice Weevil, *Sitophilus oryzae. Oikos,* 64 (3): 441-450.

Qi YT, Burkholder WE (1982). Sex pheromone biology and behavior of the cowpea weevil *Callosobruchus maculatus* (Coleoptera: Bruchidae). *J. Chem. Ecol.,* 8 (2): 527-34.

Rajput AA, Sarwar M, Ahmad N, Siddiqui QH, Tofique M (2003). Evaluation for resistance in some local and exotic chickpea genotypes against *Helicoverpa armigera* (Hubner). *Pakistan Journal of Biological Sciences,* 6 (18): 1612-1615.

Sanon A, Tricault Y, Huignard J, Ouedraogo PA (1999). Biological control of *Callosobruchus maculatus* (F.) (Bruchidae), during storage of cowpea. *Annales Societe Entomologique de France,* 35: 496-501.

Sarwar M (2004). Concept of integrated insect pests management. *Pakistan & Gulf Economists,* 23 (46 & 47): 39-41.

Sarwar M (2008). Laboratory studies on different wheat genotypes for their resistance against Khapra Beetle *Trogoderma granarium* Everts (Coleoptera: Dermestidae). *Pakistan Journal of Seed Technology,* 2 (11 & 12): 46-53.

Sarwar M (2009). Evaluating wheat varieties and genotypes for tolerance to feeding damage caused by *Tribolium castaneum* (Herbst) (Coleoptera: Tenebrionidae). *Pakistan Journal of Seed Technology,* 2 (13 & 14): 94-100.

Sarwar M (2010). Some possibilities on the effectiveness of plant powders as grain protectants against cowpea weevil, *Callosobruchus maculatus* (Fabricius) Walp (Coleoptera: Bruchidae) infestation in chickpea. *International Journal of Agronomy and Plant Production,* 1 (2): 45-50.

Sarwar M (2012). Assessment of resistance to the attack of bean beetle *Callosobruchus maculatus* (Fabricius) in chickpea genotypes on the basis of various parameters during storage. *Songklanakarin Journal of Science and Technology,* 34 (3): 287-291.

Sarwar M (2013 a). Studies on resistance discrepancy in seeds of local and exotic lentil (*Lens culinaris* Medikus) genotypes against *Callosobruchus analis* L. (Coleoptera: Bruchidae). *African Entomology,* 21 (2): 273-279.

Sarwar M (2013 b). Development and Boosting of Integrated Insect Pests Management in Stored Grains. *Journal of Agriculture and Allied Sciences,* 2 (4): 16-20.

Sarwar M (2015 a). Extermination of Insect Pests (Coleoptera: Bruchidae) and Damage of Stored Pulses by Different Methods in Market. *American Journal of Marketing Research,* 1 (3): 99-105.

Sarwar M (2015 b). Categorization of Some Advanced Local Wheat Lines against *Tribolium castaneum* (Herbst) (Coleoptera: Tenebrionidae). *International Journal of Life Science and Engineering,* 1 (3): 108-113.

Sarwar M (2015 c). Distinguishing and Controlling Insect Pests of Stored Foods for Improving Quality and Safety. *American Journal of Marketing Research,* 1 (3): 201-207.

Sarwar M. 2016. Indoor risks of pesticide uses are significantly linked to hazards of the family members. *Cogent Medicine,* 3: 1155373.

Sarwar M. (2019). Biology and Ecology of Some Predaceous and Herbivorous Mites Important from the Agricultural Perception. In: *Acarology, Mite Biology and Ecology* (ed. Hufnagel, L.). IntechOpen Ltd., London, UK. p. 29.

Sarwar M, Sattar M (2007). Varietals assessment of different wheat varieties for their resistance response to khapra beetle *Trogoderma granarium. Pakistan Journal of Seed Technology*, 1 (10): 1-7.

Sarwar M, Sattar M (2012). Appraisal of Different Plant Products against *Trogoderma granarium* Everts to Protect Stored Wheat- A Laboratory Comparison. *The Nucleus,* 49 (1): 65-69.

Sarwar M, Tofique M (2006). Resistance variability within gram seeds of different genotypes against the intrusion of cowpea weevil, *Callosobruchus analis* L. (Bruchidae: Coleoptera). *Pakistan Journal of Seed Technology,* 1 (8 & 9): 27-34.

Sarwar M, Ahmad N, Tofique M (2009). Host plant resistance relationships in chickpea (*Cicer arietinum* L.) against gram pod borer (*Helicoverpa armigera*) (Hubner). *Pakistan Journal of Botany,* 41 (6): 3047-3052.

Sarwar M, Ahmad N, Tofique M (2011). Identification of susceptible and tolerant gram (*Cicer arietinum* L.) genotypes against gram pod borer (*Helicoverpa armigera*) (Hubner). *Pakistan Journal of Botany,* 43 (2): 1265-1270.

Sarwar M, Ahmad N, Rajput AA, Tofique M (2004). Search for Varietal Resistance within Stored Wheat Genotypes against the Infestation of Red Flour Beetle, *Tribolium castaneum* Herbst (Coleoptera: Tenebrionidae). *Proce. 4th Inter. Congress of Entomological Sciences, University of Agric., Faisalabad*, September, 22-23. pp. 23-27.

Sarwar M, Ahmad N, Bux M, Tofique M (2012). Potential of Plant Materials for the Management of Cowpea Bruchid *Callosobruchus analis* (Coleoptera: Bruchidae) In Gram *Cicer arietinum* during Storage. *The Nucleus*, 49 (1): 61-64.

Sarwar M, Ahmad N, Sattar M, Tofique M (2005). Evaluating the Seed's Reaction of Certain Chickpea Genotypes against the Action of Pulse Beetle, *Callosobruchus analis* L. (Bruchidae: Coleoptera). *Pakistan Journal of Seed Technology,* 1 (6): 14-21.

Sarwar M, Ahmad N, Siddiqui QH, Mohammad R, Sattar M, Tofique M (2003). Varietal resistance in stored mungbean against the infestation of pulse beetle, *Callosobruchus analis* (Fabricius) (Coleoptera: Bruchidae). *Pakistan Journal of Zoology,* 35 (4): 301-305.

Shaheen FA, Khaliq A, Aslam M (2006). Resistance of chickpea (*Cicer arietinum* L.) cultivars against pulse beetle. *Pakistan Journal of Botany, 38* (4): *1237-1244.*

Sharmila R, Roy S (1994). Oviposition preference and development of *C. maculatus* (F.) on seeds of promising dry land legumes. *Agricultural Science Digest,* 14: 177-180.

Singh AK, Pandey RK (2016). Eco-friendly management of pulse beetle (*Callosobruchus chinensis* L.) in stored chickpea (*Cicer arietinum*). *Asian Journal of Environmental Science,* 11: 13-18.

Singh AK, Kameshwar K, Santosh K (2018). Taxonomic redescription of the rice weevil (*Sitophilus oryzae*). *Journal of Pharmacognosy and Phytochemistry,* 7 (S1): 36-39.

Somnath D, Choudhury, Chakraborty K (2014). Study on both the life cycle and morphometrics of *Sitophilus oryzae* on rice cultivar Sampa mashuri in laboratory condition. *Journal of Applied Science and Research*, 2 (6): *22-28.*

Taylor TA (1981). Distribution, Ecology and Importance of Bruchids Attacking Grain Legumes and Pulses in Africa. In: The Ecology of Bruchids Attacking Legumes

(Pulses) Labeyrie, V. (Ed.). *Proceedings of the International Symposium held at Tours (France),* April 16-19, 1980. pp.199-203.

Tuda M, Chou LY, Niyomdham C, Buranapanichpan S, Tateishi Y (2005). Ecological factors associated with pest status in *Callosobruchus* (Coleoptera: Bruchidae): high host specificity of non-pests to Cajaninae (Fabaceae). *Journal of Stored Products Research,* 4 :31-45.

Tuda M, Ronn J, Buranapanichpan S, Wasano N, Arnqvist G (2006). Evolutionary diversification of the bean beetle genus *Callosobruchus* (Coleoptera: Bruchidae): traits associated with stored-product pest status. *Molecular Ecology,* 15 (12): 3541-3551.

Ukeh DA, Sambo MJ, Isah MD (2013). Response of Potential Stored Grain Insect Pests to BFL 225 Multi-Attractant Lure in Commercial Warehouses. *Journal of Biology, Agriculture and Healthcare*, 3 (8): 87-91.

Van Emden HF, Ball, SL, Rao MR (1988). Pest disease and weed problems in pea lentil and faba bean and chickpea. p. 519-534. In: *World Crops: Cool Season Food Legumes*, ISBN 90-247-3641-2. Kluwer Academic Publishers, Dordrecht, The Netherlands.

Velasquez LPG, Faroni LRDA., Pimentel MAG, Heleno FF, Prates LHF (2017). Behavioral and physiological responses induced by ozone in five Brazilian populations of *Rhyzopertha dominica. Journal of Stored Products Research,* 72 (1): 111-116.

Via S (1999). Cannibalism facilitates the use of a novel environment in the flour beetle, *Tribolium castaneum. Heredity,* 82: 267-275.

Wade MJ (1984). The population biology of flour beetles, *Tribolium castaneum*, after interdemic selection for increased and decreased population growth rate. *Res. Popul. Ecol.,* 26: 353-366.

Wakil W, Riasat T, Ghazanfar MU, Kwon YJ, Shaheen FA (2011). Aptness of *Beauveria bassiana* and enhanced diatomaceous earth (DEBBM) for control of *Rhyzopertha dominica* F. *Entomol. Res.,* 41: 233-241.

Walter VE (1990). Stored product pests. In: Handbook of Pest Control, Story K, Moreland D. (eds.). Franzak & Foster Co., Cleveland, OH. pp. 526-529.

Waterworth PD (1986). Internal seed infesting insects. Part I e II. Beltsville: USDA. 136 p.

Weigand S (1990). Insect pests of chickpea in the Mediterranean area and possibilities for resistance. In: Saxena M.C., Cubero J. I., Wery J. (eds.). Present status and future prospects of chickpea crop production and improvement in the Mediterranean countries. Zaragoza: CIHEAM, p. 73-76.

Xu Z, Weng L, He J (2008). A New Species of Cephalonomia Westwood (Hymenoptera: Bethylidae) Parasitizing *Rhizopertha dominica* (F.) (Coleoptera: Bostrychidae. *Proceedings of the Entomological Society of Washington,* 110 (2): 477-480.

Yanagi S, Saeki Y, Tuda M (2013). Adaptive Egg Size Plasticity for Larval Competition and its Limits in the Seed Beetle *Callosobruchus chinensis*. *Entomologia Experimentalis et Applicata, 148 (2): 182-187.*

In: Handbook of Chickpeas
Editors: Albert T. Lund and Noah D. Schultz
ISBN: 978-1-53616-374-2

Chapter 12

MODULATION OF DNA DAMAGE IN HUMAN PROSTATE CANCER CELL LINES BY A CHICKPEA PROTEASE INHIBITOR CONCENTRATE

Gianfranco Bartolini[1,*], Chris I. R. Gill[1], Richard Owusu-Apenten[1], Ian R. Rowland[2] and Pamela J. Magee[1]
[1]Nutrition Innovation Centre for Food and Health,
University of Ulster, Coleraine, Northern Ireland, UK
[2]Department of Food and Nutritional Sciences, University of Reading, Reading, UK

ABSTRACT

Epidemiological studies support a role for diets rich in legumes and pulses in prostate cancer prevention. A protease inhibitor from soybean (Bowman-Birk Inhibitor; BBI) has been shown to exert anti-cancer properties in several *in vitro* and *in vivo* carcinogenesis model systems, including those of prostate cancer.

The anti-cancer effects of protease inhibitors from other legumes have been poorly investigated. BBI and other protease inhibitor concentrates (PICs) have previously been shown to decrease oxidative damage by inhibiting free radical production. This mechanism may account for BBI's potent anti-inflammatory effects and consequently BBI's anti-cancer properties.

The ability of PICs to modulate oxidatively induced DNA damage is not fully elucidated and the aim of the current study was to determine if a PIC from chickpea exerts anti-genotoxic effects on LNCaP and DU 145 human prostate cancer cells *in vitro*.

Cells were treated with PIC from chickpea (0.05, 0.1, 0.3, 1 and 3 mg/ml) for 24 hours and the Comet assay was used to assess the effect on DNA strand breaks induced

* Corresponding Author's Email: bartolini_gianfranco@yahoo.com.

by a H_2O_2 challenge (75 μM, 5 min, 4°C). Purified BBI was used as a reference. The genotoxicity of PIC was assessed using the Comet assay following exposure to PBS. The toxicity of chickpea PIC was assessed in LNCaP cells using the MTT assay.

Chickpea PIC (1, 3mg/ml) reduced H_2O_2-induced DNA damage in LNCaP cells by up to 15.4% ($P < 0.01$). Similar effects were apparent with BBI (0.1, 0.3 mg/ml).

Of note, the tested PIC showed no genotoxicity to LNCaP cells. Neither the chickpea PIC nor BBI exerted any significant anti-genotoxic effects in hydrogen peroxide challenged DU 145 cells, while no genotoxic activity was induced.

Pre-incubation with PIC (3 mg/ml) also significantly modulated the level of expression of genes linked with detoxifying or DNA repair properties.

In conclusion, chickpea PIC exerted anti-genotoxic effects on LNCaP cells, supporting previous findings that BBI and protease inhibitors have a protective effect on oxidative damage. Chickpea PIC may therefore play a role in prostate cancer prevention, however, further research is needed to fully understand the molecular mechanisms involved.

INTRODUCTION

Prostate cancer accounts for 26% of all new cases of cancer in males in the United Kingdom (2016), being the most commonly diagnosed malignancy in men, with 47,600 new cases reported in 2016 (Cancer Research UK 2019). In the United States, 174,650 new cases of prostate cancer are estimated for 2019 (American Cancer Society 2019).

The etiology of prostate cancer is not fully elucidated, however, given the high impact of this disease on the health systems of the westernised world, much effort has been invested in the identification of behavioural factors which could confer protection against it.

Epidemiological evidence highlights the clear link between the intake of legumes and pulses and a lower risk of prostate cancer incidence (Mathers 2002, Schuurman et al. 1998, Mills et al. 1989, Severson et al. 1989). Multiple mechanisms have been proposed in relation to how leguminous food sources can modulate carcinogenesis in prostatic tissue, however a wealth of research supports the role of protease inhibitors of the Bowman-Birk family as potential chemopreventive agents.

The Bowman-Birk Inhibitor (BBI), a protease inhibitor isolated from the soybean, has been shown to exert anti-cancer properties in several *in vitro* and *in vivo* cancer model systems, including those of prostate cancer (Troll, Wiesner 1983, Kennedy, Wan 2002, Tang et al. 2009, McCormick et al. 2007, Magee et al. 2012, Armstrong et al. 2013, Souza et al. 2014).

A soybean extract enriched in BBI (BBIC) has been developed as a potential anti-cancer agent for human use and has achieved Investigational New Drug status by the US Food and Drug Administration in 1992. BBIC has been successfully used in clinical trials for the treatment of benign prostatic hyperplasia (Malkowicz et al. 2001), oral leukoplakia (Armstrong et al. 2000a, Armstrong et al. 2000b) and ulcerative colitis (Lichtenstein et al. 2008).

The precise mechanism/s through which BBI exerts its anti-cancer action has not been fully elucidated to date, however its protective effects may be due to its potent anti-inflammatory activity (Kennedy 1998) and the ability to inhibit the proteolytic activity of the 26S proteasome (Chen et al. 2005) in cancer cells, the latter possibly being linked to a beneficial modulation of molecules involved in the carcinogenic pathway.

Prostate cancer occurs with higher frequency in the elderly male (Quinn, Babb 2002, Baade, Youlden & Krnjacki 2009), presumably due to the accumulation of radical-induced oxidative DNA damage (Malins et al. 2001). Reactive oxygen species (ROS), which can be produced by endogenous (e.g., mitochondria, peroxisomes, cytochrome P450, inflammatory cells) (Klaunig, Kamendulis 2004) and exogenous sources (e.g., UV light, inflammatory cytokines, pathogens), can cause a situation of oxidative stress when not efficiently counterbalanced by the cellular antioxidant defenses, with consequent damage to the genetic material and initiation of the carcinogenic process (Klaunig, Kamendulis & Hocevar 2010). An important property of BBI is the ability to prevent the release of oxygen free radicals in human polymorphonuclear cells and HL-60 human lymphocytic cells, thereby decreasing the likelihood of oxidative DNA damage (Frenkel et al. 1987).

BBI is also known to have anti-oxidant (Huang et al. 2010) and anti-inflammatory properties (Kennedy 1998).

The potential of BBI-type protease inhibitors as agents capable of inhibiting induced DNA damage in human prostate cancer cells has not been investigated to date.

Diets rich in legumes and pulses, such as those consumed in Asian countries of the Pacific basin, are rich in protease inhibitors of the Bowman-Birk family and linked with a lower incidence of prostate cancer (Mathers 2002, Schuurman et al. 1998, Mills et al. 1989, Severson et al. 1989), thus it is plausible that dietary protease inhibitor concentrates (PICs) play a key role in conferring protection against prostate cancer.

The aim of this study was to investigate the effects of a PIC from chickpea on DNA damage in prostate cancer cells and its ability to protect against DNA damage induced by H_2O_2, an important biomarker of cancer initiation, by means of the Comet assay.

The cell lines selected for the study, LNCaP and DU 145, constitute a valuable model of different stages of prostate cancer, an earlier androgen sensitive phase and a later androgen irresponsive stage, respectively.

The MTT assay was used to assess cytotoxicity in PIC-treated LNCaP cells at the tested PIC concentrations. The effects of chickpea PIC on the expression of anti-oxidative and DNA repair genes were investigated in LNCaP cells using real-time PCR.

MATERIALS AND METHODS

Tissue Culture

The prostate cancer cell lines LNCaP and DU 145 were obtained from ATCC (Rockville, MD, USA). The LNCaP cell line was derived from a left supraclavicular lymph node metastasis of a 50 year old patient with prostate cancer (Horoszewicz et al. 1983). LNCaP cells are androgen dependent and express a mutated androgen receptor (AR), in addition to being oestrogen sensitive via oestrogen receptors (ERs) α and β (Webber, Bello & Quader 1997). The LNCaP cell line constitutes a model of the early stages of prostate carcinogenesis.

The DU 145 cell line was derived from a brain metastasis of prostate carcinoma (Stone et al. 1978). DU 145 cells are androgen receptor negative (AR-) and are a model of androgen independent prostate cancer, where the AR gene is not expressed due to the methylation of its promoter region (Chlenski et al. 2001).

Cell culture media and supplements were purchased from Gibco (Paisley, UK) unless otherwise stated. LNCaP cells were cultured in Roux flasks as monolayers in RPMI 1640 medium supplemented with 2.0mM L-glutamine, 10.0mM Hepes (Sigma-Aldrich, Dorset, UK), 1% penicillin-streptomycin, 12.5 mM glucose in genomic grade water (sterilized with a 0.22 μm filter) and 10% foetal bovine serum. DU 145 cells were cultured in Roux flasks as monolayers in RPMI 1640 medium supplemented with 1% penicillin-streptomycin and 10% foetal bovine serum. Cells were incubated at 37°C with 5% CO_2 and 95% filtered air and harvested when a confluency < 75% was reached.

For cell subculturing the medium was removed from the flasks and the cells were rinsed with PBS for 2 min (LNCaP cells were not washed with PBS to avoid cell dislodgement due to the weaker attachment to the growth surface of the flask), and re-suspended by the addition of trypsin (0.25% trypsin-EDTA) at 37°C for 3-5 min. Cells were centrifuged at 1250 rpm for 3 min and the cell pellets were re-suspended in their specific growth medium.

Cell viability was assessed by the trypan blue exclusion assay using a Neubauer hemocytometer (50 μl of cell suspension + 50 μl of tryplan blue dye) (Pool-Zobel et al. 1994) and a cell count was performed under an inverted microscope using a 20x objective.

Preparation of Chickpea Protease Inhibitor Concentrate

The protease inhibitor concentrate (PIC) tested in this study was extracted from chickpeas purchased from "Neal's Yard" (Burton upon Trent, UK) and that had been grown and packed to organic standard.

PIC was prepared from acetone defatted flour and subsequently freeze-dried, according to a cold temperature method developed in house based on a modification of the one described by Kassell (1970). To prepare the legume flour, chickpeas were ground and the powder was dispensed in a beaker and covered with an approximately equal volume of acetone GPR. The flour was stirred for one hour and after two hours the acetone phase was removed. The acetone extraction was repeated three times and the flour was eventually air dried. Subsequently, 600 ml of deionized water was dispensed in a beaker and the pH was adjusted to 5.0 (a pH range of 5.0–5.5 was considered acceptable). One hundred grams of chickpea flour was weighed out and added to the previous beaker, and stirred for at least 60 minutes, so as to allow the protein fraction, having a low solubility, to precipitate. The pH was monitored throughout the process, and, if necessary, adjusted to the required value (i.e., 5.0–5.5) with HCl or NaOH. The flour / water mixture was then centrifuged (3000 rpm, 10°C, 10 minutes) and the aqueous soluble fraction was paper filtered and three times the volume of acetone was added to allow the precipitation of the Bowman-Birk Inhibitor and left overnight. The acetone was then decanted, leaving a small amount covering the powder and the residual powder / acetone was paper filtered to virtually dry. The filter was removed and put in a beaker with 100 ml of deionized water and stirred to re-suspend the powder. Following the filter removal, the powder was stirred for 30 minutes and the mixture was then dispensed in 50 ml centrifuge tubes. The tubes were centrifuged (8000 rpm, 4°C, 10 minutes) and stored at -70°C for 24 hours, and then freeze-dried. The aliquots from several preparations were mixed to make a batch of PIC that was stored at -20°C.

For genotoxicity and cytotoxicity testing (Comet assay and the MTT assay) dilutions (0, 0.05, 0.1, 0.3, 1, 3 mg/ml) were obtained from a 3 mg/ml stock solution prepared by dissolving the required amount of PIC powder in the appropriate growth medium, previously warmed in a water bath at 37°C. Briefly, the powder was weighed out and added to the appropriate amount of medium in a 25 ml universal tube. The tube was vortexed for 2 min and then the medium with PIC was transferred into a beaker and stirred for 2 hours. The mixture was subsequently centrifuged (8000 rpm, 4°C, 30 min) to remove starches and the supernatant syringe filtered (0.22 μm).

Commercially available BBI (Sigma-Aldrich) was used as a control at 0.3 and 0.1 mg/ml. A 0.3 mg/ml stock solution was prepared by dissolving the powder in previously warmed growth medium by vortexing and the 0.1 mg/ml solution was then prepared and filtered with a 0.22 μm filter.

Analysis of PIC Protein Concentration

The net protein and peptide content of PIC was determined by using the manual Ninhydrin assay (Hirst, 1967; Brewer et al. 1995). This method is based on the hydrolysis of protein and peptides into the constituent amino acids, which are then neutralized to pH

4–7 and reacted with ninhydrin (Sigma-Aldrich, Dorset, UK). A colorimeter is used to detect the colour formed at 570 ηm (E = 20,000 M-1 cm-1).

To carry out the analysis, 150 µl of a 1 mg/ml PIC solution in deionized water was pipetted into propylene tubes and oven dried (110°C, 2 hours). Then 150 µl of 10.0 M NaOH was added and the tubes were autoclaved at 121°C (15 psi) for 20 minutes. The samples were neutralized with concentrated glacial acetic acid (250 µl) until the pH was higher than 4 and transferred to glass tubes. Two ml of 15 mM PBS was added to the tubes followed by 0.5 ml of ninhydrin solution. The mixture was boiled for 15 minutes and allowed to cool, and 2.5 ml of ethanol (50% v/v) was added. The tubes were vortexed, and absorbance readings were recorded at 570 ηm. Isoleucine (1.5 mM; 50 µl) was prepared as a standard and treated as per the samples. The analysis of PIC protein concentration was determined as the average of at least three independent experiments carried out in triplicate.

Determination of PIC Chymotrypsin Inhibitor Activity (CIA)

The activity of PIC was measured as chymotrypsin inhibitor activity (CIA), where one CIA unit is defined as the amount of PIC needed for the inhibition of one unit of chymotrypsin according to the Hummel Assay (Rafiq, Bailey 1998); a unit of chymotrypsin is the amount of enzyme required for the hydrolysis of 10^{-6} moles of N-benzoyl tyrosine p-nitroanilide (BTNA) per minute.

The activity of chymotrypsin was determined using 1 mg/ml solution of commercial chymotrypsin (Sigma-Aldrich, Dorset, UK). Differing amounts of enzyme (10, 25, 50, 100 µl) were aliquoted in 1.5 ml cuvettes with 12.5 µl of BTNA (Sigma-Aldrich) and 900 µl of reaction buffer (50 mM Tris, 20 mM CaC12 pH 8.2). The mixture was incubated at constant temperature and absorbance was read at 410 ηm using a spectrophotometer. The absorbance was recorded again after ten minutes.

To determine the CIA, PIC was dissolved in reaction buffer by vortexing at a concentration of 2 mg/ml. Different aliquots of the PIC solution (0, 10, 25, 50 µl) were then added to Eppendorf tubes with 18 µl of 1 mg/ml chymotrypsin in 1 mM HCl, and the final volume was brought to 100 µl with reaction buffer. The Eppendorf tubes were left at room temperature for 10 minutes, then the solution PIC / chymotrypsin was aliquoted in cuvettes with 900 µl of reaction buffer and 12.5 µl of 10 mM BTNA in DMSO. The absorbance was read at 410 ηm with a spectrophotometer and a further reading was carried out after 10 minutes. PIC chymotrypsin activity was determined as the average of at least two independent experiments carried out in triplicate.

Comet Assay

The Comet assay exploits a single-cell gel electrophoresis (SCGE) and is a sensitive and rapid technique for measuring and analysing DNA damage leading to strand breakage in individual eukaryotic cells.

The Comet assay is widely used for testing the genotoxic potential of chemicals and for cancer research. The assay is also used for assessing potential anti-cancer properties of dietary compounds / extracts in terms of reduction of DNA damage induced by an oxidant such as hydrogen peroxide (Gill et al. 2005). The damaged DNA (single-strand or double-strand breaks) loses its supercoiling and is free to migrate towards the anode during the electrophoretic process (Collins 2004), resulting in structures resembling comets, with the "heads" containing intact DNA. The cells are observed using a fluorescence microscope and the amount of DNA displaced from the comet "head" is directly proportional to the extent of DNA damage.

The assay was carried out as described by Gill et al. (2005). Briefly, 3.8×10^5 DU145 cells and 1.5×10^6 LNCaP cells were seeded in 25 cm^2 flasks. Cells were incubated for 24 hours before pre-treatment with PIC. LNCaP and DU 145 cells were incubated with selected concentrations of PIC (0, 0.05, 0.1, 0.3, 1, 3 mg/ml for chickpea PIC; 0.1, 0.3 mg/ml for BBI) for 24 hours before being harvested. A cell suspension of 1.5×10^5 cells/ml was obtained and 450 µl was dispensed in 1.5 ml Eppendorf tubes and was challenged with 50 µl of H_2O_2 (Sigma-Aldrich) (75 µM for LNcaP cells and 200 µM for DU 145 cells) or PBS, for 5 min on ice. The Eppendorf tubes were then centrifuged for 5 min at 1250 rpm at 4°C and the supernatant removed and replaced with warm (40°C) low melting point agar (0.85% in PBS). Eighty-five µl of the mixture of low melting agar/ cell suspension was dispensed onto chilled normal melting agar gels (previously prepared on frosted slides) and covered with a coverslip. Gels were allowed to set in the fridge for approximately 15 min. The coverslips were removed and the slides were immersed for 1 hour at 4°C in lysis buffer (2.5 M NaCl, 100 mM Na_2EDTA, 10mM TRIS, pH 10.0) with 1% Triton-X added immediately prior to use. The slides were then transferred to the electrophoresis tank with the electrophoresis buffer (300 mM NaOH/ 1mM EDTA) for 20 min, to allow the DNA to unwind before running the electrophoresis at 26 V, 300 mA, 4°C, 20 min. After electrophoresis, the slides were collected and washed (3 times, 3 min each) with neutralising buffer (0.4 M TRIS, pH 7.5). Each gel was then stained with 20 µl of ethidium bromide (20 µg/ml) and coverslips added. The gels were visualised with a Nikon Eclipse 600 epi-fluorescence microscope (400x magnification) and analysed by means of Komet 3.0 image analysis software (Kinetic Imaging Ltd, Liverpool, UK).

It is important to note that the Comet assay is prone to subjectivity due to the manual input of software gating. One hundred cells were analysed for each gel and the mean DNA damage of each sample was calculated from 3 gels. The mean values from at least 3 independent experiments were used in the statistical analysis.

MTT Viability Assay

The MTT [3-(4,5-dimethylthiazol-2-yl)-2,5-diphenyltetrazolium bromide] assay is a method that detects cell viability as a response to external factors and is based on the theoretical principle of the ability of a dehydrogenase enzyme from the mitochondria of viable cells to reduce the tetrazolium rings of the pale yellow MTT, with the subsequent formation of purple formazan crystals.

The ability of cells to cleave MTT is indicative of the degree of their mitochondrial / cellular respiration. The formazan crystals accumulate in the cell, due to their inability to pass through the cell membrane. They can be solubilized by a detergent and quantified by spectrophotometric analysis (Mosmann 1983).

Fifty thousand cells in 100 µl of growth medium were seeded in each well of a 96-well plate. The cells were incubated for 24 hours to obtain a firm monolayer, then medium was tapped out onto tissue paper and the PIC treatment (0.05 – 3 mg/ml) was added (100 µl/well). Eight replicates of each treatment were performed. Growth medium was used as a negative control and commercially available soybean BBI was used as a reference. The cells were exposed to the treatment for 24 hours. At the end of the treatment period the cells were washed with PBS and 15 µl of MTT dye solution (0.5mg/ml of MTT powder, Sigma-Aldrich) in PBS and sterilized with a 0.22 µm filter was added to each well and the plates were incubated at 37°C in a humidified incubator (95% air, 5% CO_2) for 4 hours. One hundred µl of solubilizing solution (89.2% isopropanol (Sigma-Aldrich), 10% Triton-X (Sigma-Aldrich), 0.8% HCl) was then added to wells and the plates were placed back in the incubator overnight. The plates were eventually read at 570 ηm, reference 650 ηm with a VERSAmax microplate reader and results analysed with the SoftMax Pro 5.4 software. The results were presented as % viability normalized to the negative control. Each experiment was repeated on at least 3 independent occasions.

Quantitative Real-Time PCR

LNCaP cells were seeded in 25cm^2 tissue culture flasks (10^6 cells per flask) and incubated for 24 hours at 37°C, 5% CO_2, before treatment. Growth medium was then removed and the cells were treated with chickpea PIC (0 and 3 mg/ml, prepared as described for the Comet and the MTT assays) for 96 hours, with the treatment refreshed after 48 hours. At the end of the treatment period cells were harvested and counted by means of the trypan blue exclusion test using a Neubauer hemocytometer, as described previously. RNA was isolated from cells following homogenization with QIAshredder homogenizers using an RNeasy Mini Kit and treated with DNase I (Qiagen, Crawley, Germany) according to the manufacturer's guidelines. cDNA was synthesised using the

Transcriptor First Strand cDNA Synthesis Kit (Roche Diagnostics, Penzberg, Germany). Briefly, 2.5 μg total RNA was mixed with PCR-grade water and 2 μl Random Hexamer Primer to give a final volume of 13 μl. To the tube containing the template-primer mixture was then added a mix containing 4 μl Transcriptor Reverse Transcriptase Reaction Buffer, 0.5 μl Protector RNase Inhibitor, 2 μl Deoxynucleotide Mix, 0.5 μl Transcriptor Reverse Transcriptase. The resulting 20 μl mix was subjected to 25°C for 10 minutes, followed by 55°C for 30 minutes, and 85°C for 5 minutes. The reaction tubes were stored at -20°C prior to use. Gene expression quantification was carried out using RealTime ready Custom Panel Plates (Roche Diagnostics) containing function tested assays for genes with anti-oxidant or DNA repair function (see Table 1), and beta-actin (ACTB), glyceraldeide-3-phosphate dehydrogenase (GAPDH) and hypoxanthine phosphoribosyltransferase 1 (HPRT1) as reference genes.

Table 1. Genes analysed by means of Real-Time PCR and function of proteins encoded

SOD1	responsible for destroying harmful free superoxide radicals in the body, converting them to molecular oxygen and hydrogen peroxide
SOD2	encodes a mitochondrial protein that binds to the superoxide byproducts of oxidative phosphorylation and converts them to hydrogen peroxide and diatomic oxygen
GSTA1	detoxification of electrophilic compounds, including carcinogens, therapeutic drugs, environmental toxins and products of oxidative stress, by conjugation with glutathione
GSTA4	involved in cellular defense against toxic, carcinogenic, and pharmacologically active electrophilic compounds
GPX1	encodes a member of the glutathione peroxidase family functioning in the detoxification of hydrogen peroxide
GPX4	reduction of hydrogen peroxide, organic hydroperoxide, and lipid peroxides
CAT	encodes catalase, a key antioxidant enzyme converting the reactive oxygen species hydrogen peroxide to water and oxygen
ATM	encodes a protein functioning as cell cycle checkpoint kinase important for cell response to DNA damage and for genome stability
ATR	encodes a protein required for cell cycle arrest and DNA damage repair in response to DNA damage
AC103686.3	encodes the catalytic subunit of the DNA-dependent protein kinase (DNA-PK) with functions in DNA double strand break repair and recombination

Genes tested were SOD1, SOD2, GSTA1, GSTA4, GPX1, GPX4, CAT, ATM, ATR, AC103686.3, and were chosen on the basis of their specific role in protection against DNA damage and in the etiology of prostate cancer. Real-time qPCR was carried out in triplicate by means of the LightCycler® 480 instrument (Roche Diagnostics) using 10 μl LightCycler® 480 Probes Master (Roche Diagnostics), 50 ηg cDNA and genomic-grade water for a final volume of 20 μl per well. The plates, covered with an optical adhesive cover, were exposed to the following thermal cycler conditions: +95°C for 10 minutes, 45 cycles of +95°C for 10 seconds, +60°C for 30 seconds, +72°C for 1 second and a final cooling step at +40°C for 30 seconds. The relative quantification of target gene transcripts in comparison to a reference gene transcript was calculated using the mathematical model developed by Pfaffl (2001).

Statistical Analysis

Differences between means were evaluated using ANOVA with LSD post-hoc analysis or the non-parametric Kruskal-Wallis and Mann-Whitney U tests for not normally distributed datasets. SPSS statistical package was used to perform all statistical analysis.

RESULTS

Characteristics of Chickpea PIC

The crude protein (Cp) content for chickpea PIC was 25.2 ± 1.9%. CIA activity was 154.3 ± 29.1 CIA/mg. The chickpea PIC specific activity was 611.4 CIA/mg protein. Values are expressed as means ± SEM.

Comet Assay

Genotoxic effects were not observed following pre-incubation (24 hours) of LNCaP cells with chickpea PIC at any of the concentrations tested (Figure 1). Significant anti-genotoxic effects were exerted by chickpea PIC after 24 hours pre-incubation in LNCaP cells at 1 and 3 mg/ml. (Figure 1) with H_2O_2-induced damage significantly reduced by 15.4 and 14.1%, respectively ($P < 0.01$).

The anti-genotoxic effects of purified BBI in LNCaP cells are shown in Figure 2, where equal concentrations of chickpea PIC are also represented as a comparison. Pure BBI proved to have more potent effects than chickpea PIC compared to the control, with a % reduction of DNA damage of 20% and 15.6% ($P < 0.05$) for 0.1 and 0.3 mg/ml, respectively, although there was not a statistically significant difference between the effect of BBI and chickpea PIC. Pure BBI did not exert genotoxic effects in LNCaP cells (data not shown).

The effects of chickpea PIC on the DU 145 cell line are represented in Figure 3. Chickpea PIC exerted no significant anti-genotoxic effects in hydrogen peroxide challenged DU 145 cells.

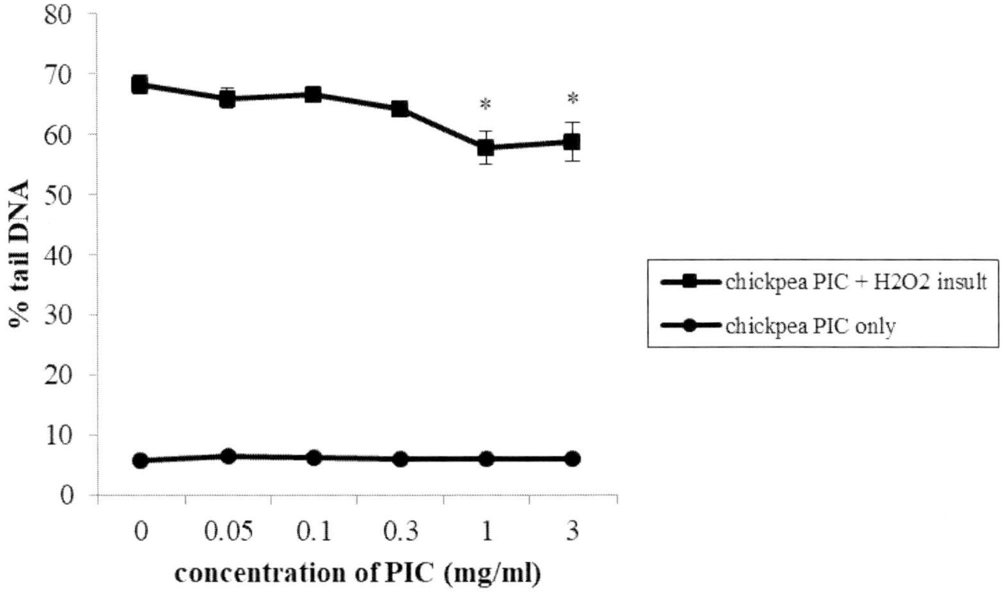

Figure 1. Genotoxic and anti-genotoxic effects of chickpea PIC (24-h incubation) in LNCaP cells. $n = 3$, 75 μM hydrogen peroxide challenge. *$P < 0.01$ vs. control; ANOVA, LSD post-hoc test. Values represent the mean ± SEM.

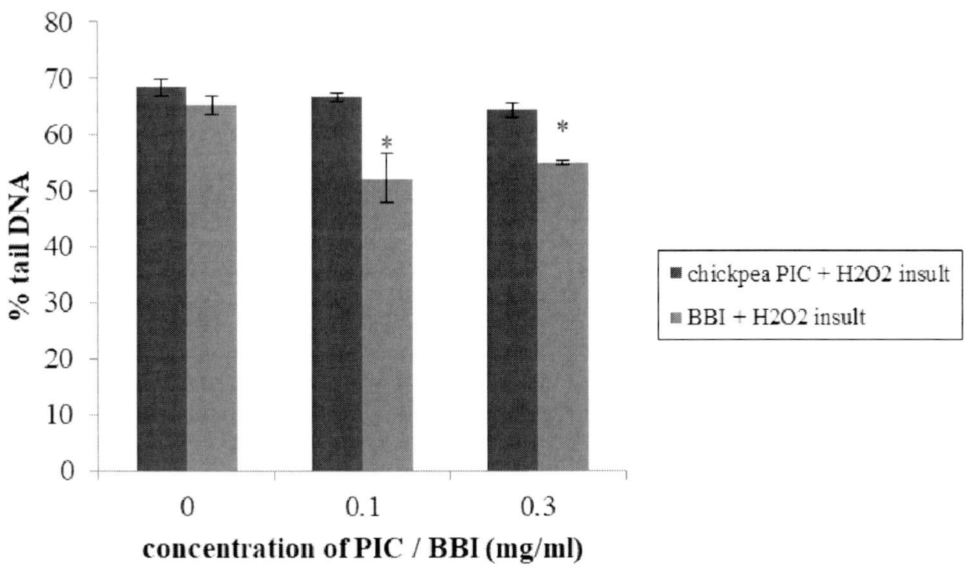

Figure 2. Anti-genotoxic effects of purified BBI and chickpea PIC (24-h incubation) in LNCaP cells. $n = 3$, 75 μM hydrogen peroxide challenge. *$P < 0.05$ vs. control; ANOVA, LSD post-hoc test . Values represent the mean ± SEM.

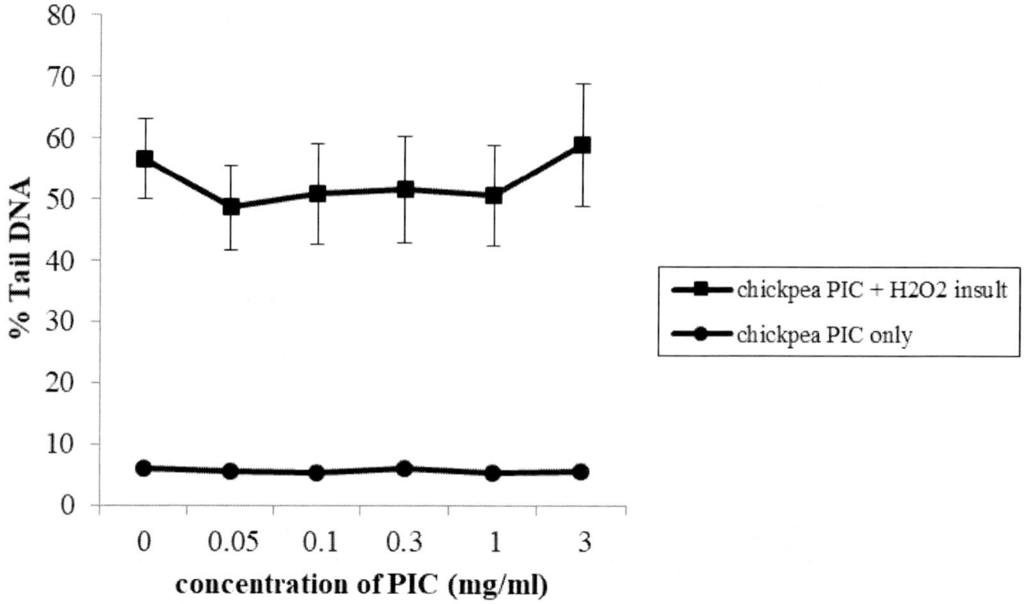

Figure 3. Genotoxic and anti-genotoxic effects of chickpea PIC (24-h incubation) in DU 145 cells. $n = 3$, 200 μM hydrogen peroxide challenge. Values represent the mean ± SEM.

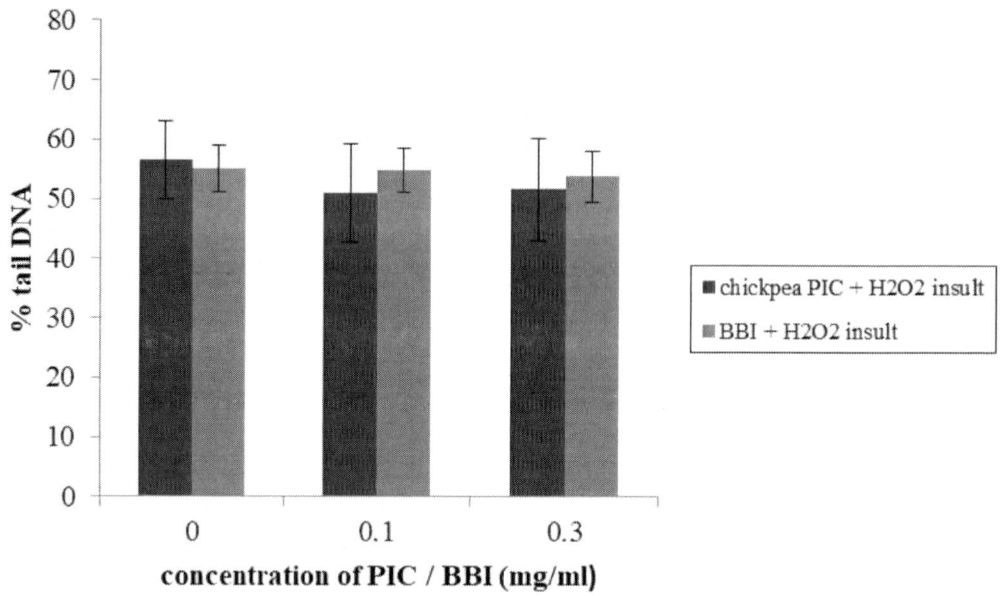

Figure 4. Anti-genotoxic effects of purified BBI and chickpea PIC (24-h incubation) in DU 145 cells. $n = 3$, 200 μM hydrogen peroxide challenge. Values represent the mean ± SEM.

In contrast to LNCaP cells, the 2 selected concentrations of purified BBI did not exert any anti-genotoxic effects on DU 145 cells, as shown in Figure 4, where equal concentrations of chickpea PIC are also represented as a comparison.

Pre-incubation of DU145 cells with chickpea PIC and BBI did not induce any genotoxic activity (Figure 3, data not shown for BBI).

MTT Viability Assay

The effects of chickpea PIC and BBI on the proliferation of LNCaP cells are shown in Figure 5. Exposure of LNCaP cells to PIC or BBI for 24 hours did not significantly inhibit cell viability at any of the concentrations tested.

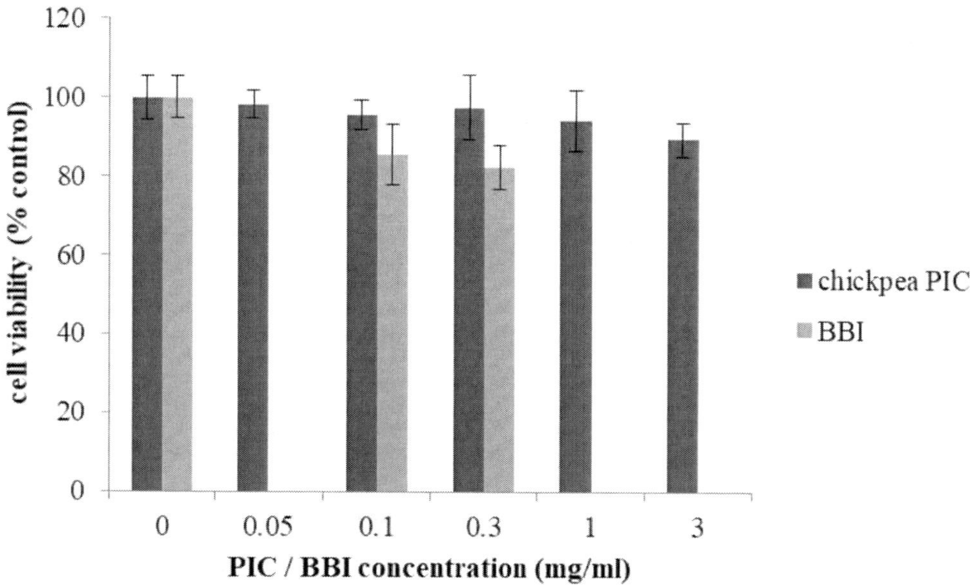

Figure 5. Effect of chickpea PIC and BBI (24-h incubation) on the viability of LNCaP prostate cancer cells as assessed by means of the MTT assay. $n = 3$, values represent the mean ± SEM normalized to untreated cells (control).

Quantitative Real-Time PCR

To investigate wether the anti-genotoxic effects exerted by chickpea PIC were mediated via the modulation of the expression of genes with anti-oxidant or DNA repair function, quantitative gene expression analysis was conducted by means of real-time PCR. The effects of 96 hours incubation with chickpea PIC on the level of expression of genes involved in anti-oxidative or DNA repair pathways are shown in Figure 6. Treatment with chickpea PIC induced a 53.4% increase (1.5 fold increase) in the expression of mitochondrial superoxide dismutase 2 (SOD 2) gene ($P < 0.01$). The expression of the ataxia telangiectasia and Rad3-related protein (ATR) gene was also significantly increased by 32.9 % (1.3 fold increase), ($P < 0.05$). Although treatment with chickpea PIC appeared to upregulate the expression of the other genes analysed, statistical significance was not reached.

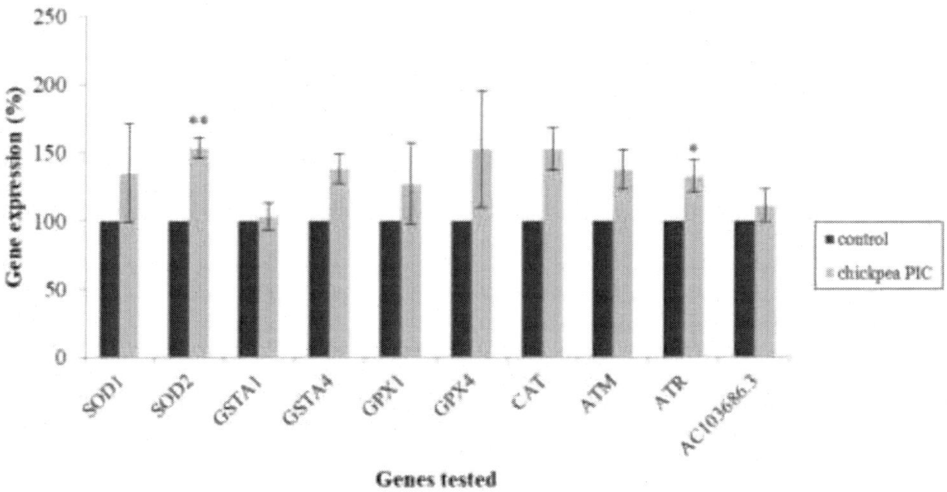

Figure 6. Effects of chickpea PIC (96-h incubation at 3 mg/ml) on the expression of genes with anti-oxidative or DNA repair properties, as determined by quantitative real-time PCR. n=3, * $P < 0.05$ vs. control, Kruskal-Wallis with Mann-Whitney U test; ** $P < 0.01$ vs. control, ANOVA, LSD post-hoc test LSD. $n = 3$, values represent the mean of relative gene expression based on untreated control ± SEM normalized to control. SOD1 = superoxide dismutase 1; SOD2 = superoxide dismutase 2; GSTA1 = glutathione S-transferase alpha 1; GSTA4 = glutathione S-transferase alpha 4; GPX1 = glutathione peroxidase 1; GPX4 = glutathione peroxidase 4; CAT = catalase; ATM = ataxia telangiectasia mutated; ATR = ataxia telangiectasia and Rad3 related; AC103686.3 = DNA-dependent protein kinase catalytic subunit.

Discussion

The present study has demonstrated, for the first time, the ability of a PIC from chickpea to inhibit DNA damage induced by H_2O_2 in LNCaP prostate cancer cells and to modulate the expression of genes linked to anti-oxidative or DNA repair processes.

To date there is a lack of scientific evidence demonstrating the modulation of oxidative DNA damage exerted by PICs in prostate cancer cells. The findings of this study show that chickpea PIC and BBI confer to LNCaP cells an increased cellular capacity to protect against oxidative damage, as demonstrated by the ability to inhibit induced DNA damage, following H_2O_2 insult. The observed anti-genotoxic activity was not associated with cytotoxic effects, as assessed by means of the MTT assay. Purified BBI exerted significant anti-genotoxic activity in LNCaP cells at lower concentrations than the chickpea PIC, likely due to a greater amount of the active compound being present in the purified BBI sample. Differences in the biochemical composition might also account for the different effects of PICs from different leguminous sources.

Chickpea PIC was characterised in terms of CIA, however several bioactive compounds might be present and responsible for its biological activity. Magee et al. (2012) report the presence of a 4000 Da peptide, possibly lunasin, determined by mass spectrometry in PIC samples from various leguminous sources. Lunasin is known for its anti-cancer properties (Hernandez-Ledesma, Hsieh & de Lumen 2013, Lam, Galvez & de Lumen 2003, Galvez et al. 2001) and the further analysis of PIC extracts by means of mass spectrometry, sodium dodecyl sulfate polyacrylamide gel electrophoresis (SDS-PAGE) or antioxidant enzyme assays would help identify the key agents responsible for the observed biological activity.

To date, there is little experimental evidence for the anti-cancer potential of chickpea (Giron-Calle et al. 2004, Sánchez-Chino et al. 2017, Sánchez-Chino et al. 2019) and chickpea PIC (Yavelow et al. 1985, Magee et al. 2012), so the choice of this legume as the source for the extract was considered important for the advancement of scientific knowledge in this area.

In contrast to LNCaP cells, chickpea PIC did not exert significant effects on DNA damage in DU145 cells. This may be due to the fact that the two cell lines model different stages of carcinogenesis. LNCaP cells are a model of an early androgen responsive prostate cancer, thus BBI might be more effective at inhibiting the initiation process triggered by DNA oxidants and might act through a different mechanism as the cancer progresses to a later androgen independent stage.

This is interesting especially in light of the potential use of BBI as a chemopreventive agent, capable of decreasing the risk of developing prostate cancer. From this perspective, it is also important to note that the tested PIC was not genotoxic at any of the concentrations tested. This finding correlates well with a study by Farombi et al. (2004) where background levels of DNA strand breaks or oxidation in liver and colon cells isolated from rats were not affected by dietary supplementation with BBI (0.25 and 0.5%), as assessed by means of the Comet assay.

LNCaP cells were more prone to H_2O_2-induced DNA damage than DU145 cells, and this is consistent with their higher sensitivity to agents that induce disruption of genetic material, such as ionising radiation (Sun et al. 2007), possibly due to functional p53 (van Bokhoven et al. 2003) and lower activity of nuclear factor-kB (NF-kB) (Josson et al. 2006); the latter is also antagonised by the androgen receptor (Palvimo et al. 1996, McKay, Cidlowski 1999). This also correlates with the development of radio-resistance of tumour cells following radiotherapy administered in early stage prostate cancer (Skvortsova et al. 2008, An et al. 2007), leading to progression to a more aggressive and metastatic stage, and supports the potential role of PICs as an adjuvant treatment to radiation therapy.

The findings of the study also support the documented anti-cancer activity of BBI in *in vitro* and *in vivo* models of prostate cancer where protective effects were observed (Kennedy, Wan 2002, Tang et al. 2009, McCormick et al. 2007, Wan et al. 1999).

Oxidative DNA damage is an important mutagenic and apparently carcinogenic factor in humans (Loft, Poulsen 1996, Valko et al. 2006, Valavanidis, Vlachogianni & Fiotakis 2009) and is involved in the initiation process of carcinogenesis. The mechanisms through which PICs exert their observed anti-genotoxic effects are not fully elucidated to date, however they can be linked to antioxidant status or stimulation of DNA repair. A study by Huang et al. (2010) demonstrated the antioxidant and thus free radical scavenging activity of BBI in a series of *in vitro* experiments, highlighting the ability to protect against hydroxyl radical-induced calf thymus DNA damage. BBI has the ability to suppress the production and release of superoxide anion radicals from purified human polymorphonuclear cells (Frenkel et al. 1987) and differentiated HL-60 human promyeloid leukemic cells *in vitro* (Ware, Wan & Kennedy 1999), thus decreasing oxidative damage to DNA, possibly resulting in malignantly transformed phenotypes. BBI is involved in the activation of the expression of genes responsible for DNA repair, as shown by Dittmann et al. (1998) in irradiated normal human skin fibroblasts. BBI also improves nucleotide excision repair *in vitro*, (Dittmann et al. 2001, Dittmann et al. 2000), however, the DNA lesions measured by the Comet assay in the present study are strand breaks, thus a different pathway may be involved.

The results obtained by real-time PCR shed light on the molecular pathways through which PICs might reduce oxidative DNA damage. It is known that the expression of antioxidant enzymes is lower in prostate cancer compared to healthy prostatic epithelium (Bostwick et al. 2000), with a shift in the prooxidant-antioxidant balance toward a more oxidative state (Ripple et al. 1997). The incidence of prostate cancer increases with age (Sakr et al. 1994) and this is consistent with an age–related decrease in the antioxidant defense mechanisms (Mo, Hom & Anderson 1995, Warner 1994) which can in turn lead to a progressive accumulation of DNA strand breaks (He, Yasumoto 1994).

Treatment of LNCaP cells with 3 mg/ml chickpea PIC significantly increased the expression of SOD2, an isoform of the superoxide dismutase family (SOD) whose transcript is a mitochondrial enzyme with the ability to scavenge superoxide anion radical (Fridovich 1997).

Chickpea PIC also increased the expression of ATR gene, encoding for a signalling molecule involved in DNA damage response (DDR) and cell cycle progression, in a statistically significant fashion. The effects of BBI on the expression of the aforementioned genes have not been investigated to date. Although treatment with PIC appeared to upregulate the expression of the other genes analysed, concurrent with a previous study demonstrating an upregualtion of anti-oxidant genes in prostate LAPC-4 cells following genistein treatment (Raschke et al. 2006), these effects were not significant.

It is worth considering, when studying the potential mechanisms through which PICs modulate H_2O_2-induced DNA damage, the experimental limitation posed by the different exposure times and doses used in the Comet and MTT assay versus quantitative real-time PCR (0.05, 0.1, 0.3, 1, 3 mg/ml for 24 hours and 3 mg/ml for 96 hours, respectively).

Conclusion

To summarize, the present study used an *in vitro* bioassay to demonstrate the anti-genotoxic activity of a PIC from chickpea against H_2O_2 insult in LNCaP prostate cancer cells. These findings support the chemopreventive potential of BBI against prostate carcinogenesis and highlight a possible target for this dietary compound, corresponding to inhibition of the initiation stage of prostate cancer, during which events leading to DNA damage take place. Further investigation is needed to address the mechanisms through which the anti-genotoxic activity of the PIC is exerted.

The investigation of the effects of chickpea PIC on H_2O_2-induced DNA damage and on genes with anti-genotoxic functions should also be carried out under the same experimental conditions with regards to dose and exposure time.

References

American Cancer Society 2019, *Cancer Statistic Center, Prostate*. Available: https://cancerstatisticscenter.cancer.org/#!/cancer-site/Prostate.

An, J., Chervin, A. S., Nie, A., Ducoff, H. S. & Huang, Z. 2007, "Overcoming the radioresistance of prostate cancer cells with a novel Bcl-2 inhibitor," *Oncogene*, vol. 26, no. 5, pp. 652-661.

Armstrong, W. B., Kennedy, A. R., Wan, X. S., Atiba, J., McLaren, E. & Meyskens, F. L. 2000a, "Single-dose administration of Bowman-Birk inhibitor concentrate in patients with oral leukoplakia," *Cancer Epidemiology Biomarkers & Prevention,* vol. 9, no. 1, pp. 43-47.

Armstrong, W. B., Kennedy, A. R., Wan, X. S., Taylor, T. H., Nguyen, Q. A., Jensen, J., Thompson, W., Lagerberg, W. & Meyskens, F. L. 2000b, "Clinical modulation of oral leukoplakia and protease activity by Bowman-Birk inhibitor concentrate in a phase IIa chemoprevention trial," *Clinical Cancer Research,* vol. 6, no. 12, pp. 4684-4691.

Armstrong, W. B., Taylor, T. H., Kennedy, A. R., Melrose, R. J., Messadi, D. V., Gu, M., Le, A. D., Perloff, M., Civantos, F., Goodwin, W. J., Wirth, L. J., Kerr, A. R. & Meyskens, F. L., Jr. 2013, "Bowman Birk Inhibitor Concentrate and Oral Leukoplakia: A Randomized Phase IIb Trial," *Cancer Prevention Research*, vol. 6, no. 5, pp. 410-418.

Baade, P. D., Youlden, D. R. & Krnjacki, L. J. 2009, "International epidemiology of prostate cancer: Geographical distribution and secular trends," *Molecular Nutrition & Food Research,* vol. 53, no. 2, pp. 171-184.

Barbosa, A. C. L., Lajolo, F. M. & Genovese, M. I. 2011, "Effect of free or protein-associated soy isoflavones on the antioxidant status in rats," *Journal of the science of food and agriculture,* vol. 91, no. 4, pp. 721-731.

Bostwick, D., Alexander, E., Singh, R., Shan, A., Qian, J., Santella, R., Oberley, L., Yan, T., Zhong, W., Jiang, X. & Oberley, T. 2000, "Antioxidant enzyme expression and reactive oxygen species damage in prostatic intraepithelial neoplasia and cancer," *Cancer,* vol. 89, no. 1, pp. 123-134.

Byun, J. S., Han, Y. S. & Lee, S. S. 2010, "The Effects of Yellow Soybean, Black Soybean, and Sword Bean on Lipid Levels and Oxidative Stress in Ovariectomized Rats," *International Journal for Vitamin and Nutrition Research,* vol. 80, no. 2, pp. 97-106.

Cancer Research UK 2019, *Prostate cancer statistics* [Homepage of Cancer Research UK], [Online]. Available: https://www.cancerresearchuk.org/health-professional/cancer-statistics/statistics-by-cancer-type/prostate-cancer#heading-Zero.

Chen, Y. W., Huang, S. C., Lin-Shiau, S. Y. & Lin, J. K. 2005, "Bowman-Birk inhibitor abates proteasome function and suppresses the proliferation of MCF7 breast cancer cells through accumulation of MAP kinase phosphatase-1.," *Carcinogenesis,* vol. 26, no. 7, pp. 1296-1306.

Chlenski, A., Nakashiro, K., Ketels, K. V., Korovaitseva, G. I. & Oyasu, R. 2001, "Androgen receptor expression in androgen independent prostate cancer cell lines," *Prostate,* vol. 47, no. 1, pp. 66-75.

Collins, A. R. 2004, "The comet assay for DNA damage and repair - Principles, applications, and limitations," *Molecular biotechnology,* vol. 26, no. 3, pp. 249-261.

DiSilvestro, R., Goodman, J., Dy, E. & LaValle, G. 2005, "Soy isoflavone supplementation elevates erythrocyte superoxide dismutase, but not plasma ceruloplasmin in postmenopausal breast cancer survivors," *Breast cancer research and treatment,* vol. 89, no. 3, pp. 251-255.

Dittmann, K., Knaus-Dittmann, D., Mayer, C. & Rodemann, H. P. 2001, "Bowman-Birk proteinase inhibitor-mediated radioprotection against UV irradiation is TP53-dependent and associated with stimulation of nucleotide excision repair," *Radiation and environmental biophysics,* vol. 40, no. 2, pp. 163-167.

Dittmann, K. H., Dikomey, E., Mayer, C. & Rodemann, H. P. 2000, "The Bowman-Birk protease inhibitor enhances clonogenic cell survival of ionizing radiation-treated nucleotide excision repair-competent cells but not of xeroderma pigmentosum cells," *International journal of radiation biology,* vol. 76, no. 2, pp. 223-229.

Dittmann, K. H., Gueven, N., Mayer, C. & Rodemann, H. P. 1998, "The radioprotective effect of BBI is associated with the activation of DNA repair-relevant genes," *International journal of radiation biology,* vol. 74, no. 2, pp. 225-230.

Farombi, E. O., Hansen, M., Ravn-Haren, G., Moller, P. & Dragsted, L. O. 2004, "Commonly consumed and naturally occurring dietary substances affect biomarkers of oxidative stress and DNA damage in healthy rats," *Food and Chemical Toxicology,* vol. 42, no. 8, pp. 1315-1322.

Frenkel, K., Chrzan, K., Ryan, C. A., Wiesner, R. & Troll, W. 1987, "Chymotrypsin-specific protease inhibitors decrease H2O2 formation by activated human polymorphonuclear leukocytes.," *Carcinogenesis,* vol. 8, no. 9, pp. 1207-1212.

Fridovich, I. 1997, "Superoxide anion radical (O-2 radical anion), superoxide dismutases, and related matters," *Journal of Biological Chemistry,* vol. 272, no. 30, pp. 18515-18517.

Galvez, A., Chen, N., Macasieb, J. & De Lumen, B. 2001, "Chemopreventive property of a soybean peptide (lunasin) that binds to deacetylated histones and inhibits acetylation," *Cancer research*, vol. 61, no. 20, pp. 7473-7478.

Gill, C. I. R., Boyd, A., McDermott, E., McCann, M., Servili, M., Selvaggini, R., Taticchi, A., Esposto, S., Montedoro, G., McGlynn, H. & Rowland, I. 2005, "Potential anti-cancer effects of virgin olive oil phenols on colorectal carcinogenesis models *in vitro*," *International Journal of Cancer,* vol. 117, no. 1, pp. 1-7.

Giron-Calle, J., Vioque, J., del Mar Yust, M., Pedroche, J., Alaiz, M. & Millan, F. 2004, "Effect of chickpea aqueous extracts, organic extracts, and protein concentrates on cell proliferation," *Journal of Medicinal Food,* vol. 7, no. 2, pp. 122-129.

Hagen, M. K., Lehenbauer-Ludke, A. R., Paludo, A. C., Schenkel, P., Goncalves, L., Fernandes, T. G., Caron, R., Llesuy, S., Mill, J. G. & Bello-Klein, A. 2009, "Diet with isolated soy protein reduces oxidative stress and preserves ventricular function in rats with myocardial infarction," *Nutrition Metabolism and Cardiovascular Diseases,* vol. 19, no. 2, pp. 91-97.

He, P. & Yasumoto, K. 1994, "Dietary Butylated Hydroxytoluene Counteracts with Paraquat to Reduce the Rate of Hepatic Dna Single-Strand Breaks in Senescence-Accelerated Mice," *Mechanisms of ageing and development,* vol. 76, no. 1, pp. 43-48.

Hernandez-Ledesma, B., Hsieh, C.-. & De Lumen, B. O. 2013, "Chemopreventive Properties of Peptide Lunasin: A Review," *Protein and Peptide Letters*, vol. 20, no. 4, pp. 424-432.

Ho, K., Li, L., Zhao, L. & Qian, Z. 2003, "Genistein protects primary cortical neurons from iron-induced lipid peroxidation," *Molecular and cellular biochemistry,* vol. 247, no. 1-2, pp. 219-222.

Horoszewicz, J. S., Leong, S. S., Kawinski, E., Karr, J. P., Rosenthal, H., Chu, T. M., Mirand, E. A. & Murphy, G. P. 1983, "Lncap Model of Human Prostatic-Carcinoma," *Cancer research,* vol. 43, no. 4, pp. 1809-1818.

Huang, G., Chiu, C., Wu, C., Huang, S., Amegaya, S., Hou, W., Sheu, M., Liao, J., Chen, Y. & Lin, Y. 2010, "Redox status of Bowman-Birk inhibitor from soybean influence its *in vitro* antioxidant activities," *Botanical Studies,* vol. 51, no. 4, pp. 431-437.

Josson, S., Xu, Y., Fang, F., Dhar, S., St Clair, D. & St Clair, W. 2006, "RelB regulates manganese superoxide dismutase gene and resistance to ionizing radiation of prostate cancer cells," *Oncogene*, vol. 25, no. 10, pp. 1554-1559.

Kassell, B. 1970, *Trypsin and Chymotrypsin Inhibitors from Soybeans-D.*

Kennedy, A. R. 1998, "The Bowman-Birk inhibitor from soybeans as an anticarcinogenic agent," *American Journal of Clinical Nutrition,* vol. 68, no. 6 Suppl, pp. 1406S-1412S.

Kennedy, A.R. & Wan, X.S. 2002, "Effects of the Bowman-Birk inhibitor on growth, invasion, and clonogenic survival of human prostate epithelial cells and prostate cancer cells.," *Prostate,* vol. 50, no. 2, pp. 125-133.

Klaunig, J. E., Kamendulis, L. M. & Hocevar, B. A. 2010, "Oxidative Stress and Oxidative Damage in Carcinogenesis," *Toxicologic pathology,* vol. 38, no. 1, pp. 96-109.

Klaunig, J. & Kamendulis, L. 2004, "The role of oxidative stress in carcinogenesis," *Annual Review of Pharmacology and Toxicology,* vol. 44, pp. 239-267.

Lam, Y., Galvez, A. & De Lumen, B. 2003, "Lunasin (TM) suppresses E1A-mediated transformation of mammalian cells but does not inhibit growth of immortalized and established cancer cell lines," *Nutrition and Cancer-an International Journal*, vol. 47, no. 1, pp. 88-94.

Lichtenstein, G. R., Deren, J. J., Katz, S., Lewis, J. D., Kennedy, A. R. & Ware, J. H. 2008, "Bowman-birk inhibitor concentrate: A novel therapeutic agent for patients with active ulcerative colitis," *Digestive diseases and sciences,* vol. 53, no. 1, pp. 175-180.

Liu, J., Chang, S. & Wiesenborn, D. 2005, "Antioxidant properties of soybean isoflavone extract and tofu *in vitro* and *in vivo*," *Journal of Agricultural and Food Chemistry,* vol. 53, no. 6, pp. 2333-2340.

Loft, S. & Poulsen, H. E. 1996, "Cancer risk and oxidative DNA damage in man," *Journal of Molecular Medicine-Jmm,* vol. 74, no. 6, pp. 297-312.

Magee, P. J., Owusu-Apenten, R., McCann, M. J., Gill, C. I. & Rowland, I. R. 2012, "Chickpea (Cicer arietinum) and other plant-derived protease inhibitor concentrates inhibit breast and prostate cancer cell proliferation *in vitro*.," *Nutrition & Cancer,* vol. 64, no. 5, pp. 741-748.

Malins, D., Johnson, P., Wheeler, T., Barker, E., Polissar, N. & Vinson, M. 2001, "Age-related radical-induced DNA damage is linked to prostate cancer," *Cancer research,* vol. 61, no. 16, pp. 6025-6028.

Malkowicz, S. B., McKenna, W. G., Vaughn, D. J., Wan, X. S., Propert, K. J., Rockwell, K., Marks, S. H. F., Wein, A. J. & Kennedy, A. R. 2001, "Effects of Bowman-Birk inhibitor concentrate (BBIC) in patients with benign prostatic hyperplasia," *Prostate,* vol. 48, no. 1, pp. 16-28.

Mathers, J. C. 2002, "Pulses and carcinogenesis: potential for the prevention of colon, breast and other cancers," *British Journal of Nutrition,* vol. 88, pp. S273-S279.

McCormick, D. L., Johnson, W. D., Bosland, M. C., Lubet, R. A. & Steele, V. E. 2007, "Chemoprevention of rat prostate carcinogenesis by soy isoflavones and by Bowman-Birk inhibitor," *Nutrition and Cancer-an International Journal,* vol. 57, no. 2, pp. 184-193.

McKay, L. & Cidlowski, J. 1999, "Molecular control of immune/inflammatory responses: Interactions between nuclear factor-kappa B and steroid receptor-signaling pathways," *Endocrine reviews,* vol. 20, no. 4, pp. 435-459.

Mills, P. K., Beeson, W. L., Phillips, R. L. & Fraser, G. E. 1989, "Cohort study of diet, lifestyle, and prostate cancer in Adventist men.," *Cancer,* vol. 64, no. 3, pp. 598-604.

Mo, J., Hom, D. & Anderson, J. 1995, "Decreases in Protective Enzymes Correlates with Increased Oxidative Damage in the Aging Mouse-Brain," *Mechanisms of ageing and development,* vol. 81, no. 2-3, pp. 73-82.

Mosmann, T. 1983, "Rapid Colorimetric Assay for Cellular Growth and Survival - Application to Proliferation and Cyto-Toxicity Assays," *Journal of immunological methods,* vol. 65, no. 1-2, pp. 55-63.

Palvimo, J., Reinikainen, P., Ikonen, T., Kallio, P., Moilanen, A. & Janne, O. 1996, "Mutual transcriptional interference between RelA and androgen receptor," *Journal of Biological Chemistry,* vol. 271, no. 39, pp. 24151-24156.

Pfaffl, M. 2001, "A new mathematical model for relative quantification in real-time RT-PCR," *Nucleic acids research,* vol. 29, no. 9, pp. e45.

Poolzobel, B. L., Lotzmann, N., Knoll, M., Kuchenmeister, F., Lambertz, R., Leucht, U., Schroder, H. G. & Schmezer, P. 1994, "Detection of Genotoxic Effects in Human Gastric and Nasal-Mucosa Cells Isolated from Biopsy Samples," *Environmental and molecular mutagenesis,* vol. 24, no. 1, pp. 23-45.

Quinn, M. & Babb, P. 2002, "Patterns and trends in prostate cancer incidence, survival, prevalence and mortality. Part I: international comparisons," *BJU international,* vol. 90, no. 2, pp. 162-173.

Rafiq, A. & Bailey, G. 1998, "A more sensitive Hummel assay for chymotrypsin," *Analytical Biochemistry,* vol. 257, no. 2, pp. 233-234.

Raschke, M., Rowland, I. R., Magee, P. J. & Pool-Zobel, B. L. 2006, "Genistein protects prostate cells against hydrogen peroxide-induced DNA damage and induces expression of genes involved in the defence against oxidative stress," *Carcinogenesis,* vol. 27, no. 11, pp. 2322-2330.

Ripple, M., Henry, W., Rago, R. & Wilding, G. 1997, "Prooxidant-antioxidant shift induced by androgen treatment of human prostate carcinoma cells," *Journal of the National Cancer Institute,* vol. 89, no. 1, pp. 40-48.

Sakr, W. A., Grignon, D. J., Crissman, J. D., Heilbrun, L. K., Cassin, B. J., Pontes, J. J. & Haas, G. P. 1994, "High grade prostatic intraepithelial neoplasia (HGPIN) and prostatic adenocarcinoma between the ages of 20-69: an autopsy study of 249 cases.," *In Vivo,* vol. 8, no. 3, pp. 439-443.

Sánchez-Chino, X. M., Jiménez Martínez, C., Vásquez Garzón, V. R., Álvarez González, I., Villa Treviño, S., Madrigal Bujaidar, E., Dávila Ortiz, G. & Baltiérrez Hoyos, R. 2017, "Cooked Chickpea Consumption Inhibits Colon Carcinogenesis in Mice Induced with Azoxymethane and Dextran Sulfate Sodium," *Journal of the American College of Nutrition,* vol. 36, no. 5, pp. 391-398.

Sánchez-Chino, X. M, Jiménez Martínez, C., León-Espinosa, E. B., Garduño-Siciliano, L., Álvarez-González, I., Madrigal-Bujaidar, E., Vásquez-Garzón, V. R., Baltiérrez-Hoyos, R. & Dávila-Ortiz, G. 2019, "Protective Effect of Chickpea Protein Hydrolysates on Colon Carcinogenesis Associated With a Hypercaloric Diet," *Journal of the American College of Nutrition,* vol. 38, no. 2, PP. 162-170.

Schuurman, A. G., Goldbohm, R. A., Dorant, E. & van den Brandt, P.A. 1998, "Vegetable and fruit consumption and prostate cancer risk: A cohort study in the Netherlands," *Cancer Epidemiology Biomarkers & Prevention,* vol. 7, no. 8, pp. 673-680.

Severson, R. K., Nomura, A. M. Y., Grove, J. S. & Stemmermann, G. N. 1989, "A Prospective-Study of Demographics, Diet, and Prostate-Cancer among Men of Japanese Ancestry in Hawaii," *Cancer research,* vol. 49, no. 7, pp. 1857-1860.

Skvortsova, I., Skvortsov, S., Stasyk, T., Raju, U., Popper, B., Schiestl, B., von Guggenberg, E., Neher, A., Bonn, G. K., Huber, L. A. & Lukas, P. 2008, "Intracellular signaling pathways regulating radioresistance of human prostate carcinoma cells," *Proteomics*, vol. 8, no. 21, pp. 4521-4533.

Souza, L. D. C., Camargo, R., Demasi, M., Santana, J. M., De Sa, C. M. And De Freitas, S. M. 2014, "Effects of an Anticarcinogenic Bowman-Birk Protease Inhibitor on Purified 20S Proteasome and MCF-7 Breast Cancer Cells," *Plos One*, vol. 9, no. 1, pp. e86600-e86600.

Stone, K. R., Mickey, D. D., Wunderli, H., Mickey, G. H. & Paulson, D. F. 1978, "Isolation of a Human Prostate Carcinoma Cell Line (Du 145)," *International Journal of Cancer,* vol. 21, no. 3, pp. 274-281.

Sun, Y., Clair, D. K. S., Fang, F., Warren, G. W., Rangnekar, V. M., Crooks, P. A. & Clair, W. H. S. 2007, "The radiosensitization effect of parthenolide in prostate cancer cells is mediated by nuclear factor-kappa B inhibition and enhanced by the presence of PTEN," *Molecular Cancer Therapeutics*, vol. 6, no. 9, pp. 2477-2486.

Tang, M., Asamoto, M., Ogawa, K., Naiki-Ito, A., Sato, S., Takahashi, S. & Shirai, T. 2009, "Induction of apoptosis in the LNCaP human prostate carcinoma cell line and prostate adenocarcinomas of SV40T antigen transgenic rats by the Bowman-Birk inhibitor.," *Pathology international,* vol. 59, no. 11, pp. 790-796.

Troll, W. & Wiesner, R. 1983, "Protease Inhibitors - Possible Anticarcinogens in Edible Seeds," *Prostate,* vol. 4, no. 4, pp. 345-349.

Valavanidis, A., Vlachogianni, T. & Fiotakis, C. 2009, "8-hydroxy-2' -deoxyguanosine (8-OHdG): A Critical Biomarker of Oxidative Stress and Carcinogenesis," *Journal of Environmental Science and Health Part C-Environmental Carcinogenesis & Ecotoxicology Reviews,* vol. 27, no. 2, pp. 120-139.

Valko, M., Rhodes, C., Moncol, J., Izakovic, M. & Mazur, M. 2006, "Free radicals, metals and antioxidants in oxidative stress-induced cancer," *Chemico-biological interactions,* vol. 160, no. 1, pp. 1-40.

van Bokhoven, A., Varella-Garcia, M., Korch, C., Johannes, W., Smith, E., Miller, H., Nordeen, S., Miller, G. & Lucia, M. 2003, "Molecular characterization of human prostate carcinoma cell lines," *Prostate*, vol. 57, no. 3, pp. 205-225.

Wan, X. S., Ware, J. H., Zhang, L., Newberne, P. M., Evans, S. M., Clark, L. C. & Kennedy, A. R. 1999, "Treatment with soybean-derived Bowman Birk inhibitor increases serum prostate-specific antigen concentration while suppressing growth of human prostate cancer xenografts in nude mice.," *Prostate,* vol. 41, no. 4, pp. 243-252.

Ware, J. H., Wan, X. S. & Kennedy, A. R. 1999, "Bowman-Birk inhibitor suppresses production of superoxide anion radicals in differentiated HL-60 cells," *Nutrition and Cancer-an International Journal,* vol. 33, no. 2, pp. 174-177.

Warner, H. 1994, "Superoxide-Dismutase, Aging, and Degenerative Disease," *Free Radical Biology and Medicine,* vol. 17, no. 3, pp. 249-258.

Webber, M. M., Bello, D. & Quader, S. 1997, "Immortalized and tumorigenic adult human prostatic epithelial cell lines: Characteristics and applications .2. Tumorigenic cell lines," *Prostate,* vol. 30, no. 1, pp. 58-64.

Yavelow, J., Collins, M., Birk, Y., Troll, W. & Kennedy, A. R. 1985, "Nanomolar concentrations of Bowman-Birk soybean protease inhibitor suppress x-ray-induced transformation *in vitro*.," *Proceedings of the National Academy of Sciences of the United States of America,* vol. 82, no. 16, pp. 5395-5399.

INDEX

A

B

C

D

E

F

G

H

I

K

L

M

N

O

P

Q

R

S

T

V

W

X